Advanced Researches in Ecology

Volume I

Advanced Researches in Ecology
Volume I

Edited by **Jeffery Clarke**

R CALLISTO REFERENCE

New York

Published by Callisto Reference,
106 Park Avenue, Suite 200,
New York, NY 10016, USA
www.callistoreference.com

Advanced Researches in Ecology: Volume I
Edited by Jeffery Clarke

International Standard Book Number: 978-1-63239-026-4 (Hardback)

Printed in the United States of America.

Contents

Preface   VII

Chapter 1   **Water Hyacinth** *Eichhornia crassipes* **(Mart.) Solms-Laubach Dynamics and Succession in the Nyanza Gulf of Lake Victoria (East Africa): Implications for Water Quality and Biodiversity Conservation**   **1**
John Gichuki, Reuben Omondi, Priscillar Boera, Tom Okorut, Ally Said Matano, Tsuma Jembe and Ayub Ofulla

Chapter 2   **Toxicity of Metals to a Freshwater Snail,** *Melanoides tuberculata*   **11**
M. Shuhaimi-Othman, R. Nur-Amalina and Y. Nadzifah

Chapter 3   **Diet Composition and Feeding Strategies of the Stone Marten (***Martes foina***) in a Typical Mediterranean Ecosystem**   **21**
Dimitrios E. Bakaloudis, Christos G. Vlachos, Malamati A. Papakosta, Vasileios A. Bontzorlos and Evangelos N. Chatzinikos

Chapter 4   **Patterns of Diversity of the Rissoidae (Mollusca: Gastropoda) in the Atlantic and the Mediterranean Region**   **32**
Sérgio P. Ávila, Jeroen Goud and António M. de Frias Martins

Chapter 5   **Quantitative Analysis of Driving Factors of Grassland Degradation: A Case Study in Xilin River Basin, Inner Mongolia**   **62**
Yichun Xie and Zongyao Sha

Chapter 6   **Caatinga Revisited: Ecology and Conservation of an Important Seasonal Dry Forest**   **76**
Ulysses Paulino de Albuquerque, Elcida de Lima Araújo, Ana Carla Asfora El-Deir, André Luiz Alves de Lima, Antonio Souto, Bruna Martins Bezerra, Elba Maria Nogueira Ferraz, Eliza Maria Xavier Freire, Everardo Valadares de Sá Barreto Sampaio, Flor Maria Guedes Las-Casas, Geraldo Jorge Barbosa de Moura, Glauco Alves Pereira, Joabe Gomes de Melo, Marcelo Alves Ramos, Maria Jesus Nogueira Rodal, Nicola Schiel, Rachel Maria de Lyra-Neves, Rômulo Romeu Nóbrega Alves, Severino Mendes de Azevedo-Júnior, Wallace Rodrigues Telino Júnior and William Severi

Chapter 7   **Application of Scenario Analysis and Multiagent Technique in Land-Use Planning: A Case Study on Sanjiang Wetlands**   **94**
Huan Yu, Shi-Jun Ni, Bo Kong, Zheng-Wei He, Cheng-Jiang Zhang, Shu-Qing Zhang, Xin Pan, Chao-Xu Xia and Xuan-Qiong Li

Chapter 8   **A Review of Surface Water Quality Models**   **104**
Qinggai Wang, Shibei Li, Peng Jia, Changjun Qi and Feng Ding

Chapter 9   **N : P Stoichiometry in a Forested Runoff during Storm Events: Comparisons with Regions and Vegetation Types**   111
Lanlan Guo, Yi Chen, Zhao Zhang and Takehiko Fukushima

Chapter 10   **Diurnal Characteristics of Ecosystem Respiration of Alpine Meadow on the Qinghai-Tibetan Plateau: Implications for Carbon Budget Estimation**   119
Yu Qin and Shuhua Yi

Chapter 11   **Shesher and Welala Floodplain Wetlands (Lake Tana, Ethiopia): Are They Important Breeding Habitats for *Clarias gariepinus* and the Migratory *Labeobarbus* Fish Species?**   124
Wassie Anteneh, Eshete Dejen and Abebe Getahun

Chapter 12   **Effluents of Shrimp Farms and Its Influence on the Coastal Ecosystems of Bahía de Kino, Mexico**   134
Ramón H. Barraza-Guardado, José A. Arreola-Lizárraga, Marco A. López-Torres, Ramón Casillas-Hernández, Anselmo Miranda-Baeza, Francisco Magallón-Barrajas and Cuauhtemoc Ibarra-Gámez

Chapter 13   **Evaluation of the Impacts of Land Use on Water Quality: A Case Study in The Chaohu Lake Basin**   142
Juan Huang, Jinyan Zhan, Haiming Yan, Feng Wu and Xiangzheng Deng

Chapter 14   **An Investigation into Occasional White Spot Syndrome Virus Outbreak in Traditional Paddy Cum Prawn Fields in India**   149
Deborah Gnana Selvam, K. M. Mujeeb Rahiman and A. A. Mohamed Hatha

Chapter 15   **A Refined Methodology for Defining Plant Communities Using Postagricultural Data from the Neotropics**   160
Randall W. Myster

Chapter 16   **World Aquaculture: Environmental Impacts and Troubleshooting Alternatives**   169
Marcel Martinez-Porchas and Luis R. Martinez-Cordova

Chapter 17   **Diversity of Woodland Communities and Plant Species along an Altitudinal Gradient in the Guancen Mountains, China**   178
Dongping Meng, Jin-Tun Zhang and Min Li

Chapter 18   **The Ecological Response of *Carex lasiocarpa* Community in the Riparian Wetlands to the Environmental Gradient of Water Depth in Sanjiang Plain, Northeast China**   185
Zhaoqing Luan, Zhongxin Wang, Dandan Yan, Guihua Liu and Yingying Xu

Chapter 19   **Impacts of Intensified Agriculture Developments on Marsh Wetlands**   192
Zhaoqing Luan and Demin Zhou

Permissions

List of Contributors

Preface

Organisms and their peculiar habitats and environments has interested the researchers for centuries. The interactions between these two is referred to as ecology. There are innumerable species of organisms on the face of the earth, thus their diversity is awe striking. Ecologists spend years studying and researching this diversity. Another important aspect of ecology is the ecosystem. Every organism belongs to a certain ecosystem. Ecosystems do not just refer to the habitat but also the different communities that might co-exist in a similar environment, are all studied under ecosystems.

Ecology finds its roots in the studies of philosophers such as Aristotle and Hippocrates when they delved deeper into the history of nature. Since then the field has been rapidly progressing. The 19th century, in particular witnessed the growth of modern ecology. Evolution became the mainstay of ecological research in this period. Ecology is often referred to as an interdisciplinary science converging the essentials of evolutionary biology and earth sciences for a better understanding of species and their environment.

Ecology is often distributed into branches such as habitat, biodiversity, hierarchal ecology, biome, biosphere and many others. It has applications in varied fields such as sustainability, environmental management, environmental protection, soil management and many other related fields.

I wish to thank all the contributing authors who have shared their researches in this book. I thank them for maintaining the time parameters set for this publication. They have also been very supportive to me throughout the editing process and guided me whenever I needed help. I also wish to thank the publisher and the publishing team for their excellent technical assistance, as and when required. Lastly, I wish to thank my friends and family for their endless support and unwavering faith in me.

Editor

Water Hyacinth *Eichhornia crassipes* (Mart.) Solms-Laubach Dynamics and Succession in the Nyanza Gulf of Lake Victoria (East Africa): Implications for Water Quality and Biodiversity Conservation

John Gichuki,[1,2] Reuben Omondi,[1] Priscillar Boera,[1] Tom Okorut,[3] Ally Said Matano,[3] Tsuma Jembe,[1] and Ayub Ofulla[4]

[1] *Kenya Marine and Fisheries Research Institute, P.O. Box 1881, Kisumu 40100, Kenya*
[2] *Big Valley Rancheria Band of Pomo Indians, 2726 Mission Rancheria Road, Lake Port, CA 95453-9637, USA*
[3] *Lake Victoria Basin Commission, P.O. Box 1510, Kisumu 40100, Kenya*
[4] *Maseno University, P.O. Box Private Bag Maseno, Kenya*

Correspondence should be addressed to John Gichuki, gichukij@yahoo.com

Academic Editors: A. Bosabalidis and B. S. Chauhan

This study, conducted in Nyanza Gulf of Lake Victoria, assessed ecological succession and dynamic status of water hyacinth. Results show that water hyacinth is the genesis of macrophyte succession. On establishment, water hyacinth mats are first invaded by native emergent macrophytes, *Ipomoea aquatica* Forsk., and *Enydra fluctuans* Lour., during early stages of succession. This is followed by hippo grass *Vossia cuspidata* (Roxb.) Griff. in mid- and late stages whose population peaks during climax stages of succession with concomitant decrease in water hyacinth biomass. Hippo grass depends on water hyacinth for buoyancy, anchorage, and nutrients. The study concludes that macrophyte succession alters aquatic biodiversity and that, since water hyacinth infestation and attendant succession are a symptom of broader watershed management and pollution problems, aquatic macrophyte control should include reduction of nutrient loads and implementing multifaceted approach that incorporates biological agents, mechanical/manual control with utilization of harvested weed for cottage industry by local communities.

1. Introduction

Macrophytes are higher plants that grow in ecosystems whose formation has been dominated by water and whose processes and characteristics are largely controlled by water. Macrophytes can be subdivided into four groups on the basis of their water requirements and habitats. Submerged macrophytes are those that are completely covered with water but rooted in the substrate, for example, *Potamogeton schweinfurthii* A. Benn.

Floating leafed macrophytes are those that are rooted but have floating leaves, for example, *Nymphaea lotus* thumb. Emergent macrophytes are rooted plants with their principal photosynthetic surfaces projecting above the water, for example, *Cyperus papyrus* L. Finally, the free-floating macrophytes are those that float on the water surface, for example, water hyacinth *Eichhornia crassipes* (Mart.) Solms-Laubach and water fern/water velvet *Azolla pinnata* Decne ex Mett. Factors that influence the establishment and distribution of macrophytes include depth, topography, and type of substrate, exposure to currents and/or wind, and water turbidity [1].

Water hyacinth *Eichhornia crassipes*, a perennial aquatic herb which belongs to the pickerelweed family (Pontederiaceae), is a native of tropical America. This assumption is based on the prevalence of other species of *Eichhornia* spp. particularly the more primitive *Eichhornia paniculata* (Spreng.) Solms and *Eichhornia paradoxa* (Mart.) Solms, in this area. It has been classified as one of the worst aquatic weeds in the world [2]. The weed has spread to many parts of the world due to its beautiful large purple and violet (lavender) flowers similar to orchids that make it a popular ornamental plant for ponds [3]. In the areas bordering Lake Victoria, the weed was first recorded in Lake Kyoga (Uganda) in May 1988 [4]. Within Lake Victoria, it was observed in the Ugandan sector in 1989 ([5–8] Tanzania in 1989 [9], and Kenya in 1990 [10, 11]). In the Kagera River of Rwanda, the water hyacinth was recorded in 1991 [12] though it was believed to have been present in this area since at least the early 1980s ([13, 14]). It was present upstream of Lake Victoria on European plantations within the Kagera catchment since at least the 1940s and reached nuisance levels here, choking riverine wetland lakes, by the 1980s. Migration via the Kagera river was the most likely point of entry for water hyacinth into Lake Victoria. The weed has flourished in Lake Victoria due to absence of natural predators as insects, fish, and other biota and due to favorable environmental conditions. It is associated with major negative economic and ecological impacts to the Lake Victoria region.

The weed forms thick mats over the infested water bodies causing obstruction to economic development activities and impacting negatively on the indigenous aquatic biodiversity. Furthermore, the weed affects the conditions of the water body and life of the flora and fauna in them. Floating mats of water hyacinth for example drastically curtail the penetration of light into the aquatic ecosystem thus inhabiting the growth of phytoplankton.

Initially, efforts to control water hyacinth in Lake Victoria during the early 1990s were of limited success and were primarily directed at manually removing water hyacinth and conducting public awareness exercises. In the mid-late 1990s, management to combat water hyacinth increased with efforts such as the Lake Victoria Environmental Management Program (LVEMP) and US Agency for International Development (USAID) funding for coordination efforts by Clean Lakes, Inc. (Martinez, CA, USA). Control actions included biocontrol using *Neochetina bruchi* and *N. eichhorniae* water hyacinth weevils and mechanical control using large harvesting and chopping boats [7]. Operational water hyacinth control through the use of herbicides was not implemented in the region.

Despite water hyacinth's invasive nature and dominance in Lake Victoria in the 1990s, water hyacinth largely disappeared from Lake Victoria by the end of 1999. For instance, no water hyacinth was found on the Gulf from April 2002 until October 2004, only appearing again at the next measurement date of December 2005. Various hypotheses have been proposed on its rapid disappearance including the introduction of water hyacinth weevils [15, 16], effects of the *El Nino* weather of 1997/1998 [17], or a combination of interacting factors involving *El Nino* of 1997/1998, and biocontrol by weevils [18].

Here, we describe a form of macrophyte ecological succession which culminated in the control of water hyacinth in the Nyanza Gulf of Lake Victoria in 2008.

2. Methodology

2.1. Study Area. The study was carried out between September and December 2008. The study sites are displayed in Figure 1. The following areas were sampled, namely, Off Kibos, Dunga beach, next to Osienala Headquarters, Sondu Miriu, Homa Bay, Oluch river mouth, Lwanda Gembe, and Asembo Bay. The study was carried out at specific areas in the Nyanza Gulf where the populations of major macrophyte species were sighted. The zones were divided into 3 zones, namely, Hippo grass *Vossia cuspidata* (Roxb.) Griff. zone, Water hyacinth *Eichhornia crassipes* (Mart.) Solms-Laubach zone, and Hippo grass/water hyacinth mixture. In most of the cases, the macrophyte mats could not be penetrated by the boat, and thus 2 zones were sampled, namely, interphase zone and open water zone. The open water zone was always located 50 metres away from the mats. Where possible, a third zone was sampled, namely, within macrophyte zone (hippo grass/water hyacinth mixture).

2.2. Physical and Chemical Parameters. Geographical coordinates were determined using a GPS Garmin GPS II Plus. Turbidity was measured with a 2100 P Hatch Turbidimeter, while pH was measured with a WTW pH 315i meter. Secchi depth was measured with a 20 cm black and white Secchi disk. Water samples were collected by a Van Dorn sampler. Nutrient analysis was carried out following the methods outlined in [19]. Total nitrogen in the samples was analyzed on unfiltered samples by digestion with concentrated sulfuric acid (by autoclave procedure) to convert organic nitrogen to ammonium nitrogen, and then analysis for total nitrogen was carried out as outlined for ammonium nitrogen. Phosphate phosphorus was measured following the ascorbic acid method as outlined in [19].

Samples for determination of phytoplankton were collected from subsurface water. The water samples (25 mLs) were preserved in acidic Lugol's solution. Phytoplankton species identification and enumeration was done using inverted microscope at 400x magnification. Phytoplankton taxa were identified using the methods of [20]. Phytoplankton densities were estimated by counting all the individuals whether these organisms were single cells, colonies, or filaments. The resulting counts were used to calculate the algal density and expressed in cells/mL. The quantification of chlorophyll *a* was performed following the methods outlined in [21]. Chlorophyll *a* content of the water was determined in ug/L and algal densities in individuals, colonies, or filaments per litre.

2.3. Macrophyte Diversity Studies. Subjective and quantitative techniques were employed in the studies of aquatic vegetation or macrophytes. Transects were taken at different zones and percentage, cover was estimated with 1×1 m quadrats. Plants with diagnostic features such as flowers, fruits, shoots, and rhizomes were collected and correctly pressed and labeled with a brief habitat description and associated taxa. Macrophyte species occurring in the various sites were recorded, and the sites at which they occurred were marked by GPS. Identification of aquatic macrophytes was carried out by use of keys of [22–25]. Photographs were also taken using an HSC- 5 50 Sony model digital camera.

Water Hyacinth Eichhornia crassipes (Mart.) Solms-Laubach Dynamics and Succession in the Nyanza Gulf of Lake Victoria (East Africa): Implications for Water Quality and Biodiversity Conservation

3

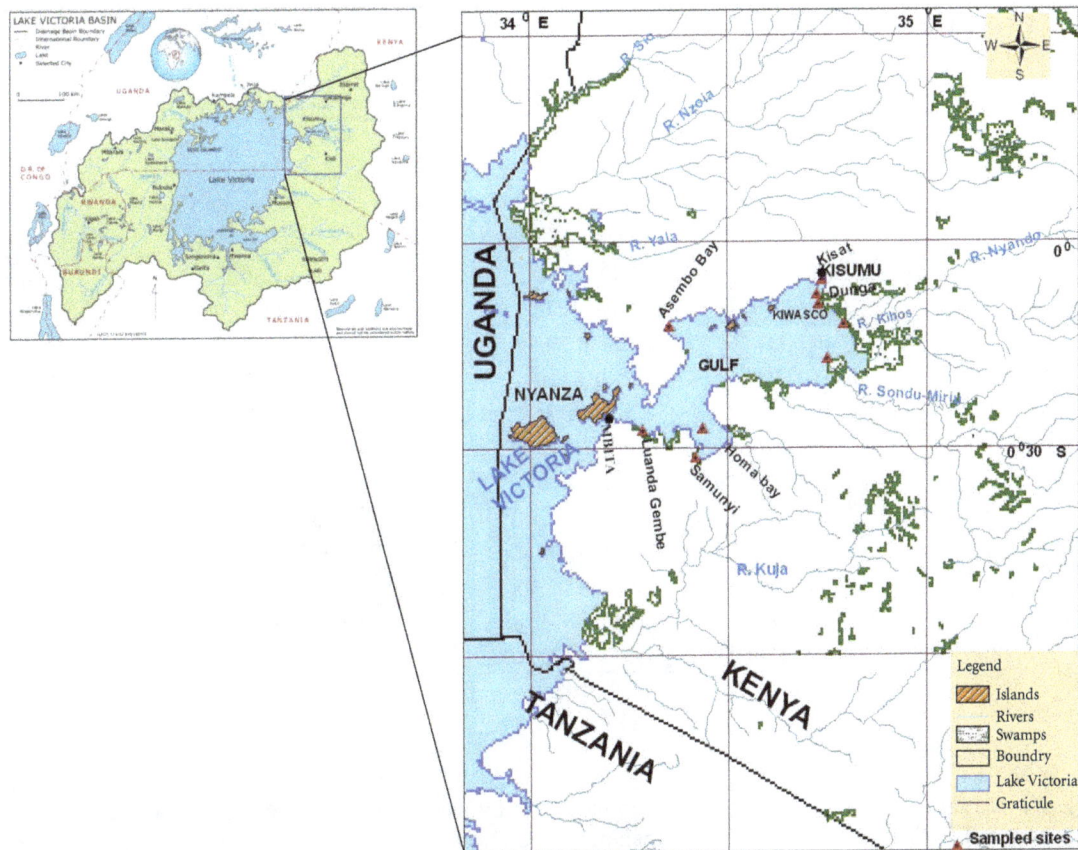

FIGURE 1: Map of the Nyanza Gulf, Lake Victoria showing the sampling sites.

2.4. Invertebrates and Fish Associated with Macrophytes. Invertebrates were sampled with a 30 × 30 cm "kick-net" with a 0.5-mm mesh size by sweeping under the hyacinth mats or hippo grass. Snails were then separated from the collected root mass by vigorously shaking each root sample in a bucket containing 10% isopropyl alcohol, causing them to detach from the roots. Samples were sorted in a white plastic tray with clear water. The snails were identified in the field using taxonomic reference and taken to the laboratory for confirmatory identification. The assessment of the effect of the relative abundance of fish associated with the macrophytes was done by sampling using an electrofisher. Electrofishing activity was carried out using a Septa model unit which discharges voltages of up to 600 volts with accompanying Amperes of between 5 to 30 Amps. A pulsed mode of discharge was adopted for electrocution lasting 10 minutes at each attempt. Species identification followed descriptions given by [26].

3. Results and Discussion

3.1. Physicochemical Parameters. The physical and chemical characteristics of the sampled sites are given in Table 1.

Results of the environmental parameters showed that dissolved oxygen ranged from $2.0\,mgL^{-1}$ at Samunyi (Homa Bay) to $9.5\,mgL^{-1}$ at Asembo Bay. Conductivity ranged from $156.6\,\mu Scm^{-1}$ at Lwanda Gembe to $176.0\,\mu Scm^{-1}$ at Samunyi (Homa Bay). The sampled sites from the different habitats had low transparency but elevated turbidity, total phosphorus (TP), total nitrogen (TN), chlorophyll a, and algal counts. The high nutrient concentrations provoked the proliferation of algal blooms in the hyacinth and hippo grass habitats. There was significantly higher turbidity in the hippo grass habitats compared to the water hyacinth and open water habitats ($P = 0.009$, one way ANOVA).

3.2. Macrophyte Inventory. Results of the inventory of macrophytes are shown in Table 2. An inventory of the macrophytes during the succession revealed that the dominant macrophytes were hippo grass *Vossia cuspidata* and water hyacinth *Eichhornia crassipes*. Other important aquatic plants identified were *Ipomoea aquatica*, *Enydra fluctuans*, *Cyperus papyrus* L., and *Aeschynomene elaphroxylon* (Guill. and Perr.) Taub. *Azolla pinnata* Decne ex Mett. and *Lemna* sp. were only observed in areas where hippo grass had been cut while fishing for *Clarias gariepinus* juvenile, used as bait for Nile perch *Lates niloticus* fishery.

TABLE 1: Variations (Mean ± SD) in the physicochemical parameters between the habitats of water hyacinth complex. Values in parenthesis indicate the range.

Habitat parameter	Water hyacinth	Hippo grass	Mixed (water hyacinth and hippo grass)	Open water
Secchi depth (m)	0.35 ± 0.13	0.23 ± 0.10	0.31 ± 0.14	0.42 ± 0.12
	(0.10–0.45)	(0.15–0.35)	0.15–0.45	(0.25–0.60)
Temp (°C)	28.50 ± 2.39	26.80 ± 2.33	28.00 ± 1.77	27.70 ± 2.10
	(25.70–31.90)	(24.20–28.70)	(25.90–30.20)	(25.50–31.8)
pH	7.80 ± 0.82	7.40 ± 1.36	7.60 ± 0.40	7.80 ± 0.47
	(6.60–8.80)	(5.90–8.50)	(7.20–8.10)	(7.20–8.50)
Turbidity (NTU)	112.20 ± 111.79	461.30 ± 360.13	152.80 ± 121.89	71.1 ± 24.51
	(54.00–339.00)	(77.00–791.00)	(78.30–335.00)	(47.30–115.00)
TP (μgP/L)	538.70 ± 823.37	486.9 ± 444.77	391.90 ± 147.04	208.10 ± 47.78
	(146.90–2216.90)	(179.70–996.90)	(179.70–511.10)	(149.70–284.00)
TN (μgN/L)	975.20 ± 796.49	542.4 ± 139.30	587.80 ± 157.18	633.30 ± 118.73
	(750.40–2584.20)	(517.30–692.50)	(396.30–726.70)	(485.70–834.60)
Chloro-a (μg/L)	9.90–5989.40	231.90 ± 367.83	301.80 ± 245.82	89.95 ± 76.81
	(1045.90 ± 1045.00)	(19.40–656.70)	(11.20–612.30)	(9.90–200.20)
Algal densities (individuals/cells/colonies/L)	12172.5 ± 16177.38)	6914.67 ± 9340.42	8058.25 ± 5731.59	4772.12 ± 4264.31
	(700.00–36120.00)	(1493.00–17700.00)	(1652.00–15522.00)	(1116.00–13638.00)

3.3. *Macrophyte Succession.* Several stages of macrophyte succession were noted during the study. These were then categorized as pure water hyacinth population, early, mid, late, and climax stages of succession. During pure water hyacinth stage, the plant community was entirely a population of water hyacinth (Figure 2). At the time of the study, such zones were rare and were only found at Homa Bay, Lwanda Gembe, and Asembo Bay on the shoreline in hotspots with high nutrient concentrations. Areas covered were also limited with each population cover rarely exceeding 500 m^{-2}.

In the early stages of water hyacinth infestation, the weed takes foothold on the shoreline in the areas where native aquatic plants thrive. The early stages of water hyacinth succession occur in this zone and start when the pure water hyacinth mats were invaded by a plethora of opportunistic (usually emergent macrophytes) native invaders. The first common invaders were observed to be *Ipomoea aquatica* followed by *Enydra fluctuans* and an unidentified macrophyte of the Commelinaceae family. These are emergent runners which venture into the lake by creeping on the water hyacinth plants. In some areas, it is at this stage that scanty shoots of *Vossia cuspidata* started to appear among the other plants (Figure 3).

During the midstages of succession, the invader native aquatic plants were found to coexist within water hyacinth. It is at this stage that we observed hippo grass shoots within the macrophyte community consisting of the water hyacinth, *I. aquatica*, *E. fluctuans*, and a macrophyte of Commelinaceae family. With the increase of the opportunistic emergent invaders, there was an observed decrease in the proportion of the water hyacinth in the mat. During this stage, hippo grass had established itself within the community. Although the hippo grass is an emergent plant, its survival on water is by use of the water hyacinth as a substrate while proliferation within the community is because of the nutrients and detritus of the decaying water hyacinth. The buoyance of the hippo grass is provided by the water hyacinth biomass. The results of this succession are that the proportion of the water hyacinth decreases further in the mat. At the climax stage the water hyacinth is fully covered by the hippo grass owing to the fact that the hippo grass grows to height of 1.5 meters, while the water hyacinth grows to a height of 0.5 m (Figure 4).

The taller hippo grass shades the water hyacinth, *Ipomoea aquatic*, and *Enydra fluctuans* from sunlight. The *Ipomoea aquatica*, however, evades the shading effects of the hippo grass by climbing/twinning itself around the hippo grass. The shaded water hyacinth and *Enydra fluctuans* die off due to lack of sunlight contributing significantly to the organic matter (rich in nutrients) which fuels more proliferation of the hippo grass.

During heavy storms and wind activity, the population of the hippo grass is sloughed off the sheltered bays into the lake resulting into floating islands. The sloughing is aided by the compact mass of the hippo grass and its height (Figure 5).

After the nutrient-rich heterotrophic layer substrate of the dying water hyacinth is exhausted, the hippo grass (now existing as floating islands in the lake) starts to die off since it cannot extract nutrients from the water as it is an emergent plant living on the shoreline extracting the nutrients from the substrate in the littoral zones of the lake (Figure 6).

The few mats of water hyacinth existing under the mat do not sink with the mat but float out into the open water to start the new colonies of the water hyacinth mats. Observations from trawl surveys indicated that large fragments of fleshly sunk hippo grass had sunk at the bottom of the lake.

Water Hyacinth Eichhornia crassipes (Mart.) Solms-Laubach Dynamics and Succession in the Nyanza Gulf of Lake Victoria (East Africa): Implications for Water Quality and Biodiversity Conservation

5

TABLE 2: An inventory of the macrophyte vegetation and the proportions (%) observed at the sampled sites.

Station	Vegetation	Proportion (%)
Dunga beach	Hippo grass *Vossia cuspidata* (Roxb.) Griff.	98
	Water hyacinth *Eichhornia crassipes* (Mart.) Solms-Laubach	1.5
	Nile cabbage/water lettuce *Pistia stratiotes* L.	0.5
Kibos	Water hyacinth *Eichhornia crassipes*	50
	Hippo grass *Vossia cuspidata*	50
Sondu Miriu at interface zone	Water hyacinth *Eichhornia crassipes*	50
	Hippo grass, *Vossia cuspidata*	50
Homa Bay at Samunyi—A	Water hyacinth, *Eichhornia crassipes*	90
	Swamp cabbage/water spinach *Ipomoea aquatica var. aquatica* Forsk.	8
	Enydra fluctuans Lour.	1.5
	Hippo grass *Vossia cuspidata*	0.5
Homa Bay at Samunyi—B	Water hyacinth *Eichhornia crassipes*	90
	Swamp cabbage/water spinach *Ipomoea aquatica var. aquatica*	2
	Enydra fluctuans Lour.	8
Homa Bay at Samunyi—C	Water hyacinth *Eichhornia crassipes*	60
	Hippo grass *Vossia cuspidata*	30
	Enydra fluctuans Lour.	8
	Swamp cabbage/water spinach *Ipomoea aquatica var. aquatica*	2
Homa Bay (Floating mat)	Hippo grass *Vossia cuspidata*	85
	Water hyacinth *Eichhornia crassipes*	10
	Enydra fluctuans	2
	Swamp cabbage/water spinach *Ipomoea aquatica var. aquatica*	1
	Common papyrus *Cyperus papyrus* L.	1
	Ambach tree *Aeschyonomene elaphroxylon* (Guill. and Perr.) Taub.	1
Homa Bay offshore (floating mat)	Water hyacinth *Eichhornia crassipes*	99
	Swamp cabbage/water spinach *Ipomoea aquatica var. aquatica*	1
Oluch river mouth	Hippo grass *Vossia cuspidata*	99
	Water hyacinth *Eichhornia crassipes*	1
Lwanda Gembe—A	Water hyacinth *Eichhornia crassipes Ipomoea aquatica var. aquatic*	95
	Swamp cabbage/water spinach	1
	Commelinaceae	2
	Enydra fluctuans	2
Lwanda Gembe—B	Unidentified macrophyte—B	90
	Water hyacinth *Eichhornia crassipes*	5
	Polygonum setosulum Meisn.	5
Lwanda Gembe—C	Water hyacinth *Eichhornia crassipes*	80
	Enydra fluctuans	15
	Unidentified macrophyte—B	4
	Polygonum setosulum	1
Lwanda Gembe—D	Water hyacinth *Eichhornia crassipes*	90
	Hippo grass *Vossia cuspidata*	5
	Enydra fluctuans	5
Asembo Bay—A	Hippo grass *Vossia cuspidata*	100
Asembo Bay—B	Hippo grass *Vossia cuspidata*	80
	Swamp cabbage/water spinach *Ipomoea aquatica var. aquatica*	20
Asembo Bay—C	Hippo grass *Vossia cuspidata*	80
	Water hyacinth *Eichhornia crassipes*	20

TABLE 2: Continued.

Station	Vegetation	Proportion (%)
Asembo Bay—D	Hippo grass *Vossia cuspidata*	90
	Water hyacinth *Eichhornia crassipes*	9
	Swamp cabbage/water spinach *Ipomoea aquatica var. aquatica*	1
Asembo Bay—E	Water hyacinth *Eichhornia crassipes*	50
	Hippo grass *Vossia cuspidata*	20
	Water fern, Water velvet *Azolla pinnata*	15
	Enydra fluctuans	14
	Duckweed *Lemna* sp.	1
Asembo Bay	Water hyacinth *Eichhornia crassipes*	90
	Hippo grass *Vossia cuspidata*	10

FIGURE 2: A pure population of the water hyacinth.

FIGURE 4: The late stages of macrophyte succession showing hippograss replacing the water hyacinth.

FIGURE 3: Water hyacinth mat showing encroachment by *Ipomoea aquatica* and *Enydra fluctuans*.

FIGURE 5: The climax stages of the macrophyte succession showing a single population of hippograss.

Initial estimates revealed that more than 3,000 hectares of hippo grass could have sunk to the bottom of the lake (Figure 7).

Results on the distribution of invertebrates are shown in Figure 8. The study found a strong association of snails associated with aquatic macrophytes including *Biomphalaria sudanica* and *Bulinus africanus,* the two most common hosts for schistosomiasis in the Nyanza Gulf of Lake Victoria.

The inventory and abundance composition of fish species from areas covered by water hyacinth, hippo grass, or both varied from that of the open water obtained using trawls.

In this habitat, *Clarias gariepinus* populations dominated, contributing up to 48.6% of the biomass. *Oreochromis niloticus* is the second most abundant species, contributing 40.6% of the biomass (Figure 9). An endemic species like

Water Hyacinth Eichhornia crassipes (Mart.) Solms-Laubach Dynamics and Succession in the Nyanza Gulf of
Lake Victoria (East Africa): Implications for Water Quality and Biodiversity Conservation

7

FIGURE 6: The sinking mats of hippograss after collapse of the water hyacinth substrate.

FIGURE 7: Fragments of the hippograss at the bottom of the lake as obtained from the bottom trawls of the sampled sites.

Labeo victorianus was also found under the mats with a composition of 4.3% of the total catch.

The most abundant species in the open lake was *Lates niloticus* which accounted for 51.4% of the biomass. *O. niloticus* contributed 30.9% of the total trawl biomass (Figure 10). *Clarias gariepinus* is the 3rd most abundant fish species except for *R. argentea* and *Caridina niloticus* (both captured in the codend 10 mm mesh). From the aforementioned, it is apparent that macrophyte succession has the capacity to alter aquatic biodiversity in the lake. The most important factor contributing to these changes was dissolved oxygen.

The main anoxic tolerant species are *C. gariepinus*, *O. niloticus*, *Synodontis* spp., and *P. aethiopicus*, while *Lates niloticus* prefers areas with high oxygen levels.

Before 1970s, the shoreline wetlands of Lake Victoria were dominated by the emergent *C. papyrus*. During this period, establishment of submerged plants in the inshore areas was hindered by the constant disturbance of the substrate by the once dominant haplochromine fish species. For instance, during the peak biomass of haplochromines in Lake

Victoria, before 1960s, the fishes hindered the establishment of macrophytes in the inshore areas through enhancement of turbidity by disturbing the bottom substrate and sediments [27]. These papyrus dominated mats frequently sloughed off river mouths during the rainy seasons and formed floating islands in the offshore areas. *Vossia cuspidata* was mainly found at the banks of potamon sections of major affluent rivers. Before the invasion of water hyacinth in the 1990s, the only other free-floating macrophytes in the lake were *Pistia stratiotes, Azolla pinnata, A. nilotica*, and *Lemna* sp. The common floating leafed species included *Trapa natans* and *Nymphaea lotus* while the submerged were *Potamogeton schweinfurthii, Vallisneria spiralis* L., *Ceratophyllum demersum* L., and *Najas horrida*. According to [28] the aquatic macrophyte, communities of the wetlands are a part of a vegetation continuum from land to below water. Frequently, though there are distinct zones of vegetation along the continuum. In this study, approaching the wetland from dry land, the dominant macrophytes were found to consist of *Phragmites australis* and stands of *Typha domingensis*. Immediately after this, the *Cyperus papyrus* was the most dominant consisting mainly of mono specific stands intertwined with creepers such as *Vigna lutea* (Del.) Hook. and *Ipomoea aquatica var. aquatica*. Next in line from the papyrus zone was a thin strip of hippograss *Vossia cupsidata* interspersed with *Echinochloa pyramidalis*. The lake-swamp interface zone was colonized by the floating mats of *Eichhornia crassipes*. On establishment, the massive mats of the hyacinth overwhelmed these other macrophytes by competing for nutrients, smothering, and cutting out sunlight. The impact of water hyacinth could have led to the extinction of *Azolla nilotica* which was common at the mouth of River Nyando [11, 29]. Results from this study showed that water hyacinth is the genesis of plant succession phenomenon in the Nyanza Gulf of Lake Victoria. Water hyacinth established itself in the Nyanza Gulf after the short rains of 2006 (Kusewa pers. comm). The main control of water hyacinth is through biological method using *Neochetina eichhorniae* and *Neochetina bruchi*. The dying plants of the water hyacinth provide a substrate for emergent macrophytes, and this results in rapid increase of these especially *Vossia cuspidata, Ipomoea, aquatica, Enydra fluctuans*, and unidentified macrophyte of the family Commelinaceae. Our observation revealed that the macrophytes that initiate the colonization on the water are *Enydra fluctuans* and *Ipomoea aquatica* these plants are closely followed by the hippograss. Initially, the hippograss seems to coexist with the water hyacinth in a mutually beneficial association such that both plants extend their range along the shore and out into the open water. The mats that develop are mainly dominated by opportunistic native plants such as *Vossia cuspidata, Ipomoea aquatica var. aquatica, Enhydra fluctuans*, and a macrophyte of the Commelinaceae family. The flourishment of the floating islands is nourished by the nutrients from the water hyacinth substrate, and the cohesion is provided by the underwater biomass of the water hyacinth. According to [30], freshwater hyacinth plant contains high levels of nutrients. These nutrients enhance the proliferation of the hippo grass and its associated vegetation. According to [17], light is an

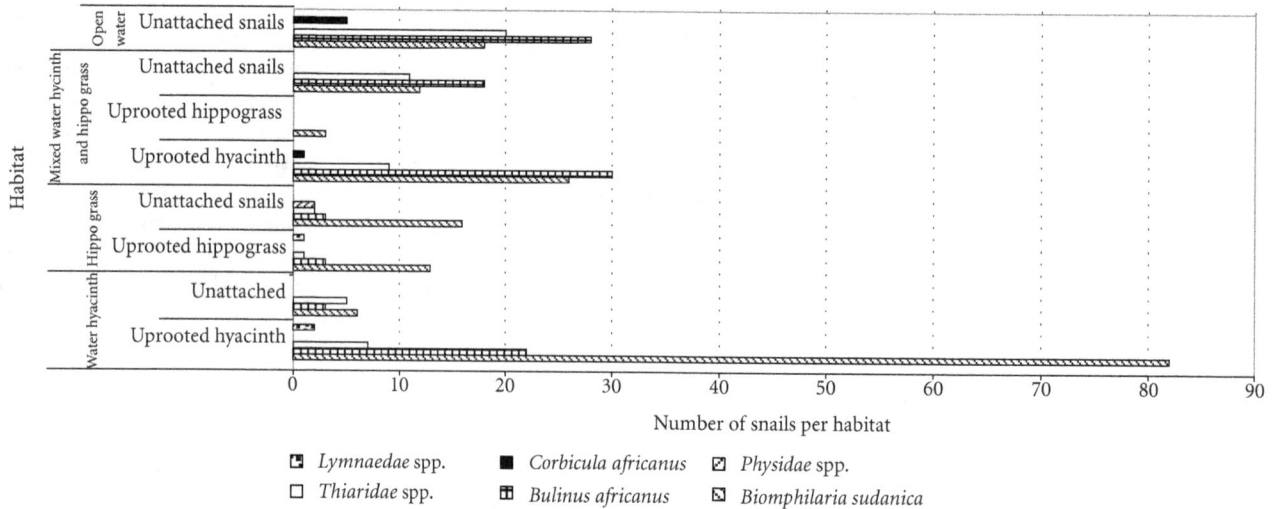

FIGURE 8: Relative abundance of different invertebrate species collected in the habitats of water hyacinth and hippo grass and in open water at different sites within the Nyanza Gulf of Lake Victoria.

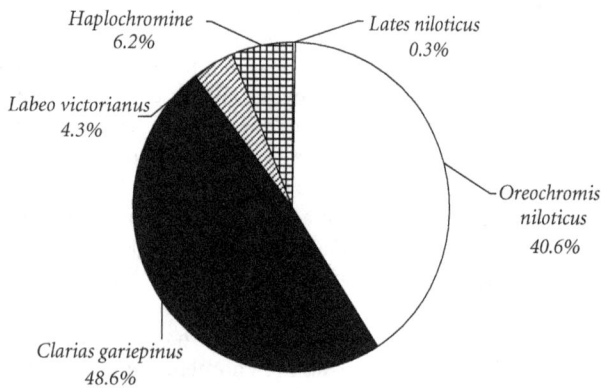

FIGURE 9: % composition of the fish species in the areas covered by water hyacinth complex.

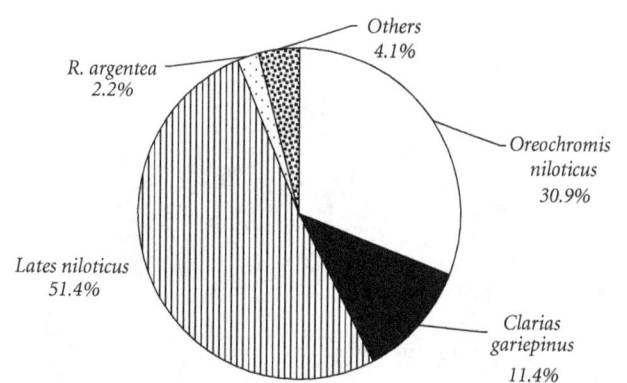

FIGURE 10: Percent (%) composition of the fish species in the open water areas.

important limiting factor to water hyacinth growth. Light becomes nonlimiting to CO_2 uptake at a photosynthetic active radiation (PAR) of 2000 $\mu E\ m^{-2}\ s^{-1}$. In Lake Victoria, this light level occurs for about 6 h around midday. [15, 17], associated the decline of the water hyacinth in 1998 in Lake Victoria to climatic perturbations leading to the decreased 9 light intensity occasioned by the cloudy, wet El Nino weather phenomenon of 1997-1998 [31]. Prolonged suboptimal light will reduce growth and reproduction rates and relatively increase the effect of other debilitating influences such as other weather-related factors, for example, water level, wave action, water quality, temperature, and humidity as well as weevil herbivory and phytopathogenic attack [17, 18, 32]. In this study, we observed that the reduced light conditions were majorly caused by the shading effects of the water hyacinth by the hippo grass. Whereas the water hyacinth is a perennial aquatic herb; rhizome and stems normally floating, rooting at the nodes, with long black pendant roots, it only grows generally to a height of 0.5 m [33]. The hippo grass is perennial with submerged or floating culms.

It has a two to six on a short axis, or solitary 15–22 cm long; sessile spikelets up to 10 mm long, lower glume of both spikelets with a winged tail 5–30 mm long, rarely shorter. The plant can grow to a total height of 1.5 metres. The taller hippo grass shades the water hyacinth, Ipomoea aquatica, Enydra fluctuans, and a macrophyte of the Commelinaceae family from sunlight, heat, and UV radiation. The Ipomoea aquatica, however, evades the shading effects of the hippo grass by climbing/twinning itself around the hippo grass. The shaded water hyacinth and Enydra fluctuans die off due to lack of sunlight contributing significantly to the organic matter (rich in nutrients) which fuels the more proliferation of the hippo grass. During heavy storms, strong wind activity, or currents, the population of the hippo grass is sloughed off the sheltered bays into the lake resulting into floating islands. The sloughing is aided by the compact mass of the hippo grass and its height. The fate of the floating islands is determined by the time; it takes for total disintegration of the heterotrophic substrate. After the nutrient-rich heterotrophic layer substrate of the dying water

Water Hyacinth Eichhornia crassipes (Mart.) Solms-Laubach Dynamics and Succession in the Nyanza Gulf of
Lake Victoria (East Africa): Implications for Water Quality and Biodiversity Conservation

9

hyacinth is exhausted, the hippo grass (now existing as floating islands in the lake) starts to die and sink to the bottom of the lake.

4. Conclusion

This study provides compelling evidence to show that macrophyte succession has the capacity to alter aquatic biodiversity in the lake. For instance, areas dominated by water hyacinth and hippo grass were found to be anoxic and dominated by anoxic tolerant species such as *C. gariepinus*, *O. niloticus*, *Synodontis* spp., and *P. aethiopicus*. Areas without water hyacinth were dominated by *Lates niloticus*. In addition, the succession threatened endemic macrophytes such as the free-floating *Pistia stratiotes* and *Azolla pinnata*.

Water hyacinth infestation is a symptom of broader watershed management and pollution problems in the Nyanza Gulf of Lake Victoria. Methods for water hyacinth control should include reduction of nutrients in the water bodies. This can be achieved through treatment of waters flowing from sewage works and factories at least up to tertiary level using constructed wetlands and reducing the diffuse loadings by changing land use practices as well as development of holistic management of pollution involving land, water, and air shed. There is need to implement an integrated approach to water hyacinth management in which biological control agents play the central role with leverages on manual/mechanical controls in a multifaceted approach. Research indicates that the present control programmes are expensive and do not provide any returns on investments. Most are self-propagating and return nutrient back into the system. The paradigm shift is emphasized for water hyacinth control through harvesting and utilization which is eco-friendly. Communities have expressed willingness to participate in community-based water hyacinth control strategies. The most practical approach is to involve them in manual and biological control activities, for example, in rearing weevils. Some little incentives may be necessary to leverage for lost time. The most sustainable method however includes organizing communities into cottage industrial production units with water hyacinth as raw material, and this is value added because an economic use is found for an unwanted plant. In this way, water hyacinth can provide a substitute for bulking agents, reducing the requirements for expensive goods, and generating income (through creation of employment, generating income) and improve standards of living from sale of byproducts providing alternative livelihoods and relieving the pressure from the fishery, forestry, and energy.

Conflict of Interests

The authors declare that they have no conflict of interests.

Acknowledgments

The authors would like to acknowledge the contributions of a number of scientists and technical personnel who contributed in this work namely Dr. Henry Lungaiya of Masinde Muliro University of Technology, Ms. Mwende Kusewa of Kenya Agricultural Research Institute, Carolyne Lwenya and John Ouko of Kenya Marine and Fisheries Research Institute. They would also like to thank the fisher folk from Usoma beach, Rakwaro beach, Sango Rota and Asembo Bay who took time of their busy schedules to participate in study. Finally, they wish to thank profusely the Russel Fund EFN Professional Development Grant for accepting to fund the first author to attend and make a presentation at the 13th World Lakes Conference in Wuhan, China from November 1–5, 2009. Thanks are due to Dr. Mercedes Gonzalez for this kind gesture. The study was funded by the Lake Victoria Basin Commission based in Kisumu, KENYA.

References

[1] C. D. Sculthorpe, *The Biology of Aquatic Vascular Plants*, Arnold, London, UK, 1967.

[2] L. G. Holm, D. L. Plucknett, J. V. Pancho, and J. P. Herberger, *The World's Worst Weeds, Distribution and Biology*, Krieger Publishing Co., Malabar, Fla, USA, 1991.

[3] W. T. Penfound and T. T. Earle, "The biology of water hyacinth," *Ecological Monographs*, vol. 18, pp. 447–472, 1948.

[4] T. Twongo, "Status of water hyacinth in Uganda," in *Control of Africa's Floating Water Weeds*, A. Greathead and P. de Groot, Eds., pp. 55–57, Commonwealth Science Council, Zimbabwe, 1991.

[5] T. Twongo and O. K. Ondongkara, "Invasive water weeds in Lake Victoria basin: proliferation, impacts, and control, 2000," Lake Victoria, Lake Victoria Fisheries Organization, 2000.

[6] T. Twongo and J. S. Balirwa, "The water hyacinth problem and the biological control option in the highland lake region of the upper Nile basin—Uganda's experience. 1995," in *Proceedings of the Nile 2002 Conference: Comprehensive Water Resources Development of the Nile Basin—Taking off*, Arusha, Tanzania, 2002.

[7] T. Twongo and T. O. Okurut, "Control of water hyacinth in Lake Victoria, challenges in the resurgence succession," in *Proceedings of the Science Policy Linkages Workshop*, Imperial Botanical Gardens, Entebbe, Uganda, September 2008, Sponsored by the UNU-INWEH in collaboration with the LVBC.

[8] A. B. Freilink, "Water hyacinth in Lake Victoria: a report on an aerial survey," in *Proceedings of the a National Workshop on Water Hyacinth in Uganda*, K. Thompson, Ed., Ecology, Distribution, Problems and Strategy for Control, FAO, Kampala, Uganda, October, 1991, TCP/UGA/9153/A.

[9] P. Bwathondi and G. Mahika, "A report of the infestation of water hyacinth (*Eichhornia crassipes*) along the Tanzanian side of Lake Victoria," National Environment Council, Dar es Salaam, Tanzania, 1994.

[10] A. M. Mailu, G. R. S. Ochiel, W. Gitonga, and S. W. Njoka, "Water hyacinth: an environmental disaster in the Winam Gulf of Lake Victoria and its control," in *Proceedings of the 1st IOBC Global Working Group Meeting for the Biological and Integrated Control of Water Hyacinth*, M. P. Hill, M. H. Julien, and T. D. Center, Eds., pp. 101–105, Weeds Research Division, ARC, South Africa, 1999.

[11] A. Othina, R. Omondi, J. Gichuki, D. Masai, and J. Ogari, "Impact of water hyacinth, *Eichhornia crassipes*, (Liliales: Ponteridiaceae) on other macrophytes and Fisheries in the Nyanza Gulf of Lake Victoria," in *Lake Victoria Fisheries: Status, Biodiversity and Management*, M. van der Knaap and M. Munawar,

Eds., Aquatic Ecosystem health and Management Society, 2003, http://www.aehms.org/glow_lake_victoria.html.

[12] A. R. D. Taylor, "Floating water-weeds in East Africa," in *Control of Africa's Floating Water Weeds*, A. Greathead and P. de Groot, Eds., pp. 111–124, Commonwealth Science Council, Zimbabwe, 1991.

[13] T. Twongo, F. W. B. Bugenyi, and F. Wanda, "The potential for further proliferation of water hyacinth in Lakes Victoria, Kyoga and Kwania and some urgent aspects for research," *African Journal of Tropical Hydrobiology and Fisheries*, vol. 6, pp. 1–10, 1995.

[14] A. R. D. Taylor, "Floating water-weeds in East Africa, with a case study in Northern Lake Victoria," in *Proceedings of the Control of Africa's Floating Water Weeds*, A. Greathead and P. J. de Groot, Eds., vol. 295, no. 93, AGR-18, p. 187, Commonwealth Science Council, 1993.

[15] A. E. Williams, R. E. Hecky, and H. C. Duthie, "Water hyacinth decline across Lake Victoria—was it caused by climatic perturbations or biological control? A reply," *Aquatic Botany*, vol. 87, no. 1, pp. 94–96, 2007.

[16] M. H. Julien, M. W. Griffiths, and A. D. Wright, *Biological Control of Water Hyacinth. The Weevils Neochetina Bruchi and N. Eichhorniae: Biologies, Host Ranges and Rearing, Releasing and Monitoring Techniques for Biological Control of Eichhornia Erassipes*, ACIAR Monograph no. 60, Australian Centre for International Agricultural Research, 1999.

[17] A. E. Williams, H. C. Duthie, and R. E. Hecky, "Water hyacinth in Lake Victoria: why did it vanish so quickly and will it return?" *Aquatic Botany*, vol. 81, no. 4, pp. 300–314, 2005.

[18] T. P. Albright, T. G. Moorhouse, and T. J. McNabb, "The rise and fall of water hyacinth in Lake Victoria and the Kagera River basin, 1989–2001," *Journal of Aquatic Plant Management*, vol. 42, pp. 73–78, 2004.

[19] R. G. Wetzel and G. E. Likens, *Limnological Analyses*, Springer, New York, NY, USA, 3rd edition, 2000.

[20] G. Huber-Pestalozzi, *Cryptophyceae, Chloromonadophyceae, Dinophyceae*, Das Phytoplankton des Susswassers, 1968.

[21] J. D. H. Strickland and T. R. Parsons, *A Practical Handbook of Seawater Analysis*, Canadian Government, Publishing Centre, Ottawa, Canada, 1972.

[22] A. D. Q. Agnew and S. Agnew, *Upland Kenya Wild Flowers: A Flora of the Ferns and Herbaceous Flowering Plants of Upland Kenya*, East African Natural History Society, 2nd edition, 1994.

[23] C. D. K. Cook, E. M. Gut, E. M. Rix, J. Schneller, and M. Seitz, *Water Plants of the World: A Manual of the Identification of the Genera of Freshwater Plants*, Dr. W. Junk Publishers, The Hague, The Netherlands, 1974.

[24] J. O. Kokwaro and T. Johns, *Luo Biological Dictionary*, East African Educational Publishers, Nairobi, Kenya, 1998.

[25] G. R. Sainty and S. W. L. Jacobs, *Water Plants of Australia*, Royal Botanical Gardens, Sydney, Australia, 3rd edition, 1994.

[26] F. Witte and M. J. P. van Oijen, "Taxonomy, ecology and fishery of Lake Victoria haplochromine trophic groups," *Zool Verh Leiden*, vol. 262, no. 1, pp. 1–47, 1990.

[27] F. Witte, T. Goldschmidt, P. C. Goudswaard, W. Ligtvoet, M. J. P. van Oijen, and J. H. Wanink, "Species extinction and concomitant ecological changes in Lake Victoria," *Netherlands Journal of Zoology*, vol. 42, no. 2-3, pp. 214–232, 1991.

[28] D. Denny, "African wetlands," in *Wetlands*, M. Finalyson and M. Moser, Eds., pp. 115–148, International Waterfowl and Wetlands Research Bureau (IWRB), Facts on File Ltd, Oxford, UK, 1991.

[29] R. Omondi and M. Kusewa, "Macrophytes of Lake Victoria, Kenya and succession after invasion of water hyacinth," in

Proceedings of the 11th World Lake Conference, pp. 600–602, October 2005.

[30] S. Matai and D. K. Bagchi, "Water hyacinth: a plant with prolific bioproductivity and photosynthesis," in *Proceedings of the Proceedings of the International Symposium on Biological Applications of Solar Energy*, A. Gnanam, S. Krishnaswamy, and J. S. Kahn, Eds., pp. 144–148, Macmillan, India, 1980.

[31] A. Anyamba, C. J. Tucker, and J. R. Eastman, "NDVI anomaly patterns over Africa during the 1997/98 ENSO warm event," *International Journal of Remote Sensing*, vol. 22, no. 10, pp. 1847–1859, 2001.

[32] J. R. U. Wilson, O. Ajuonu, T. D. Center et al., "The decline of water hyacinth on Lake Victoria was due to biological control by *Neochetina* spp," *Aquatic Botany*, vol. 87, no. 1, pp. 90–93, 2007.

[33] G. Gopal, *Water Hyacinth: Aquatic Plant Studies*, Elsevier Science, Amsterdam, The Netherlands, 1987.

2

Toxicity of Metals to a Freshwater Snail, *Melanoides tuberculata*

M. Shuhaimi-Othman, R. Nur-Amalina, and Y. Nadzifah

School of Environmental and Natural Resource Sciences, Faculty of Science and Technology, National University of Malaysia (UKM), Selangor, 43600 Bangi, Malaysia

Correspondence should be addressed to M. Shuhaimi-Othman, shuhaimi@ukm.my

Academic Editors: T. Brock, K. Kannan, J. Ruelas-Inzunza, and B. C. Suedel

Adult freshwater snails *Melanoides tuberculata* (Gastropod, Thiaridae) were exposed for a four-day period in laboratory conditions to a range of copper (Cu), cadmium (Cd), zinc (Zn), lead (Pb), nickel (Ni), iron (Fe), aluminium (Al), and manganese (Mn) concentrations. Mortality was assessed and median lethal times (LT_{50}) and concentrations (LC_{50}) were calculated. LT_{50} and LC_{50} increased with the decrease in mean exposure concentrations and times, respectively, for all metals. The LC_{50} values for the 96-hour exposures to Cu, Cd, Zn, Pb, Ni, Fe, Al, and Mn were 0.14, 1.49, 3.90, 6.82, 8.46, 8.49, 68.23, and 45.59 mg L^{-1}, respectively. Cu was the most toxic metal to *M. tuberculata*, followed by Cd, Zn, Pb, Ni, Fe, Mn, and Al (Cu > Cd > Zn > Pb > Ni > Fe > Mn > Al). Metals bioconcentration in *M. tuberculata* increases with exposure to increasing concentrations and Cu has the highest accumulation (concentration factor) in the soft tissues. A comparison of LC_{50} values for metals for this species with those for other freshwater gastropods reveals that *M. tuberculata* is equally sensitive to metals.

1. Introduction

Metals are released from both natural sources and human activity. The impact of metals on the environment is an increasing problem worldwide. The impact of metals on aquatic ecosystems is still considered to be a major threat to organisms health due to their potential bioaccumulation and toxicity to many aquatic organisms. Although metals are usually considered as pollutants, it is important to recognize that they are natural substances. Zinc, for example, is an essential component of at least 150 enzymes; copper is essential for the normal function of cytochrome oxidase; iron is part of the haemoglobin in red blood cells; boron is required exclusively by plants [1]. Malaysia, as a developing country, is no exception and faces metals pollution caused especially by anthropogenic activities such as manufacturing, agriculture, sewage, and motor vehicle emissions [2–5]. Metals are nonbiodegradable. Unlike some organic pesticides, metals cannot be broken down into less harmful components. Managing metal contamination requires an understanding of the concentration dependence of toxicity. Dose-response relationships provide the basis for the assessment of hazards and risks presented by environmental chemicals. Toxicity

testing is an essential tool for assessing the effect and fate of toxicants in aquatic ecosystems and has been widely used as a tool to identify suitable organisms as a bioindicator and to derive water quality standards for chemicals. There are many different ways in which toxicity can be measured, and most commonly the measure (end point) is death [1, 6, 7]. Metals research in Malaysia, especially using organisms as a bioindicator, is still scarce. Therefore, it is important to conduct studies with local organisms that can be used to gain data on metal toxicity, to determine the organism's sensitivity and to derive a permissible limit for Malaysian's water that can protect the local aquatic communities.

The freshwater molluscs of the Malaysian region are common, and most extant species are relatively easy to collect. The snails are rich fauna, while bivalve are the second. More than 150 aquatic nonmarine mollusc species have been recorded from the Malaysian region. *Melanoisdes tuberculata* (Müller 1774) is from class Gastropoda with shells higher than wide (elongate), conical, usually light brown in colour, and it is a cosmopolitan species [8]. *M. tuberculata* is a species of freshwater snail with an operculum, a parthenogenetic, aquatic gastropod mollusc in the family Thiaridae. The average shell length is about

20–27 mm and this species is native to subtropical and tropical northern Africa and southern Asia (Indo-Pacific region, Southern Asia, Arabia, and northern Australia), but they have established populations throughout the globe. The snail has an operculum that can protect it from desiccation and can remain viable for days on dry land [9]. It is a warm-climate species, prefers a temperature range of 18 to 32°C, and is primarily a burrowing species that tends to be most active at night. This snail feeds primarily on algae (microalgae) and acts as an intermediate host for many digenetic trematodes. *M. tuberculata* is a viviparous, gonochoric species with polyploid strains that reproduces by apomictic parthenogenesis. Because meiosis usually does not occur, offspring are identical to their mother. Females can be recognized by their greenish coloured gonads while males have reddish gonads. Under good conditions, females will produce fertilized eggs that are transferred to a brood pouch where they remain until they hatch. *M. tuberculata* will begin reproducing at a size as small as 5 to 10 mm in length and broods may contain over seventy offspring embryos which develop in the mother [10–12].

Molluscs have long been regarded as promising bioindicator and biomonitoring subjects. They are abundant in many terrestrial and aquatic ecosystems, being easily available for collection. They are highly tolerant to many pollutants and exhibit high accumulations of them, particularly heavy metals [13, 14]. Little information exists in the literatures concerning the toxic effects of metals for this snail. So far, only a few studies have been reported on metal toxicity to *M. tuberculata* [15, 16] and most of the studies were on the accumulation of metals [14, 17, 18]. Therefore, the purpose of this study was to determine the acute toxicity of eight metals (Cu, Cd, Zn, Pb, Ni, Fe, Al, and Mn) to the freshwater mollusc *M. tuberculata* and to examine the bioconcentration of these metals in the body after four days of exposure.

2. Materials and Methods

Snails *M. tuberculata* were collected from canals in the university in Bangi, Selangor, Malaysia. Identification of the species was based on Panha and Burch [8]. Prior to toxicity testing, the snails were acclimatized for one week under laboratory conditions (28–30°C with 12 h light : 12 h darkness) in 50-L stocking tanks using dechlorinated tap water (filtered by several layers of sand and activated carbon; T.C. Sediment Filter (TK Multitrade, Seri Kembangan, Malaysia)) aerated through an air stone. During acclimation the snails were fed on lettuce. The standard stock solution ($100 \, \text{mg} \, \text{L}^{-1}$) of Cu, Cd, Zn, Pb, Ni, Fe, Al, and Mn was prepared from analytical grade metallic salts of $CuSO_4 \cdot 5H_2O$, $CdCl_2 \cdot 2.5H_2O$, $ZnSO_4 \cdot 7H_2O$, $Pb(NO_3)_2$, $NiSO_4 \cdot 6H_2O$, $FeCl_3$, $Al_2(SO_4)_3 \cdot 18H_2O$, and $MnSO_4 \cdot H_2O$, respectively (Merck, Darmstadt, Germany). The stock solutions were prepared with deionized water in 1 L volumetric flasks. Acute Cu, Cd, Zn, Pb, Ni, Fe, Al, and Mn toxicity experiments were performed for a four-day period using adult snails (shell length approximately 1.5–2.0 cm, mean wet weight 22.5 ± 1.6 mg) obtained from stocking tanks.

Following a range finding test, five Cu, Cd, Zn, Pb, Ni, Fe, Al, and Mn nominal concentrations were chosen (Table 1). Metal solutions were prepared by dilution of a stock solution with dechlorinated tap water. A control with dechlorinated tap water only was also used. The tests were carried out under static conditions with renewal of the solution every two days. Control and metal-treated groups each consisted of two replicates of five randomly allocated snails in a 500 mL glass beaker containing 400 mL of the appropriate solution. No stress was observed for the snails in the solution, indicated by 100% survival for the snails in the control water until the end of the study. A total of 10 animals per treatment/concentration were used in the experiment and a total of 410 animals were employed in the investigation [42, 43]. Samples of water for metal analysis taken before and immediately after each solution renewal were acidified to 1% with ARISTAR nitric acid (65%) (BDH Inc, VWR International Ltd., England) before metal analysis by flame or furnace Atomic Absorption Spectrophotometer (AAS-Perkin Elmer model AAnalyst800, Massachusetts, USA) depending on the concentrations.

During the toxicity test, the snails were not fed. The experiments were performed at room temperature of 28–30°C with photoperiod 12 h light : 12 h darkness, using fluorescent lights (334–376 lux). Water quality parameters (pH, conductivity, and dissolved oxygen) were measured every two days using portable meters (model Hydrolab Quanta, Hach, Loveland, USA) and water hardness samples were fixed with ARISTAR nitric acid and measured by flame atomic absorption spectrophotometer (AAS—Perkin Elmer model AAnalyst 800). Mortality was recorded every 3 to 4 hours for the first two days and then at 12 to 24 hour intervals throughout the rest of the test period. The criterion used to determine mortality were failure to respond to gentle physical stimulation. The death was further confirmed by putting the snail on the glass petri dish for few minutes and if it did not show any movement, it was considered dead. Any dead animals were removed immediately.

At the end of day four, the live snails were used to determine bioconcentration of the metals in the whole body (soft tissues) according to the concentrations used. The snails were cleaned with dechlorinated tap water, and soaked in boiling water for approximately 3 min. Tissues of the molluscs were removed from the shell, rinsed with deionized water, and each sample contained three replicates of three to five animals in a glass test tube (depending on how many live animals were left) and was oven-dried (80°C) for at least 48 hours before being weighed [14]. Each replicate was digested (whole organism) in 1.0 mL ARISTAR nitric acid (65%) in a block thermostat (80°C) for 2 hours. Upon cooling, 0.8 mL of hydrogen peroxide (30%) was added to the solutions. The test tubes were put back on the block thermostat for another 1 hour until the solutions became clear. The solutions were then made up to 25 mL with the addition of deionized water in 25 mL volumetric flasks. Efficiency of the digestion method was evaluated using mussel and lobster tissue reference material (SRM 2976 and TORT-2, National Institute of Standard and Technology, Gaithersburg, USA and National Research Council Canada, Ottawa, Ontario,

TABLE 1: Median lethal times (LT$_{50}$) for *M. tuberculata* exposed to different concentrations for Cu, Cd, Zn, Pb, Ni, Fe, Al, and Mn.

Nominal (and measured) concentration (mg L^{-1})	LT$_{50}$ (h)	95% Confidence limits
Cu		
0.075 (0.081)	163.42	63.60–419.91
0.1 (0.145)	134.97	53.22–342.28
0.32 (0.292)	98.89	44.35–220.50
0.56 (0.549)	75.87	26.88–214.15
0.87 (0.915)	55.42	25.36–121.11
Cd		
0.56 (0.611)	283.44	85.46–940.12
1.0 (1.21)	114.89	52.78–250.09
5.6 (4.87)	57.21	30.29–108.05
10 (10.82)	22.34	11.03–45.27
32 (33.49)	7.82	4.63–13.23
Zn		
1.0 (1.09)	216.96	630.53–1541.63
5.6 (5.30)	96.71	52.82–177.09
10 (8.19)	61.83	39.85–95.94
32 (32.45)	32.44	22.13–47.55
56 (49.60)	12.34	8.25–18.45
Pb		
1.0 (1.02)	250.72	430.55–2057.69
5.6 (5.42)	179.32	38.42–837.02
10 (10.95)	88.25	30.24–257.71
18 (17.16)	40.36	13.91–117.15
32 (31.18)	11.17	6.71–18.57
Ni		
5.6 (5.51)	105.96	59.28–189.40
10 (9.02)	92.38	48.25–176.87
32 (31.53)	58.11	39.82–84.80
75 (67.11)	36.84	21.66–62.67
100 (97.84)	16.26	12.50–21.17
Fe		
5.6 (5.27)	134.97	53.22–342.28
8.7 (8.86)	102.06	40.17–259.29
10 (11.76)	79.72	28.18–225.53
32 (33.47)	34.71	13.36–90.15
56 (58.17)	20.04	7.5–53.51
Al		
56 (88.38)	80.87	111.48–58.66
100 (160.83)	57.91	87.09–38.50
320 (362.83)	42.75	65.99–27.70
560 (884.34)	18.57	40.94–8.42
1000 (1229.91)	8.40	18.52–3.81
Mn		
10 (12.98)	119.53	62.44–228.82
32 (31.60)	67.81	35.06–131.15
56 (57.81)	35.07	19.04–64.60
87 (85.61)	16.97	9.00–32.00
100 (97.01)	8.35	5.05–13.79

Canada, resp.). Efficiencies obtained were within 10% of the reference values. To avoid possible contamination, all glassware and equipment used were acid-washed (20% HNO$_3$) (Dongbu Hitek Co. Ltd., Seoul, Korea, 68%), and the accuracy of the analysis was checked against blanks. Procedural blanks and quality control samples made from

standard solutions for Cu, Cd, Zn, Pb, Ni, Fe, Al, and Mn (Spectrosol, BDH, England) were analyzed in every ten samples in order to check for sample accuracy. Percentage recoveries for metals analyses were between 85–105%.

Median lethal times (LT$_{50}$) and concentrations (LC$_{50}$) for the snails exposed to metals were calculated using measured metal concentrations. FORTRAN programs based on the methods of Litchfield [44] and Litchfield and Wilcoxon [45] were used to compute the LT_{50} and LC_{50}. Data were analyzed using time/response (TR) and concentration/response (CR) methods by plotting cumulative percentage mortality against concentration and time, respectively, on logarithmic-probit paper. Concentration factors (CFs) were calculated for whole animals as the ratio of the metals concentrations in the tissues to the metals concentration measured in the water.

3. Results and Discussion

In all data analyses, the actual (measured concentration) rather than nominal Cu, Cd, Zn, Pb, Ni, Fe, Al, and Mn concentrations were used (Table 1). The mean water quality parameters measured during the test were pH 6.68 ± 0.22, conductivity $180.0 \pm 46.0\,\mu S\,cm^{-1}$, dissolved oxygen $6.1 \pm 0.27\,mg\,L^{-1}$, and total hardness (Mg^{2+} and Ca^{2+}) $18.72 \pm 1.72\,mg\,L^{-1}$ as $CaCO_3$.

One hundred percent of control animals maintained in dechlorinated tap water survived throughout the experiment. The median lethal times (LT_{50}) and concentrations (LC_{50}) increased with a decrease in mean exposure concentrations and times, respectively, for all metals (Tables 1 and 2). However, the lethal threshold concentration could not be determined since the toxicity curves (Figures 1 and 2) did not become asymptotic to the time axis within the test period. Figures 1 and 2 show that Cu was the most toxic metal to *M. tuberculata*, followed by Cd, Zn, Pb, Ni, Fe, Mn, and Al. Other studies show different trends of toxicity with different snails. According to Luoma and Rainbow [7] the rank order of toxicity of metals will vary between organisms. With *Lymnaea luteola*, Khangarot and Ray [28, 30] showed that the order of toxicity was Cd > Ni > Zn; with *Viviparus bengalensis*, Gupta et al. [27] and Gadkari and Marathe [34] found that the order of toxicity was Zn > Cd > Pb > Ni; and with *Juga plicifera*, Nebeker et al. [20] found that Cu was more toxic than Ni.

The present study showed that LC_{50}s for 48 and 96 hours of Cu, Cd, Zn, Pb, Ni, Fe, Al, and Mn were 0.39, 11.85, 13.15, 10.99, 36.46, 21.78, 306.89, and 120.43 mg L^{-1}, and 0.14, 1.49, 3.90, 6.82, 8.46, 8.49, 68.23 and 45.59 mg L^{-1}, respectively (Table 1). A few studies had reported on the acute toxicity of metals to *M. tuberculata*. Bali et al. [15] and Mostafa et al. [16] showed that 96 h-LC_{50} of Cu to *M. tuberculata* were 0.2 and 3.6 mg L^{-1}, respectively, which were higher than the present study. In comparison with other freshwater gastropods (Table 3), this study showed that in general LC_{50}s for *M. tuberculata* were lower or similar compared to other freshwater snails. Direct comparisons of toxicity values obtained in this study with those in the literature were difficult because of differences in the characteristics (primarily water hardness, pH, and temperature) of the test waters. With similar water hardness (soft water) and using adult snails, Nebeker et al. [20] reported that 96 h-LC_{50} of Cu for *Fluminicola virens* was 0.08 mg L^{-1}, and of Zn for *Physa Gyrina* was 1.27 mg L^{-1}, which was lower than the present study. The toxicity reported by other studies (Table 3)

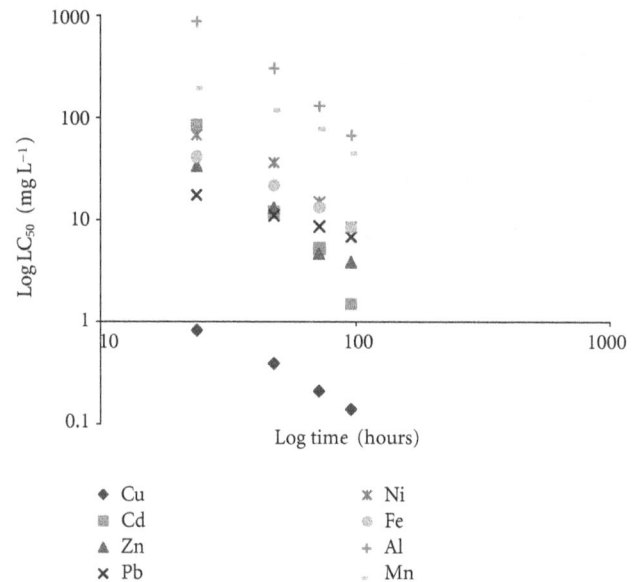

FIGURE 1: The relationship between median lethal concentration (LC_{50}) and exposure times for *M. tuberculata*.

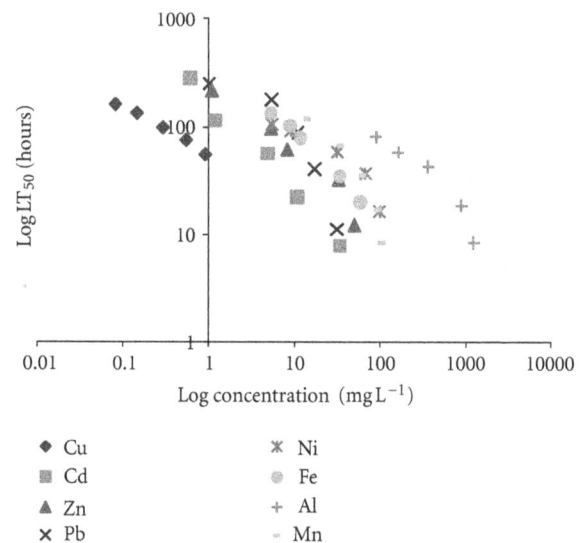

FIGURE 2: The relationship between median lethal time (LT_{50}) and exposure concentrations for *M. tuberculata*.

differs from that reported in this study owing to the different species, ages, and sizes of the organisms as well as varied test methods (water quality and water hardness) as this can affect toxicity [46–49]. In the present study, the water hardness used was considered low (18.7 mg L^{-1} $CaCO_3$), and the water was categorized as soft water (<75 mg L^{-1} as $CaCO_3$).

In comparison with other taxa, *M. tuberculata* shows less sensitivity to metals. LC_{50}s reported for other taxa from this laboratory such as Crustacea (prawn *Macrobrachium lanchesteri* [50] and ostracod *Stenocypris major* [51]), fish (*Rasbora sumatrana* and *Poecilia reticulata* [52]), and Annelida (*Nais elinguis* [53]) were lower than the LC_{50} values of *M. tuberculata* in the present study. Von Der Ohe and Liess [54] showed that 13 taxa belonging to Crustacea were among

TABLE 2: Median lethal concentrations (LC$_{50}$) for *M. tuberculata* at different exposure times for Cu, Cd, Zn, Pb, Ni, Fe, Al, and Mn.

Time (hour)	LC$_{50}$ (mg L^{-1})	95% Confidence limits
Cu		
24	0.82	0.49–4.21
48	0.39	0.23–0.88
72	0.21	0.12–0.33
96	0.14	0.09–0.20
Cd		
24	85.03	13.94–518.57
48	11.85	2.70–51.99
72	5.24	0.96–28.43
96	1.49	0.34–6.53
Zn		
24	33.97	21.59–65.31
48	13.15	6.93–26.06
72	4.73	2.28–8.10
96	3.90	1.81–6.67
Pb		
24	17.39	12.06–29.47
48	10.99	6.04–19.68
72	8.57	4.37–14.79
96	6.82	2.89–12.67
Ni		
24	68.35	48.18–102.23
48	36.46	20.76–70.91
72	15.04	5.23–28.97
96	8.46	3.53–14.02
Fe		
24	42.12	25.74–133.99
48	21.78	10.52–88.85
72	13.29	4.09–29.47
96	8.49	1.58–15.25
Al		
24	880.78	553.91–2147.55
48	306.89	184.29–487.20
72	130.22	35.51–226.38
96	68.23	2.24–123.87
Mn		
24	194.52	112.85–335.27
48	120.43	58.08–249.72
72	78.35	36.20–169.56
96	45.59	20.17–103.04

the most sensitive to metal compounds and concluded that taxa belonging to Crustacea are similar to one another and to *Daphnia magna* in terms of sensitivity to organics and metals and that Molluscs have an average sensitivity to metals. Mitchell et al. [9] reported that the snail has a tightly sealing operculum that allows it to withstand desiccation and apparently also increases its tolerance to chemicals.

Bioconcentration of Cu, Cd, Zn, Pb, Ni, Fe, Al, and Mn in surviving *M. tuberculata* is as shown in Figure 3. Bioconcentration data for live snails were obtained from five Cd (0.61, 1.21, 4.87, 10.82 and 33.49 mg L^{-1}), Fe (5.27, 8.86, 11.76, 33.47, and 58.17 mg L^{-1}), and Mn (12.98, 31.60, 57.81, 85.61 and 97.01 mg L^{-1}) concentration exposures;

four Pb (1.02, 5.42, 10.95 and 17.16 mg L^{-1}) concentration exposures; three Cu (0.081, 0145 and 0.292 mg L^{-1}), Zn (1.09, 5.30 and 8.19 mg L^{-1}), Ni (5.51, 9.02 and 31.53 mg L^{-1}), and Al (88.38, 160.83 and 362.83 mg L^{-1}) concentration exposures. In general, the Cu, Cd, Pb, Zn, Ni, Fe, Al, and Mn bioconcentration in *M. tuberculata* increases with increasing concentration exposure. Similar results were reported by Moolman et al. [18] on Cd and Zn accumulation by two freshwater gastropods (*M. tuberculata* and *Helisoma duryi*). Hoang and Rand [55] showed that whole body Cu concentration of juvenile apple snails (*Pomacea paludosa*) was significantly correlated with soil and water Cu concentrations. In other experiments, Hoang et al. [56] showed that

TABLE 3: Comparison of LC$_{50}$ values of freshwater gastropod *M. tuberculata* with other freshwater mollusc.

Metal	Species	Water hardness (mg L^{-1})	Live stage	Test duration	LC$_{50}$ (mg L^{-1})	Reference
Copper	*M. tuberculata*	18.7	Adult	96 h	0.14	This study
	M. tuberculata			48 h	3.6	[16]
	M. tuberculata		Juvenile	24 h	0.2	[15]
	B. glabrata	44	Adult	48 h	0.18	[19]
	F. virens	21	Adult	96 h	0.08	[20]
	J. plicifera	21	Adult	96 h	0.015	[20]
	B. glabrata	100	—	96 h	0.04	[21]
	P. paludosa	68	60 d	96 h	0.14	[22]
	P. jenkinsi	—	Adult	96 h	0.08	[23]
Cadmium	*M. tuberculata*	18.7	Adult	96 h	1.49	This study
	Amnicola sp.	50	Adult	96 h	8.4	[24]
	P. fontinalis	—	—	96 h	0.08	[25]
	A. hypnorum	45	Adult	96 h	0.09	[26]
	B. glabrata	100	—	96 h	0.3	[21]
	V. bengalensis	180	—	96 h	1.2	[27]
	L. luteola	195	Adult	96 h	1.5	[28]
Zinc	*M. tuberculata*	18.7	Adult	96 h	3.90	This study
	P. gyrina	36	Adult	96 h	1.27	[20]
	L. acuminata	375	—	96 h	10.49	[29]
	L. luteola	195	Adult	96 h	11.0	[30]
	V. bengalensis	180	—	96 h	0.64	[27]
	P. heterostropha	20	Adult	96 h	1.11	[31]
	P. heterostropha	100	Adult	96 h	3.16	[31]
Lead	*M. tuberculata*	18.7	Adult	96 h	6.82	This study
	L. emarginata	150	—	48 h	14.0	[32]
	E. livescens	150	—	48 h	71.0	[32]
	Filopaludina sp.	—	Adult	96 h	190	[33]
	V. bengalensis	165	—	96 h	2.54	[34]
	A. hypnorum	60.9	—	96 h	1.34	[35]
Nickel	*M. tuberculata*	18.7	Adult	96 h	8.46	This study
	Amnicola sp.	50	Adult	96 h	14.3	[24]
	J. plicifera	59	Adult	96 h	0.24	[20]
	L. luteola	195	Adult	96 h	1.43	[28]
	V. bengalensis	180	—	96 h	9.92	[27]
	L. acuminata	375	—	96 h	2.78	[29]
Iron	*M. tuberculata*	18.7	Adult	96 h	8.49	This study
	P. gyrina	109	—	96 h	12.09	[36]
	Planorbarius sp.	—	—	48 h	7.32	[37]
	S. libertina	—	—	48 h	76.0	[38]
Aluminium	*M. tuberculata*	18.7	Adult	96 h	68.23	This study
	Physa sp.	47	—	96 h	55.5	[39]
	A. limosa	PH 3.5	—	96 h	1.0	[40]
	A. limosa	PH 4.5	—	96 h	0.40	[40]
Manganese	*M. tuberculata*	18.7	Adult	96 h	45.59	This study
	B. globosus	53	—	96 h	100.0	[41]

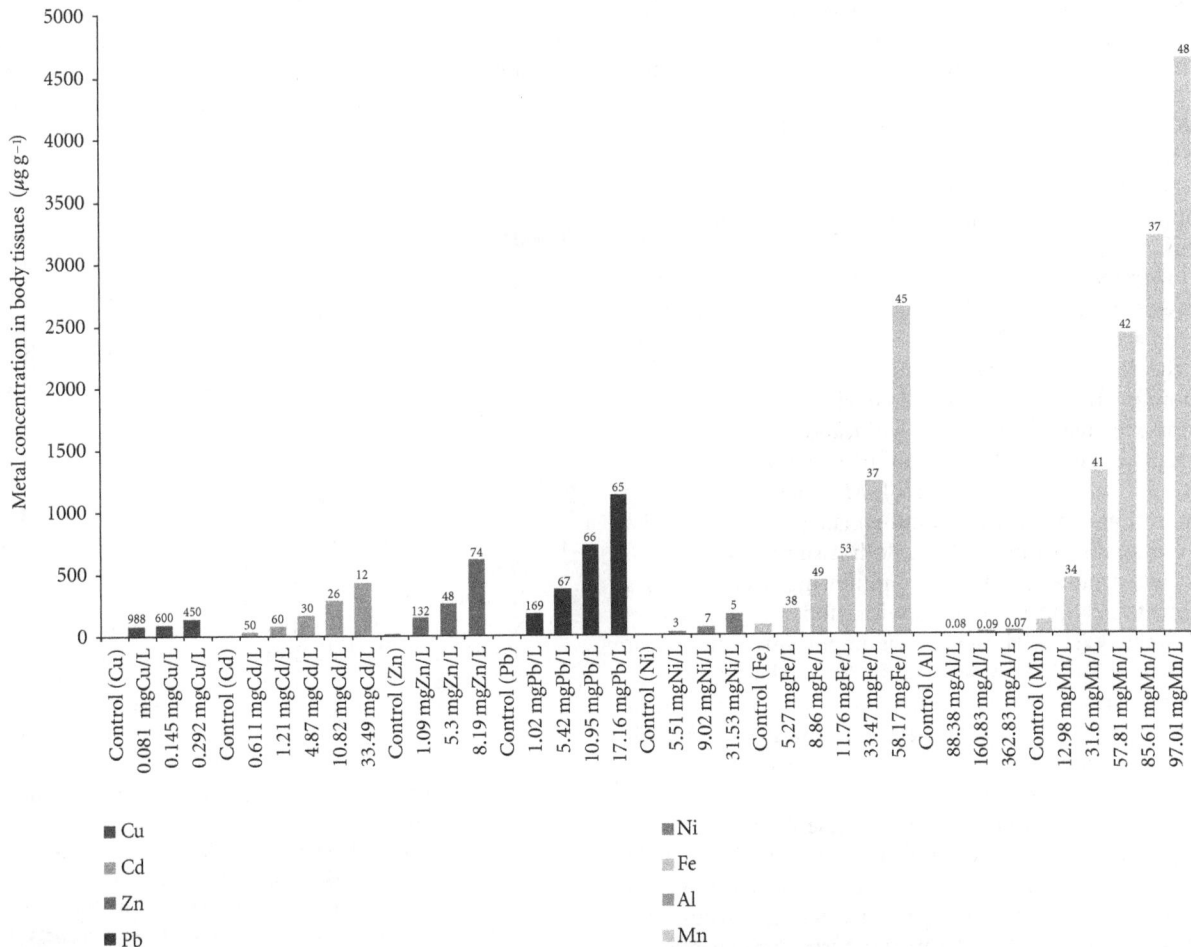

FIGURE 3: Bioconcentration of Cu, Cd, Zn, Pb, Ni, Fe, Al, and Mn (mean) in *M. tuberculata* soft tissues ($\mu g\,g^{-1}$ dry weight) after a four-day exposure to different concentrations of Cu, Cd, Zn, Pb, Ni, Fe, Al, and Mn. Concentration factor (CF) is indicated at the top of each bar.

whole body Cu concentrations of juvenile snails (*P. paludosa*) increased with exposure time and concentration and reached a plateau (saturation) after 14 days of exposure. These results are in agreement with the statement of Luoma and Rainbow [7] who state that the uptake of trace metals from solution by an aquatic organism is primarily concentration dependent. The higher the dissolved concentration of the trace metal, the higher the uptake of the metal from solution into the organism will be, until the uptake mechanism becomes saturated.

The present study shows that in general the highest concentration factor (CF) was noted for Cu (988), Pb (169), and Zn (132), and the lowest CF was for Al (0.07) (Figure 3). Similar results were reported by Lau et al. [14] who reported that *M. tuberculata* collected from the wild (Sarawak River) accumulated higher amounts of Cu, Zn, and As in the soft tissues compared to other metals. Adewunmi et al. [17] showed that Cu, Pb, and Cd were the highest metal accumulated in tissues of freshwater snails in dams and rivers in southwest Nigeria, and metal concentrations in the snails were varied with the seasons, especially for Cu which was higher in the dry season compared to the rainy season. According to Luoma and Rainbow [7] the factors

that affect the rate of uptake of metals affect the toxicity of metal. This is in agreement with the results from the present study which shows that Cu, which was the most toxic to the snail, also has the highest CF in the soft tissues of *M. tuberculata*. In explaining the toxicity of Cu, Hoang and Rand [55] demonstrate that the potential toxicity of Cu carbonate to snails may be explained by the carbonate content in the snails. The carbonate requirement for snails is more than for fish because snails require it for shell development. Copper may enter snails as Cu carbonate. After entering snails, Cu carbonate may be disassociated through biological and chemical reactions. Carbonate would be available for shell development and Cu would be accumulated in soft tissue. Hoang et al. [56] also reported that with the juvenile apple snail (*Pomacea paludosa*), most of the accumulated Cu was located in soft tissue (about 60% in the viscera and 40% in the foot) and the shell contained <4% of the total accumulated copper. However, a comparison of the uptake rate in aquatic organisms showed that in general the order of the uptake rate constant is Ag > Zn > Cd > Cu > Co > Cr > Se [7]. This discrepancy is probably due to short time of exposure (four days) to metals in this study. Other factors which may influence the bioaccumulation of heavy metals in aquatic organisms has been suggested, such as their feeding habit [57], growth rate and age of the organism [14, 58], and the bioavailability of the metals, which greatly depends on hardness of water, pH, and the acid-volatile sulphide of the water [59]. Hoang and Rand [55] showed that the apple snails (*Pomacea paludosa*) accumulated more Cu from soil-water than from water-only treatments and this suggests that apple snails accumulate Cu from soil (-sediment)/water systems. Organisms with higher growth rates also usually have lower metal concentrations in their bodies as the rate of increase in the weight of its tissue and shell will be higher than the accumulated metals [14]. According to Lau et al. [14], the shell of *M. tuberculata* would be most suitable for monitoring Cu in the aquatic environment, which has an approximately thirtyfold magnification capability and with standard errors of less than 10%. Zn would be best monitored by using the shell of *M. tuberculata*, whose magnification capability was approximately 35 times and its error was at approximately 15%. Both tissue and shell of *M. tuberculata* could also be used for monitoring arsenic as it has good magnification capabilities with moderate irregularity approximately 23%. However, it is important to note that the Lau et al. [14] study was conducted in the field (long-term exposure), while the present study was conducted in the laboratory with short-term exposure, and differences in accumulation trend and strategies (higher accumulation in soft tissues or shell) may exist.

Aquatic molluscs possess very diverse strategies in the handling and storage of accumulated metals, which include being in the forms of metal-rich granules metallothioneins (MT) or metallothionein-like proteins [60–62]. Accumulation strategies of invertebrates vary intraspecifically between metals and interspecifically for the same metal in closely related organisms [62, 63]. Moolman et al. [18] showed that *M. tuberculata* had a much higher uptake of Zn in the Zn and in the mixed Cd/Zn exposures compared to *Helisoma duryi*, and Zn was readily accumulated with increasing metal concentrations. Lau et al. [14] also demonstrated that Zn concentrations in *M. tuberculata* were significantly higher than those in the molluscs *Brotia costula* and *Clithon* sp. The present study shows that the CF of Zn was higher than the Cd in the soft tissues of *M. tuberculata*. With the juvenile apple snail, Hoang et al. [56] showed that the snails accumulated Cu during the exposure phase and eliminated Cu during the depuration phase. Metals accumulated in animals can be stored without excretion leading to high body concentrations (accumulators), or the metal levels in the body can be maintained at a low constant body concentration (regulators) by balancing the uptake with controlled rates of excretion [64].

4. Conclusions

This study showed that *M. tuberculata* was equally sensitive to metals compared to other freshwater gastropods. Cu was the most toxic metal to *M. tuberculata* followed by Cd, Zn, Pb, Ni, Fe, Mn, and Al. A comparison of the bioconcentration of metals in soft tissues of *M. tuberculata* showed that among the eight metals studied; Cu, Pb, and Zn

were the most accumulated and Al was least accumulated. *M. tuberculata* is widely distributed in urban and suburban areas which makes it easy to sample and very useful in ecotoxicology studies. This study indicates that *M. tuberculata* could be a potential bioindicator organism of metals pollution and in toxicity testing.

Acknowledgments

This study was funded by the Ministry of Science and Technology, Malaysia (MOSTI) under e-Science Fund code nos. 06-01-02-SF0217 and 06-01-02-SF472. The authors do not have any direct financial relation with the commercial identity mentioned in this paper.

References

[1] C. H. Walker, S. P. Hopkin, R. M. Silby, and D. B. Peakall, *Principles of Ecotoxicology*, CRC Press, Boca Raton, Fla, USA, 3rd edition, 2006.

[2] N. A. M. Shazili, K. Yunus, A. S. Ahmad, N. Abdullah, and M. K. A. Rashid, "Heavy metal pollution status in the Malaysian aquatic environment," *Aquatic Ecosystem Health and Management*, vol. 9, no. 2, pp. 137–145, 2006.

[3] DOE (Department of Environment, Malaysia), *Malaysia Environment Quality Report 2008*, Department of Environment, Ministry of Natural Resources and Environment, Kuala Lumpur, Malaysia, 2009.

[4] S. Z. Zulkifli, F. Mohamat-Yusuff, T. Arai, A. Ismail, and N. Miyazaki, "An assessment of selected trace elements in intertidal surface sediments collected from the Peninsular Malaysia," *Environmental Monitoring and Assessment*, vol. 169, no. 1–4, pp. 457–472, 2010.

[5] C. K. Yap and B. H. Pang, "Assessment of Cu, Pb, and Zn contamination in sediment of north western Peninsular Malaysia by using sediment quality values and different geochemical indices," *Environmental Monitoring and Assessment*, vol. 183, no. 1–4, pp. 23–39, 2011.

[6] W. J. Adams and C. D. Rowland, "Aquatic toxicology test methods," in *Handbook of Ecotoxicology*, D. J. Hoffman, B. A. Rattner, G. A. Burton Jr., and J. Cairns Jr., Eds., CRC Press, Boca Raton, Fla, USA, 2nd edition, 2003.

[7] S. N. Luoma and P. S. Rainbow, *Metal Contamination in Aquatic Environment: Science and Lateral Management*, Cambridge University Press, New York, NY, USA, 2008.

[8] S. Panha and J. B. Burch, "Mollusca," in *Freshwater Invertebrates of the Malaysian Region*, C. M. Yule and Y. H. Sen, Eds., pp. 225–253, Academy of Science Malaysia, Kuala Lumpur, Malaysia, 2004.

[9] A. J. Mitchell, M. S. Hobbs, and T. M. Brandt, "The effect of chemical treatments on red-rim melania Melanoides tuberculata, an exotic aquatic snail that serves as a vector of trematodes to fish and other species in the USA," *North American Journal of Fisheries Management*, vol. 27, no. 4, pp. 1287–1293, 2007.

[10] G. Pererea and J. G. Walls, *Apple Snails in the Aquarium*, 1996.

[11] A. J. Benson, "*Melanoides tuberculatus*," USGS Nonindigenous Aquatic Species Database, Gainesville, Fla, USA, 2008, http://nas.er.usgs.gov/queries/FactSheet.aspx?speciesID=1037.

[12] C. C. Appleton, A. T. Forbes, and N. T. Demetriades, "The occurrence, bionomics and potential impacts of the invasive freshwater snail *Tarebia granifera* (Lamarck, 1822)

(Gastropoda: Thiaridae) in South Africa," *Zoologische Med-edelingen*, vol. 83, pp. 525–536, 2009.

[13] U. Gardenfors, T. Westermark, U. Emanuelsson, H. Mutvei, and H. Walden, "Use of land-snail shells as environmental archives: preliminary results," *AMBIO*, vol. 17, no. 5, pp. 347–349, 1988.

[14] S. Lau, M. Mohamed, A. Tan Chi Yen, and S. Su'Ut, "Accumulation of heavy metals in freshwater molluscs," *Science of the Total Environment*, vol. 214, no. 1–3, pp. 113–121, 1998.

[15] H. S. Bali, S. Singh, and D. P. Singh, "Trial of some molluscicides on snails *Melanoides tuberculatus* and *Vivipara bengalensis* in laboratory," *Indian Journal of Animal Sciences*, vol. 54, no. 4, pp. 401–403, 1984.

[16] B. B. Mostafa, F. A. El-Deeb, N. M. Ismail, and K. M. El-Said, "Impact of certain plants and synthetic molluscicides on some fresh water snails and fish," *Journal of the Egyptian Society of Parasitology*, vol. 35, no. 3, pp. 989–1007, 2005.

[17] C. O. Adewunmi, W. Becker, O. Kuehnast, F. Oluwole, and G. Dörfler, "Accumulation of copper, lead and cadmium in freshwater snails in southwestern Nigeria," *Science of the Total Environment*, vol. 193, no. 1, pp. 69–73, 1996.

[18] L. Moolman, J. H. J. Van Vuren, and V. Wepener, "Comparative studies on the uptake and effects of cadmium and zinc on the cellular energy allocation of two freshwater gastropods," *Ecotoxicology and Environmental Safety*, vol. 68, no. 3, pp. 443–450, 2007.

[19] E. C. De Oliveira-Filho, R. Matos Lopes, and F. J. Roma Paumgartten, "Comparative study on the susceptibility of freshwater species to copper-based pesticides," *Chemosphere*, vol. 56, no. 4, pp. 369–374, 2004.

[20] A. V. Nebeker, A. Stinchfield, C. Savonen, and G. A. Chapman, "Effects of copper, nickel and zinc on three species of oregon freshwater snails," *Environmental Toxicology and Chemistry*, vol. 5, no. 9, pp. 807–811, 1986.

[21] C. Bellavere and J. Gorbi, "A comparative analysis of acute toxicity of chromium, copper and cadmium to *Daphnia magna*, *Biomphalaria glabrata*, and *Brachydanio rerio*," *Environmental Technology Letters*, vol. 2, no. 3, pp. 119–128, 1981.

[22] E. C. Rogevich, T. C. Hoang, and G. M. Rand, "The effects of water quality and age on the acute toxicity of copper to the Florida apple snail, *Pomacea paludosa*," *Archives of Environmental Contamination and Toxicology*, vol. 54, no. 4, pp. 690–696, 2008.

[23] A. J. Watton and H. A. Hawkes, "The acute toxicity of ammonia and copper to the gastropod *Potamopyrgus jenkinsi* (Smith)," *Environmental Pollution Series A*, vol. 36, no. 1, pp. 17–29, 1984.

[24] R. Rehwoldt, L. Lasko, C. Shaw, and E. Wirhowski, "The acute toxicity of some heavy metal ions toward benthic organisms," *Bulletin of Environmental Contamination and Toxicology*, vol. 10, no. 5, pp. 291–294, 1973.

[25] K. A. Williams, D. W. J. Green, and D. Pascoe, "Studies on the acute toxicity of pollutants to freshwater macroinvertebrates. I. Cadmium," *Archiv fur Hydrobiologie*, vol. 102, no. 4, pp. 461–471, 1985.

[26] G. W. Holcombe, G. L. Phipps, and J. W. Marier, "Methods for conducting snail (*Aplexa hypnorum*) embryo through adult exposures: effects of cadmium and reduced pH levels," *Archives of Environmental Contamination and Toxicology*, vol. 13, no. 5, pp. 627–634, 1984.

[27] P. K. Gupta, B. S. Khangarot, and V. S. Durve, "Studies on the acute toxicity of some heavy metals to an Indian freshwater pond snail *Viviparus bengalensis* L," *Archiv für Hydrobiologie*, vol. 91, no. 2, pp. 259–264, 1981.

[28] B. S. Khangarot and P. K. Ray, "Sensitivity of freshwater pulmonate snails, *Lymnaea luteola* L., to heavy metals," *Bulletin of Environmental Contamination and Toxicology*, vol. 41, no. 2, pp. 208–213, 1988.

[29] B. S. Khangarot, S. Mathur, and V. S. Durve, "Comparative toxicity of heavy metals and interaction of metals on a freshwater pulmonate snail *Lymnaea acuminata* (Lamarck)," *Acta Hydrochimica et Hydrobiologica*, vol. 10, no. 4, pp. 367–375, 1982.

[30] B. S. Khangarot and P. K. Ray, "Zinc sensitivity of a freshwater snail, *Lymnaea luteola* L., in relation to seasonal variations in temperature," *Bulletin of Environmental Contamination and Toxicology*, vol. 39, no. 1, pp. 45–49, 1987.

[31] C. B. Wurtz, "Zinc effects on fresh-water Mollusks," *Nautilus*, vol. 76, pp. 53–61, 1962.

[32] J. Cairns Jr., D. I. Messenger, and W. F. Calhoun, "Invertebrate response to thermal shock following exposure to acutely sub lethal concentrations of chemicals," *Archiv fur Hydrobiologie*, vol. 77, no. 2, pp. 164–175, 1976.

[33] S. Lantataeme, M. Kruatruchue, S. Kaewsawangsap, Y. Chitramvong, P. Sretarugsa, and E. S. Upatham, "Acute toxicity and bioaccumulation of lead in the snail, Eilopaludina (Siamopaludina) *Martensi martensi* (Frauenfeldt)," *Journal of the Science Society of Thailand*, vol. 22, no. 3, pp. 237–247, 1996.

[34] A. S. Gadkari and V. B. Marathe, "Toxicity of cadmium and lead to a fish and a snail from two different habitats," *Indian Association Water Pollution Control Technology*, vol. 5, pp. 141–148, 1983.

[35] D. J. Call, L. T. Brooke, N. Ahmad, and D. D. Vaishnav, *Aquatic Pollutant Hazard Assessments and Development of a Hazard Prediction Technology by Quantitative Structure-Activity Relationships*, U.S. EPA Cooperation Agreement No. CR 809234-01-0, Centre for Lake Superior Environmental Studies, University of Wisconsin, Superior, Wis, USA, 1981.

[36] W. J. Birge, J. A. Black, A. G. Westerman et al., *Recommendations on Numerical Values for Regulating Iron and Chloride Concentrations for the Purpose of Protecting Warmwater Species of Aquatic Life in the Commonwealth of Kentucky*, University of Kentucky, Lexington, Ky, USA, 1985.

[37] M. Furmanska, "Studies of the effect of copper, zinc, and iron on the biotic components of aquatic ecosystems," *Polskie Archiwum Hydrobiologii*, vol. 26, no. 1-2, pp. 213–220, 1979.

[38] Y. Nishiuchi and K. Yoshida, "Toxicities of pesticides to some fresh water snails," *Bulletin of the Agricultural Chemicals Inspection Station*, vol. 12, pp. 86–92, 1972.

[39] D. J. Call, L. T. Brooke, C. A. Lindberg, T. P. Markee, D. J. McCauley, and S. H. Poirier, *Toxicity of Aluminum to Freshwater Organisms in Water of pH 6.5-8.5*, Technical Report Project No.549-238-RT-WRD, Center for Lake Superior Environmental Studies, University of Wisconsin, Superior, Wis, USA, 1984.

[40] G. L. Mackie, "Tolerances of five benthic invertebrates to hydrogen ions and metals (Cd, Pb, Al)," *Archives of Environmental Contamination and Toxicology*, vol. 18, no. 1-2, pp. 215–223, 1989.

[41] P. Tomasik, C. M. Magadza, S. Mhizha, A. Chirume, M. F. Zaranyika, and S. Muchiriri, "Metal-metal interactions in biological systems. Part IV. Freshwater snail *Bulinus globosus*," *Water, Air, and Soil Pollution*, vol. 83, no. 1-2, pp. 123–145, 1995.

[42] APHA (American Public Health Association), *Standard Method for the Examination of Water and Wastewater*, part

8000, toxicity, Washington American Public Health Association, 18th edition, 1992.

[43] J. D. Cooney, "Freshwater test," in *Fundamental of Aquatic Toxicology: Effects, Environmental fate and Risk Assessment*, G. M. Rand, Ed., pp. 71–102, Taylor & Francis, 2nd edition, 1995.

[44] J. T. Lichfield, "A method for the rapid graphic solution of time-percentage effect curves," *Journal of Pharmacology and Experimental Therapeutics*, vol. 97, pp. 399–408, 1949.

[45] J. T. Lichfield and F. Wilcoxon, "A simplified method of evaluating dose-effect experiments," *Journal of Pharmacology and Experimental Therapeutics*, vol. 96, pp. 99–113, 1949.

[46] C. P. McCahon and D. Pascoe, "Use of *Gammarus pulex* (L.) in safety evaluation tests: culture and selection of a sensitive life stage," *Ecotoxicology and Environmental Safety*, vol. 15, no. 3, pp. 245–252, 1988.

[47] I. Landner and R. Reuther, *Metals in Society and in the Environment*, Kluwer Publisher, Dordrecht, The Netherlands, 2004.

[48] J. Gorski and D. Nugegoda, "Sublethal toxicity of trace metals to larvae of the blacklip abalone, *Haliotis rubra*," *Environmental Toxicology and Chemistry*, vol. 25, no. 5, pp. 1360–1367, 2006.

[49] M. Ebrahimpour, H. Alipour, and S. Rakhshah, "Influence of water hardness on acute toxicity of copper and zinc on fish," *Toxicology and Industrial Health*, vol. 26, no. 6, pp. 361–365, 2010.

[50] M. Shuhaimi-Othman, Y. Nadzifah, R. Nur-Amalina, and A. Ahmad, "Sensitivity of the freshwater prawn, *Macrobrachium lanchesteri* (Crustacea: Decapoda), to heavy metals," *Toxicology and Industrial Health*, vol. 27, no. 6, pp. 523–530, 2011.

[51] M. Shuhaimi-Othman, Y. Nadzifah, R. Nur-Amalina, and A. Ahmad, "Toxicity of metals to a freshwater ostracod, Stenocypris major," *Journal of Toxicology*, vol. 2011, pp. 1–8, 2011.

[52] M. Shuhaimi-Othman, N. Yakub, N. A. Ramle, and A. Abas, "Comparative metal toxicity to freshwater fish," submitted to *Journal of Toxicology and Industrial Health*, In press.

[53] N. S. Umirah, *Toxicity of metals to aquatic worm Nais elinguis and midge Chironomus javanus*, Unpublished M.S. thesis, Universiti Kebangsaan Malaysia (UKM), Bangi, Malaysia, 2009.

[54] P. C. Von Der Ohe and M. Liess, "Relative sensitivity distribution of aquatic invertebrates to organic and metal compounds," *Environmental Toxicology and Chemistry*, vol. 23, no. 1, pp. 150–156, 2004.

[55] T. C. Hoang and G. M. Rand, "Exposure routes of copper: short term effects on survival, weight, and uptake in Florida apple snails (*Pomacea paludosa*)," *Chemosphere*, vol. 76, no. 3, pp. 407–414, 2009.

[56] T. C. Hoang, E. C. Rogevich, G. M. Rand, and R. A. Frakes, "Copper uptake and depuration by juvenile and adult Florida apple snails (*Pomacea paludosa*)," *Ecotoxicology*, vol. 17, no. 7, pp. 605–615, 2008.

[57] G. Mance, *Pollution Threat of Heavy Metal in Aquatic Environments*, Elsevier Applied Science, New York, NY, USA, 1990.

[58] R. J. Pentreath, "The accumulation of organic mercury from sea water by the plaice, *Pleuronectes platessa* L," *Journal of Experimental Marine Biology and Ecology*, vol. 24, no. 2, pp. 121–132, 1976.

[59] J. M. Besser, C. G. Ingersoll, and J. P. Giery, "Effects of spatial and temporal variation of acid-volatile sulphide on the bioavailability of copper and zinc in freshwater sediment," *Environmental Technology and Chemistry*, vol. 15, pp. 286–293, 1996.

[60] G. Roesijadi, "Metallothioneins in metal regulation and toxicity in aquatic animals," *Aquatic Toxicology*, vol. 22, no. 2, pp. 81–114, 1992.

[61] A. Z. Mason and K. D. Jenkin, "Metal detoxification in aquatic organisms," in *Metal Speciation and Bioavailability in Aquatic Systems*, A. Tessier and D. R. Turner, Eds., pp. 479–608, Wiley, Chichester, UK, 1995.

[62] P. S. Rainbow, "Trace metal concentrations in aquatic invertebrates: why and so what?" *Environmental Pollution*, vol. 120, no. 3, pp. 497–507, 2002.

[63] P. S. Rainbow and R. Dallinger, "Metal uptake, regulation, and excretion in freshwater invertebrates," in *Ecotoxicology of Metals in Invertebrates*, R. Dallinger and P. S. Rainbow, Eds., pp. 119–131, Lewis Publishers, Boca Raton, Fla, USA, 1993.

[64] P. S. Rainbow, "Heavy metals in marine invertebrates," in *Heavy Metals in the Marine Environment*, R. W. Furness and P. S. Rainbow, Eds., pp. 68–75, CRC Press, Boca Raton, Fla, USA, 1990.

3

Diet Composition and Feeding Strategies of the Stone Marten (*Martes foina*) in a Typical Mediterranean Ecosystem

Dimitrios E. Bakaloudis,[1, 2] **Christos G. Vlachos,**[2] **Malamati A. Papakosta,**[2]
Vasileios A. Bontzorlos,[2, 3] **and Evangelos N. Chatzinikos**[2, 4]

[1] *Laboratory of Wildlife Ecology and Management, Department of Forestry and Natural Environment Management,*
Technological Educational Institute of Kavala, 1st km Drama-Mikrohori, 661 00 Drama, Greece
[2] *Department of Wildlife and Freshwater Fisheries, Faculty of Forestry and Natural Environment, Aristotle University of Thessaloniki,*
P.O. Box 241, 540 06 Thessaloniki, Greece
[3] *Hunting Confederation of Greece, 8 Fokionos Street, 105 63 Athens, Greece*
[4] *4th Hunting Federation of Sterea Hellas, 8 Fokionos Street, 105 63 Athens, Greece*

Correspondence should be addressed to Dimitrios E. Bakaloudis, dimbak@teikav.edu.gr

Academic Editors: R. Julliard and B. Tóthmérész

Stone martens (*Martes foina*) are documented as generalist throughout their distributional range whose diet composition is affected by food availability. We tested if this occurs and what feeding strategies it follows in a typical Mediterranean ecosystem in Central Greece by analysing contents from 106 stomachs, seasonally collected from three different habitats during 2003–2006. Seasonal variation in diet and feeding strategies was evident and linked to seasonal nutritional requirements, but possibly imposed by strong interference competition and intraguild predation. Fleshy fruits and arthropods predominated in the diet, but also mammals and birds were frequently consumed. An overall low dietary niche breadth ($B_A = 0.128$) indicated a fruit specialization tendency. A generalised diet occurred in spring with high individual specialisation, whereas more animal-type prey was consumed than fruits. A population specialization towards fruits was indicated during summer and autumn, whereas insects were consumed occasionally by males. In those seasons it switched to more clumped food types such as fruits and insects. In winter it selectively exploited both adult and larvae insects and partially fruits overwinter on plants. The tendency to consume particular prey items seasonally reflected both the population specialist behaviour and the individual flexibility preyed on different food resources.

1. Introduction

The stone marten (*Martes foina* Erxleben, 1777) is one of the most widely distributed mustelid in the Eurasian region, ranging west from Central and Southern Europe to the East in Mongolia, Afghanistan, and Tibet [1]. It is a strictly nocturnal animal, but it could be diurnal during summer [2] living on deciduous woodlands, wooded margins [3], and commonly reported to be found in towns and villages [4–7]. Its population is stable across its range [8]; however, the legal persecution remains a possible threat for its number.

The stone marten has been referred to as a generalist, and its diet is well known in many European countries, mainly in the central [9–11] and southwestern parts of its distribution

[12–15]. Its wide spectrum of food types exploited allows the species to occur in variable environments, from undisturbed forests to human settlements. In addition, the numerous studies carried out on the feeding habits of the stone marten note its opportunistic feeding behaviour and that it feeds on fruits, small mammals, insects, birds, reptiles, carrion, and domestic garbage [4, 9, 12, 16–19], but its diet composition is likely affected by regional and seasonal food-type availability [1, 15, 19, 20] and abiotic factors [10], as well as by interspecific competition [21]. Furthermore, there are many studies about its biology and feeding habits suggesting the importance of fruit consumption in its diet [2, 14, 21, 22], as well as reporting the significant contribution of this mammal at the potential improvement of forested areas by enhancing the flora composition with seed dispersal [5, 23, 24]. The stone marten has a terrestrial life, but it is documented as an arboreal species searching for prey on shrubs and/or trees. Its preference for both terrestrial and arboreal prey suggests a relationship with morphological adaptations, such as carnivore dentition, small body size, and long and powerful talons [3]. Its ability to climb allows it to use this wide spectrum of food resources, ranging from fruits to arboreal small mammals, birds, and their eggs.

The study of food habits is applied in wildlife species to describe the dietary composition, to compare diets among geographical regions or among seasons, and to assess the nutritional value of the diet [25]. This information is important to determine the ecological dietary breadth or the "niche" of an animal in the ecosystem and to understand its foraging behaviour, habitat use, and population dynamic, as well as being crucial in order to assess its likely impact on species with ecological and/or sport hunting interest [26]. In most food habit studies the numerical percentage (%N) and the percentage of frequency of occurrence (%F) in combination with the percentage volume or weight of prey items were used in analyses of faecal or stomach contents in mammals [26–28]. Recently, a technique based on stomach contents which includes two of the aforementioned parameters (%F and %N) has been used to explore prey importance, feeding strategy, and the inter- and intraindividual components of niche breadth in predatory fishes [29].

Most of the stone marten's diet studies have analyzed faecal using the percentage of frequency of occurrence. Here, we attempted to estimate the seasonal food habits analyzing stomach contents from Central Greece, where no other dietary study had been carried out before. This approach facilitates investigation into dietary differences between sexes. Thus, the main purposes of this study were (a) to investigate if variation over habitats, seasons, and sexes in prey types taken by the stone marten exist and (b) to gain a better understanding of its food niche characteristics both at the individual and at the population level in its southern distributional range. Both of these can help us to ascertain whether the species is a carnivore or its diet turns to frugivory in a typical Mediterranean ecosystem.

2. Materials and Methods

2.1. Study Area.
Our study area, covering 495,000 Ha, is situated in Central Greece (38°44′–38°59′ N, 22°02′–22°37′ E). Elevations range from 180 to 1,826 m, and the climate is characterized by cold, wet winters and hot, dry summers, with mean annual precipitation ranging between 542 and 1,100 mm and mean annual ambient temperatures over most of the study area averaging 6–17°C. Most of the study area is nonforested land. The dominant habitat type is agricultural land (56.17%) which occurs primarily in extended plains on low altitudes. In the field margins, there are many shrub species such as *Rubus* spp., *Prunus* spp., *Pyrus* spp., and so forth. Shrublands and grasslands (28.33%) contain a variety of plants (*Quercus coccifera*, *Juniperus* spp., *Fragaria vesca*, *Brachypodium sylvaticum*, etc.) mainly on low hills with

a mid-relief terrain. Oak forest (14.59%) contains various *Quercus* spp. which are common dominants on higher altitudes with high-relief topography. Large population not only of the red fox (*Vulpes vulpes*) but also of game species such as the European hare (*Lepus europaeus*), the wild boar (*Sus scrofa*), and the rock partridge (*Alectoris graeca*) occupy the study area. Most of the study area has heavy livestock grazing by goat (*Capra hircus*) and sheep (*Ovis aries*).

2.2. Stomach Analysis.
A total of 106 stomachs mainly hunted animals collected between April of 2003 and March of 2006. Stomachs were sorted according to year, season (spring: March–May, summer: June–August, autumn: September–November, and winter: December–February), gender, and habitat type (farmland, shrubland, and oak forest). Fourteen stomachs were found empty, and they were not included in the dietary analysis. The content of each stomach was analyzed under a dissecting scope and sorted into one of the following prey groups: mammals, birds, reptiles, amphibians, arthropods, molluscs, other invertebrates (molluscs and earthworms), plants, and others (i.e., paper, plastic, string, etc.). All prey items in the stomachs were identified to the lowest taxon possible. The identification was conducted by comparing hairs, teeth, feathers, scales, bones, and seeds by reference collection [30, 31]. We used two common techniques to analyze the diet composition, the percentage of frequency of occurrence (%F = number of stomachs containing prey *i* / total number of stomachs × 100) and the percentage of numerical abundance (%N = number of prey *i* / total number of prey items × 100) [1]. Furthermore, we evaluated the feeding strategy and prey importance using the %F of different prey types plotted against the percentage of prey-specific abundance (%P = number of prey *i* / total number of prey items only in stomachs with prey *i* × 100) (Figure 2(f)) (see Amundsen et al. [29] for detailed description).

We calculated dietary breadth of the stone marten using the Levins standardized equation for food niche [32] $B_A = [(1/\sum p_i^2) - 1]/(n - 1)$, where p_i = proportion of occurrence of each prey category in marten's diet and n = number of prey categories in stone marten's diet. B_A values range between 0 and +1, indicating narrow food niche (specialist) when value is close to 0 and broad diet niche (generalist) when the value is close to +1. We used numerical data based on pooled prey categories for the seasonal and overall assessment of the stone marten dietary niche breadth.

We calculated dietary overlap (O) between the two sexes using Pianka's [33] modification to the MacArthur and Levin measure of niche overlap [32] $O = \sum p_{ij} p_{ik}/\sqrt{\sum p_{ij}^2 \sum p_{ik}^2}$, where p_{ij} and p_{ik} are the proportions prey class *i* comprised of the diets of the *j* (male) and *k* (female) stone marten genders. Niche overlap values range from 0, for no overlap, to +1, for complete overlap.

2.3. Statistical Analysis.
We used the log-likelihood ratio G-test to analyze the frequency of occurrence of each food group according to years, seasons, and habitat types, as this test has more advantages over the chi-square [34]. Because

prey groups did not differ among the three study years ($P >$ 0.05) and among the three habitat types ($P > 0.05$), values of the frequency of occurrence were pooled for further analyses. We used log-linear analysis to test for overall interactions among six prey groups, four seasons, and two sexes. Frequencies of the occurrence of six prey groups were used in the log-linear analysis, because reptiles and amphibians were pooled together as well as other invertebrates and others. We tested for the interaction effect among terms of categorical variables using the 95% of confidence intervals criterion. When values of parameter estimation did not contain between the lower and upper 95% of confidence interval, we assumed that the contribution of the parameter (λ) to the model was significant [35]. In addition, we tested for differences within each food group, when possible, using the chi-square test for contingency tables.

All statistical analyses were performed using the statistical package SPSS (release 15.0 for windows), and statistical tests were significant if $P < 0.05$.

3. Results

3.1. Diet Composition. In total 14 stomachs were empty, and the proportion of empty stomachs varied significantly between sex and season (2×4 contingency table: $\chi^2 = 10.611$, d.f. $= 3, P = 0.014$). The highest number of empty stomachs was found in spring (50%), and it corresponded positively to that of males (64.3%) than that of females (35.7%).

A total of 1,025 prey items were recognised in the stomachs of the stone marten, including 21 species, 7 genera, 12 families/orders, and unidentified prey items. Nine major prey groups were identified in the stomach contents (Table 1). Arthropods constitute the most frequently consumed food group, which was observed in 60.9% of the stone marten stomachs, followed by fruits (%F = 55.4), mammals (%F = 30.4), birds and birds' eggs (%F = 20.7), reptiles (%F = 13.01), and molluscs (%F = 7.6). Other prey groups, such as amphibians and earthworms, were almost scarcely consumed. Among arthropods, the Orthoptera (mainly species from families of Acrididae, Gryllotalpidae and Tettigoniidae), the Myriapoda, the Coleoptera, and the Lepidoptera were best represented in terms of %F, and, among plants, fruits of mulberries (*Morus alba*), wild pears (*Pyrus amygdaliformis*), vegetable remains, and grapes (*Vitis vinifera*) were observed in most stomachs analyzed. Among mammals, the southern vole (*Microtus levis*) and the white-toothed shrew (*Crocidura suaveolens*) occurred in the diet more frequently than other small mammals, while the European hare was a rare prey. In addition, the domestic sheep and the edible dormouse (*Glis glis*) accounted for a relatively high proportion of the stone marten diet.

3.2. Seasonal Variation. In terms of seasonal diet composition, there emerged a significant variation using the log-linear analysis (Table 2). First, the log-linear analysis revealed a significant interaction between season and gender ($P = 0.0115$). That is, greater numbers of stomachs of males were collected during summer and winter in comparison

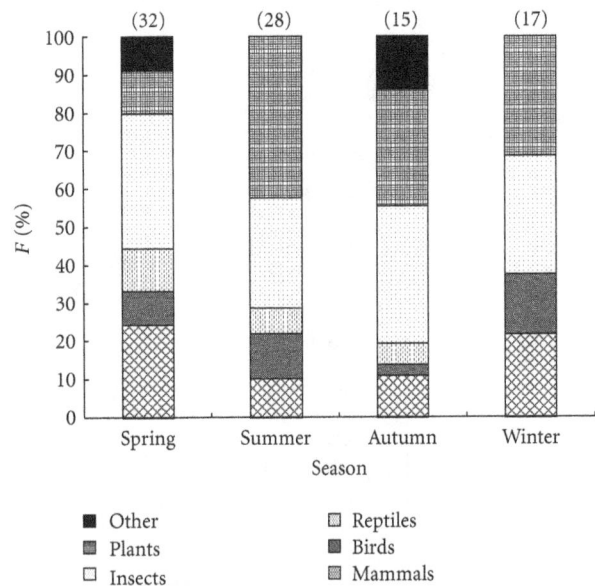

FIGURE 1: Seasonal percentage of frequency of occurrence (%F) of prey groups in the stone marten diet, in Central Greece during 2003–2006. Numbers above columns are sample sizes.

to females, there were similar proportions during spring, whereas there were fewer male stomachs in autumn than females'. Second, there was observed a significant seasonal variation in the proportion of food groups in the diet of stone martens ($P = 0.0003$) (Figure 1). The animal groups were especially dominant in the diet during spring (83.3%), they reduced to the lowest proportion during the summer (51.9%), and then increased gradually from 56% in autumn to 63% in winter in terms of frequency of occurrence. Furthermore, the significant contribution of the prey groups was different among the four seasons (Table 3). Insects occurred evenly throughout the year in the stone marten diet. Fruits were observed less frequently than expected during spring ($\lambda = -0.990$) while they were represented with higher frequencies than expected during summer ($\lambda = 0.570$). Mammals were present in higher proportions of stomachs during spring ($\lambda = 0.518$). Birds were found less frequently during autumn ($\lambda = -0.851$) but in higher frequencies during winter ($\lambda = 0.666$). Finally, both reptiles and amphibians were not observed in the stone marten diet during winter ($\lambda = -1.027$). There was also a variation in diet composition within food groups. Within insects, Coleoptera were consumed in similar proportions throughout the year ($\chi^2 = 0.254$, d.f. $= 3, P = 0.968$), but adult insects were found in the stomachs during spring and summer while larvae were found during autumn and winter. Myriapoda were uniformly found in stomachs throughout the year. Lepidoptera were observed in high proportions during spring ($\chi^2 = 22.745$, d.f. $= 3, P < 0.001$), whereas Orthoptera were consumed in high proportions during summer. Similarly, there was observed a seasonal fluctuation within fruits. Mulberries were the most consumed food in summer, grapes in autumn and early winter ($\chi^2 = 84.09$, d.f. $= 3, P < 0.001$), plums (*Prunus spinosa*) in winter, while

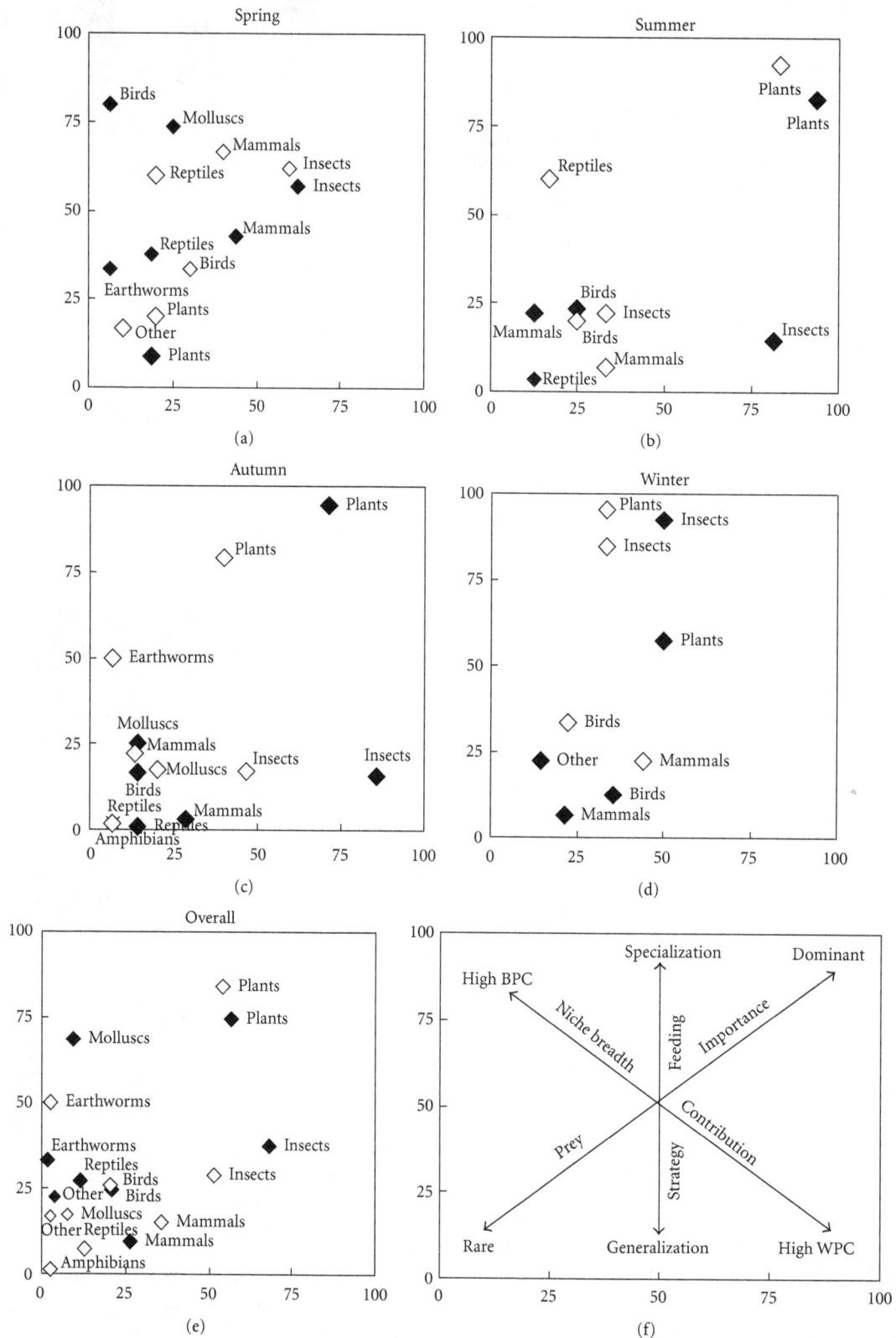

FIGURE 2: Seasonal (a)–(d) and overall (e) feeding strategy of the male (filled symbols) and the female (open symbols) stone martens in Central Greece, (f) graph redrawn from Amundsen et al. [29]. X-axis represents %F (frequency of occurrence), and Y-axis %P (prey-specific abundance).

TABLE 1: Diet composition of the stone marten in Central Greece, during 2003–2006.

Group	Order/family	Species/item	%F	%N
Mammals	Lagomorpha	*Lepus europaeus*	1.09	0.10
	Rodentia	*Apodemus sylvaticus*	1.09	0.10
		Apodemus mystacinus	2.17	0.20
		Glis glis	3.26	0.29
		Micromys minutus	1.09	0.20
		Clethrionomys glareolus	1.09	0.10
		Rattus rattus	1.09	0.10
		Microtus levis	4.35	0.49
	Soricomorpha	*Crocidura leucodon*	2.17	0.20
		Crocidura suaveolens	4.35	0.39
	Artiodactyla	*Ovis aries*	4.35	0.39
		Capra hircus	1.09	0.10
		Nonidentified mammals	4.35	0.39
		Total mammals	**30.4**	**3.0**
Birds		Nonidentified birds	13.04	1.66
		Eggs	7.61	0.59
		Total birds	**20.7**	**2.2**
Reptiles	Sauria	*Lacerta viridis*	2.17	0.29
		Podarcis muralis	1.09	0.10
		Nonidentified lizards	7.61	0.68
	Ophidia	Nonidentified snakes	1.09	0.10
Amphibians		*Rana* spp.	1.09	0.10
		Total herptiles	**14.1**	**1.3**
Arthropods	Coleoptera		13.04	1.56
	Hymenoptera		3.26	0.29
	Lepidoptera		11.96	4.29
	Orthoptera		30.43	3.51
	Myriapoda		15.22	2.34
	Trichoptera		1.09	0.20
	Libellulidae		1.09	0.10
	Arachnida		2.17	0.20
		Nonidentified insects	15.22	11.51
		Total insects	**60.9**	**24.0**
Molluscs		*Helix* spp.	1.09	0.10
		Arion spp.	7.61	2.83
Earthworms	Lumbricidae		2.17	0.39
		Total other invertebrates	**9.8**	**3.3**
Plants		*Morus alba*	13.04	17.27
		Pyrus amygdaliformis	10.87	1.56
		Prunus spinosa	3.26	4.20
		Prunus spp.	3.26	0.39
		Rubus spp.	1.09	0.10
		Rosa canina	2.17	0.20
		Ficus spp.	3.26	0.39
		Amygdalus communis	1.09	0.10
		Actinidia polygama	1.09	0.10
		Vitis vinifera	8.70	40.00
		Vegetable remains	9.78	0.98

TABLE 1: Continued.

Group	Order/family	Species/item	%F	%N
		Hordeum spp.	1.09	0.10
		Nonidentified plants	6.52	0.49
		Total plants	**55.4**	**65.6**
Other		other items	**3.3**	**0.3**

TABLE 2: Log-linear model for frequency of occurrence of prey items in the stone marten diet in Central Greece, during 2003–2006.

Source of variation	d.f.	χ^2	P value
sex × season × food	15	15.74	0.3994
sex × season	3	11.04	0.0115
sex × food	5	3.83	0.5744
season × food	15	41.43	0.0003
sex	1	7.92	0.0049
season	3	8.69	0.0336
food	5	109.03	<0.001

TABLE 3: Parameters (λ) of the interaction term season × food in the log-linear model. Bold number indicates significant contribution of the parameter to the model by using the 95% confidence interval criterion.

Prey group	Spring	Summer	Autumn	Winter
Mammals	**0.518**	−0.082	−0.333	−0.103
Birds	−0.187	0.372	**−0.851**	**0.666**
Herptiles[a]	0.545	0.219	0.263	**−1.027**
Arthropods	−0.067	0.008	0.253	−0.194
Plants	**−0.990**	**0.570**	0.262	0.158
Other[b]	0.181	−1.087	0.406	0.500

[a] Reptiles and amphibians.
[b] Molluscs, earthworms, and other food items.

vegetable remains and wild pears were found in the stomachs in similar proportions throughout the year ($P > 0.05$). Finally, according to log-linear analysis of diet composition, it was found to be relatively homogeneous between the two sexes ($P = 0.57$).

Dietary niche breadth of the stone marten pooled across the study years was relatively low ($B_A = 0.128$). It was higher in spring ($B_A = 0.317$), then decreased gradually in summer ($B_A = 0.101$) and in autumn ($B_A = 0.058$), and increased in winter ($B_A = 0.303$). Similar pattern was observed both for male and female dietary niche breadth. Females had higher values than males in spring (0.450 versus 0.336), in autumn (0.103 versus 0.056), and in winter (0.520 versus 0.299), whilst males had slightly higher values than females only in summer (0.109 versus 0.086). However, the overall dietary niche breadth was higher in male ($B_A = 0.156$) than in female ($B_A = 0.107$).

Dietary niche overlap between the two genders was extremely high ($O = 0.986$). However, their food niche overlap was relatively low in spring ($O = 0.856$).

3.3. Feeding Strategy. Both male and female stone martens exhibited a similar pattern in their feeding strategies (Figure 2(e)). Furthermore, both sexes showed an overall specialisation on fruits. At the individual level, there was

observed a tendency towards a specialised feeding strategy for both sexes, as some prey points were located in the upper half of the diagrams, and this pattern was found for all seasons (Figures 2(a)–2(d)). At the population level, two discernible feeding strategies were observed for both sexes, firstly, a relative population generalisation in spring (Figure 2(a)) and, secondly, a population specialisation in summer (Figure 2(b)), autumn (Figure 2(c)), and winter (Figure 2(d)). The stone marten was relatively generalised as a whole population in spring, as all prey points were located below the diagonal from the lower right to the upper left corner. However, some prey types (e.g., birds and molluscs) were consumed by a few males displaying specialization (high interphenotype component). These types of prey consumed had a high prey-specific abundance value, but they appeared in low frequency of occurrence in the diet, resulting in a relatively narrow niche breadth only in a limited fraction of the population during the spring.

The population specialisations of both sexes were demonstrated by the prey points being positioned on the upper right part of the graph. In summer and autumn, the population specialisation of both sexes was directed to fruits, whereas in winter fruits and insects were the dominant prey taxa of the population specialisation for both male and female stone martens. The specialisation in those three seasons was almost more pronounced for male than for female stone martens. However, there was observed a high intraphenotype contribution to the male's niche breadths during summer and autumn, as insects appeared in the lower right part of the diagrams. Insects were consumed occasionally by most males during summer and autumn, reflecting a relatively wide niche breadth for this gender.

4. Discussion

4.1. Diet Composition and Feeding Strategies. In our study the stone marten is a polyphagous mesopredator that consumes a wide spectrum of food types ranging from fruits to invertebrates and small vertebrates and occasionally carrion, as are most of the species of the genus Martes [3, 18, 36]. However, our results suggest that the stone marten in our study area principally fed on insects (%F = 60.9) and to a lesser degree on fruits (%F = 55.4), and this pattern was consistent both interannually (among the three study years) and spatially (among the three main habitat types studied). Small mammals, birds, and birds' eggs constituted a significant part of the diet, whereas reptiles, amphibians, molluscs, and earthworms were of minor importance and thereby may be considered as occasional food. Although insects represent the most important food type in our area, with a Mediterranean character in climate and a hetero-geneous landscape, only few studies in the Mediterranean basin have shown similar findings (Portugal: [37], Spain: [16], Italy: [13]), but others have shown that this prey group was not always the case (Portugal: [38], Spain: [12, 21, 39], France: [10, 22], Italy: [19, 20]). Most of the latter authors found the mammalian prey group to be the dominant one in the stone martens' diet, similar to what has been reported in other dietary studies in some Central and Northern European countries (Romania: [9], Czech Republic: [40], Germany: [17], Luxemburg: [41]). In our study area, which is characterised by a mosaic of natural habitats with oak forests, shrublands, and grasslands, insects are favoured both in numbers and diversity due to lack of agrochemicals. Furthermore, insects and especially grasshoppers were found in high densities in agricultural farmlands mainly in non-intensively cultivated crops, possibly due to the low use of insecticides (personal observations). In addition, it is evident that, as the stone marten can utilise a wide variety of habitats [42], it is not surprisingly that insects, which occurred in high numbers in most of natural and seminatural habitats in the study area, had both a consistent (Table 3) and a high presence in its diet (Table 1).

The second most important food group was fruits mainly from wild trees and shrubs, but also cultivated fruits were included in the stone marten's diet. Fruits have been reported as the main food type of the marten's diet in most studies carried out in Central and Northern European countries (see Clevenger [18]), as well as in a few cases in the Mediterranean [2, 21]. In our study area, wild fruits are available from early summer (e.g., mulberries) until late winter. Trees and shrubs which produce fruits are common species in the understory of broadleaved forest, but they especially occur along rain water gullies, while shrubs comprise the main component in

shrubland and grassland habitat types in the Mediterranean region. Furthermore, the human-altered agricultural envi-ronment studied here was dominated by wild shrubs (e.g., *Rubus* spp., *Prunus* spp., *Ficus* spp., etc.) and trees (*Pyrus* spp., *Morus* spp.) along the field margins, whereas extended agricultural areas on hillsides are covered by vineyards. Thus, fruits are almost always available in high numbers across the study area and thereby may constitute the food for a wide range of animals including the stone marten [12, 15, 43].

Our results demonstrate that a seasonal variation in the stone marten's diet apparently exists. This pattern could be related to the species' nutritional requirements throughout the year, although inter- and intraspecific competition could be involved [44]. In spring, the stone marten consumed a high disproportionate percentage (up to 84%F) of small-sized animals to fulfil its highly energy requirements that season, including insects, mammals, birds, reptiles, while in contrast, fruits and vegetable remains were not encountered frequently in its diet. In particular, small mammals seemed to constitute an important component in the diet of stone martens, as this was revealed by the log-linear analysis (Table 3 and Figure 1). Most of small mammals consumed were shrews, voles, and wood mice, but also carrion from dead domesticated animals was taken. During that season adults and juveniles rodents are encountered at a high rate and are easy to capture [3]. In addition, arthropods, included Myriapoda, Lepidoptera, and Coleoptera, composed a high proportion of the marten's diet [10]. Fruits made up a negligible portion of the diet in spring, and most of those found in stomachs were probably collected from human refuse. In spring the stone marten exhibited a generalist feeding strategy as was expected, consuming a wide range of prey types. In our study this was suggested by both the broad diet niche breadth index (B_A = 0.317) and the graph-ical representation of prey points (Figure 2(a)). However, in spring a high interindividual phenotype specialization emerged [45], as birds, reptiles, and molluscs have been eaten by relatively few individuals. It has been suggested that specialization of a generalist species could be attributed to interspecific competition as the result of a facultative behavioural change in certain resource use [46]. Further-more, asymmetric intraguild predation among mammalian carnivores can have effects analogous to those of competition [46–48]. In our study, the stone marten population suffered a high predation rate from the red fox year-round, but especially in spring [49]. Similar findings were reported in other studies where the stone marten was found to be a prey of other mesopredators, like the red fox [39]. Another explanation for the high interphenotype component could be the interference competition (intrapopulation compe-tition) that limits the range of food resources utilised by territorial individuals within their breeding space [50, 51]. Indeed, this may occur in our stone marten population due to the highly heterogeneous landscape in the study area [52], and; thus, different individuals are specializing on those food resources that are abundant within their home ranges [53], reflecting a variation in behavioural or physiological traits of individuals that determine resource-use efficiencies and different preferences [45, 54]. Furthermore, it has been suggested that strong interference competition leads to decreased rates of resource (food) intake per individual [51], and probably in our study this was the reason why more male stone martens were found with empty stomachs during spring, the critical breeding season [50].

In summer and autumn, fruits and insects became the most important foods in the stone marten diet [55]. In addition, reptiles, amphibians, and birds were consumed during these seasons but to a lesser degree. Both fruits and insects are abundant in the study area during summer. Due to seasonal ripeness of fruits in our study area, the stone marten shifts its diet seasonally to the most abundant species. Early summer mulberries composed the principal food items, whereas during the summer other abundant fruits (e.g., *Ficus* spp., *Pyrus* spp., *Rubus* spp., *Prunus* spp.) were taken, with grapes being dominant during autumn. Insects were also an important prey type taken [1], but Orthoptera predominate over other arthropods in the stone marten's diet during that period [10]. Crickets and grasshoppers (Orthoptera) are very abundant in the central part of Greece, and usually they appear to experience population explosions especially during the summer. In particular, mammals and partly birds did not contribute to the stone marten diet during this period. Although the food resource diversity increases in Mediterranean ecosystems during summer and autumn [52], the dietary niche breadth in our studied stone marten population decreased in both seasons. In addition, both sexes of the stone marten exhibited a specialized feeding strategy, consuming mainly fruits during summer and autumn [1, 18] (Figures 2(b) and 2(c)). However, an increased intraphenotype component to the niche breadth of males demonstrated a generalised diet on insects at the individual level [29]. Although there are no data on the availability of small mammals (rodents) and birds in our study area, we assumed that these animal groups were abundant during summer as they are in environments similar to those in our area [10, 38] and easy to capture by a predator as juveniles and nestlings appeared in high numbers during that season [19, 20, 56]. These prey groups could be considered optimal prey types for stone marten in energetic terms during summer, as the species has to breed due to delayed implantation and; thus, it has to fulfil its high energy demands by the more profitable prey [3, 57]. Surprisingly, we found a high proportion of fruits and insects in the stone marten's diet during summer. Fleshy fruits could be considered suboptimal food for stone marten from a nutritional point of view [44, 51], although they are nutritious and digestible [3]. Similarly, insects could be rated as suboptimal prey as they provide less energy and require a great deal of time for searching and capturing [51]. However, according to the optimal foraging theory, specialization on a less profitable food type can be optimal if the food type is sufficiently clumped [46]. The dietary switching to less profitable food types, such as fruits and insects [21, 37], could arise again by the intense interspecific competition between the red fox and the stone marten in our study area [58]. Therefore, two different scenarios could be associated with specialising the diet of the stone marten during summer. Strong interspecific competition may, on one hand, switch

to suboptimal prey types resulting in decreased dietary niche breadth [58]. Alternatively, stone marten displays a specialised feeding strategy at the population level as it expends a great deal of time and energy selectively searching for suboptimal food types [44, 51].

In winter arthropods and fruits were the dominant food types in the stone marten's diet, but also birds were taken in higher proportions than expected (Table 3 and Figure 1). In contrast with other studies where mammals dominated in diet during winter [2, 10, 13, 19, 38], in our studied stone marten population it showed an apparent selectivity for arthropods and fruits [55]. Other studies have also shown that the stone marten fed on birds in winter [2, 20], as did other mustelids [59]. Although insects were not abundant during that season, the stone marten exploited high numbers of this prey items both as adult and larvae beetles (Coleoptera). Furthermore, even when the period of ripe fruit had passed, stone marten consumed high numbers of fleshy fruits overwinter on the plant and those which had remained intact until late season, such as plums and wild pears [40]. In winter, both sexes displayed a relative specialisation for fruits and arthropods (Figure 2(d)), whereas neither intra nor inter-phenotype component at the niche breadth was detected during that season. Even in that season, the specialisation is not pronounced ($B_A = 0.303$) as in summer and in autumn; the stone marten could be considered a relative specialist, due to the high contribution of insects and fruits and to the occasional participation of small mammals, birds, and other food items in its diet (Figure 2(d)). Although a broader diet during unproductive environments, as winter, was revealed [51], the stone marten seemed to spend more time and energy searching for insects' larvae digging from fallen woods, demonstrating specialist behaviour during that season.

Finally, in our study the impact of stone marten on economically important wildlife species (i.e., the European hare) or domestic animals could be regarded negligible. In stone marten stomachs, there were occasionally found items of food of unexpected size, such as domestic sheep and goat, and these were nearly taken as carrion.

4.2. Specialist or Generalist Mustelid? A specialist is an animal which exploits efficiently a narrow prey spectrum regardless of its availability [36, 44]. On the other hand, a generalist is an animal which can exploit several alternative prey types according to their availability. In our study, both classic niche breadth indices ($B_{Am} = 0.156$, $B_{Af} = 0.107$) and graphical representation of prey-specific abundance against the frequency of occurrence of prey types suggest that stone marten exhibited a specialised feeding strategy. Furthermore, at the individual and at the population level, it showed a mixed feeding strategy according to seasons. During three out of the four seasons, stone marten indicated a pronounced population specialization while a clear generalization of male individuals was observed during summer and autumn. An evident generalised diet was revealed only during spring, but again a specialised tendency for few food groups was observed in the diet of some individuals. Unfortunately, only one study of the stone marten food habits has shown individual specialization within a generalist population [1]. However, individual specialization has been detected in other species of the family Mustelidae (pine marten *Martes martes*: [60], American marten *Martes americana*: [53], genet *Genetta genetta*: [61], badger *Meles meles*: [62]). Most of the studies conducted on food habits of stone marten have used analysis of faeces, and probably they failed to detect dietary specialization neither at the individual nor at the population level. With the results of our study, we suppose that stone marten exploited heavily one or two food groups year-round [18], and; thus, its tendency towards specialization than generalization is more evident [36].

5. Conclusion

In conclusion, this work is the first attempt to investigate both the diet composition and the feeding strategy of stone marten by analysing its stomach contents from mainland Greece. Stone marten shows seasonal differences in diet as well as mixed feeding strategies at least at the local level. The tendency of consumption of particular prey items seasonally, which is not always associated with an increased abundance in the environment, reflects, on one hand, population specialist behaviour, while on the other hand, it shows the individual flexibility on different food resources (intra and interindividual specialization) of this medium-sized mustelid. Possible mechanisms which have driven the stone marten to a more specialised diet both at the individual and at the population level were interference competition and intraguild predation. However, the extent to which stone marten behaves as a specialist, at least locally, under a strong inter- and/or intraspecific competition in mainland Greece could be better clarified by comparing the food habits from similar areas where no intraguild predation occurs.

Acknowledgments

This paper was a part of the Project entitled the *"Predation of red fox and stone marten on wildlife species in Sterea Hellas"* which was funded by the 4th Hunting Federation of Sterea Hellas. The authors are most grateful to G. Liagas, P. Panagitsas, J. Psomas, and G. Kourelas for their help in the fieldwork and especially to Margaret Gallacher-Koletsou for the linguistic assistance. They thank the Direction of the National Parks, Aesthetic Forests and Hunting of the Ministry of Agriculture for the permission to collect samples during the three years of the project. The authors declare that the experiments comply with the Greek and the EU laws.

References

[1] P. Genovesi, M. Secchi, and L. Boitani, "Diet of stone martens: an example of ecological flexibility," *Journal of Zoology*, vol. 238, no. 3, pp. 545–555, 1996.

[2] M. Posillico, P. Serafini, and S. Lovari, "Activity patterns of the stone marten *Martes foina* Erxleben, 1777, in relation to some environmental factors," *Hystrix*, vol. 7, no. 1-2, pp. 79–97, 1995.

[3] D. Macdonald, *European Mammals. Evolution and Behaviour*, Harper Collins Publishers, London, UK, 1993.

[4] A. M. Rasmussen and A. B. Madsen, "The diet of the Stone Marten *Martes foina* in Denmark," *Natura Jutlandica*, vol. 21, no. 8, pp. 141–144, 1985.

[5] M. Lucherini and G. Crema, "Diet of urban stone martens in Italy," *Mammalia*, vol. 57, pp. 274–277, 1993.

[6] A. Tóth, "Data to the diet of the urban stone marten (*Martes foina*) in Budapest," *Opuscula Zoologica Budapest*, vol. 31, pp. 113–118, 1998.

[7] J. Lanszki, "Feeding habits of stone martens in a Hungarian village and its surroundings," *Folia Zoologica*, vol. 52, no. 4, pp. 367–377, 2003.

[8] A. Tikhonov, P. Cavallini, T. Maran et al., "*Martes foina*. IUCN red list of threatened species," version 2009.2, 2008, http://www.iucnredlist.org/.

[9] J. Romanowski and G. Lesinski, "A note of the diet of stone marten in southeastern Romania," *Acta Theriologica*, vol. 36, pp. 201–204, 1991.

[10] T. Lode, "Feeding habits of the stone marten *Martes foina* and environmental factors in western France," *Zeitschrift für Säugetierkunde*, vol. 59, no. 3, pp. 189–191, 1994.

[11] P. Tryjanowski, "Food of the Stone marten (*Martes foina*) in Nietoperek Bat Reserve," *Zeitschrift für Säugetierkunde*, vol. 62, no. 5, pp. 318–320, 1997.

[12] M. Delibes, "Feeding habits of the Stone Marten, *Martes foina* (Erxleben, 1777), in northern Burgos, Spain," *Zeitschrift für Säugetierkunde*, vol. 43, pp. 282–288, 1978.

[13] P. Serafini and S. Lovari, "Food habits and trophic niche overlap of the red fox and the stone marten in a Mediterranean rural area," *Acta Theriologica*, vol. 38, no. 3, pp. 233–244, 1993.

[14] M. Pandolfi, A. M. De Marinis, and I. Petrov, "Fruit as a winter feeding resource in the diet of stone marten (*Martes foina*) in east-central Italy," *Zeitschrift für Säugetierkunde*, vol. 61, no. 4, pp. 215–220, 1996.

[15] T. Bermejo and J. Guitian, "Fruit consumption by foxes and martens in NW Spain in autumn: a comparison of natural and agricultural areas," *Folia Zoologica*, vol. 49, no. 2, pp. 89–92, 2000.

[16] F. Amores, "Feeding habits of the stone martens, *Martes foina*, in south western Spain," *Säugetierkundliche Mitteilungen*, vol. 28, pp. 316–322, 1980.

[17] K. Skirnisson, "Untersuchungen zum Raum-Zeit-System freilebender Steinmarder (*Martes foina* Erxleben, 1777)," *Beiträge zur Wildbiologie*, vol. 6, pp. 1–200, 1986.

[18] A. Clevenger, "Feeding ecology of Eurasian pine marten *Martes martes* and stone marten *Martes foina* in Europe," in *Martens, Sables, and Fishers: Biology and Conservation*, S. W. Buskirk, A. S. Harestad, M. G. Raphael, and R. A. Powell, Eds., pp. 326–340, Cornell University Press, Ithaca, NY, USA, 1994.

[19] A. Martinoli and D. G. Preatoni, "Food habits of the stone marten (*Martes foina*) in the upper Aveto Valley (northern Apennines, Italy)," *Hystrix*, vol. 7, no. 1-2, pp. 137–142, 1995.

[20] S. Bertolino and B. Dore, "Food habits of the stone marten *Martes foina* in "La Mandria" Regional Park (Piedmont region, north-western Italy)," *Hystrix*, vol. 7, no. 1-2, pp. 105–111, 1995.

[21] R. Barrientos and E. Virgós, "Reduction of potential food interference in two sympatric carnivores by sequential use of shared resources," *Acta Oecologica*, vol. 30, no. 1, pp. 107–116, 2006.

[22] G. Cheylan and P. Bayle, "Le regime alimentaire de quatre especes de mustelides en Provence: la fouine *Martes foina*,

le blaireau *Meles meles*, la belette *Mustela nivalis* et le putois *Putorius putorius*," *Faune de Provence*, vol. 9, pp. 14–26, 1988.

[23] A. Hernandez, "The role of birds and mammals in the dispersal ecology of *Rhamnus alpinus* (Rhamnaceae) in the Cantabrian Mts," *Folia Zoologica*, vol. 42, no. 2, pp. 105–109, 1993.

[24] E. Andresen, "Ecological roles of mammals: the case of seed dispersal," in *Priorities for the Conservation of Mammalian Diversity. Has the Panda had its Day?* A. Entwistle and N. Dunstone, Eds., pp. 11–25, Cambridge University Press, Cambridge, UK, 2000.

[25] L. J. Korschgen, "Procedures for food-habit analyses," in *Wildlife Management Techniques Manual*, S. D. Schemnitz, Ed., pp. 113–127, The Wildlife Society, Washington, DC, USA, 1980.

[26] J. A. Litvaitis, "Investigating food habits of terrestrial vertebrates," in *Research Techniques in Animal Ecology: Controversies and Consequences*, L. Boitani and T. K. Fuller, Eds., pp. 165–190, Columbia University Press, New York, NY, USA, 2000.

[27] J. C. Reynolds and N. J. Aebischer, "Comparison and quantification of carnivore diet by faecal analysis: a critique, with recommendations, based on a study of the fox *Vulpes vulpes*," *Mammal Review*, vol. 21, no. 3, pp. 97–122, 1991.

[28] P. Ciucci, L. Boitani, E. R. Pelliccioni, M. Rocco, and I. Guy, "A comparison of scat-analysis methods to assess the diet of the wolf *Canis lupus*," *Wildlife Biology*, vol. 2, no. 1, pp. 37–48, 1996.

[29] P. A. Amundsen, H. M. Gabler, and F. J. Staldvik, "A new approach to graphical analysis of feeding strategy from stomach contents data-modification of the Costello (1990) method," *Journal of Fish Biology*, vol. 48, no. 4, pp. 607–614, 1996.

[30] N. K. Papageorgiou, C.G. Vlachos, A. Sfougaris, and D. E. Bakaloudis, *Identification of Reptiles by Scale Morphology*, University Studio Press, Thessaloniki, Greece, 1993.

[31] N. K. Papageorgiou, C. G. Vlachos, and D. E. Bakaloudis, *Identification of Mammals by Skull and Dental Morphology*, University Studio Press, Thessaloniki, Greece, 1997.

[32] C. J. Krebs, *Ecological Methodology*, Benjamin-Cummings, Menlo Park, Calif, USA, 2nd edition, 1999.

[33] E. R. Pianka, "The structure of lizard communities," *Annual Review of Ecology and Systematics*, vol. 4, pp. 53–74, 1973.

[34] R. R. Sokal and J. F. Rohlf, *Biometry. The Principles and Practice of Statistics in Biological Research*, W.H. Freeman, New York, NY, USA, 3rd edition, 1995.

[35] S. E. Fienberg, *The Analysis of Cross-Classified Categorical Data*, MIT Press, Cambridge, UK, 2nd edition, 1994.

[36] S. Erlinge, "Specialists and generalists among the mustelids," *Lutra*, vol. 29, pp. 5–11, 1986.

[37] M. J. Santos, B. M. Pinto, and M. Santos-Reis, "Trophic niche partitioning between two native and two exotic carnivores in SW Portugal," *Web Ecology*, vol. 7, pp. 53–62, 2007.

[38] J. C. Carvalho and P. Gomes, "Feeding resource partitioning among four sympatric carnivores in the Peneda-Gerês National Park (Portugal)," *Journal of Zoology*, vol. 263, no. 3, pp. 275–283, 2004.

[39] J. M. Padial, E. Avila, and J. M. Sanchez, "Feeding habits and overlap among red fox (*Vulpes vulpes*) and stone marten (*Martes foina*) in two Mediterranean mountain habitats," *Mammalian Biology*, vol. 67, no. 3, pp. 137–146, 2002.

[40] M. Ryšavá-Nováková and P. Koubek, "Feeding habits of two sympatric mustelid species, European polecat Mustela putorius and stone marten *Mattes foina*, in the Czech Republic," *Folia Zoologica*, vol. 58, no. 1, pp. 66–75, 2009.

[41] A. Baghli, E. Engel, and R. Verhagen, "Feeding habits and trophic niche overlap of two sympatric Mustelidae, the polecat *Mustela putorius* and the beech marten *Martes foina*," *Zeitschrift fur Jagdwissenschaft*, vol. 48, no. 4, pp. 217–225, 2002.

[42] M. J. Santos and M. Santos-Reis, "Stone marten (*Martes foina*) habitat in a Mediterranean ecosystem: effects of scale, sex, and interspecific interactions," *European Journal of Wildlife Research*, vol. 56, no. 3, pp. 275–286, 2010.

[43] C. M. Herrera, "Frugivory and seed dispersal by carnivorous mammals, and associated fruit characteristics, in undisturbed Mediterranean habitats," *Oikos*, vol. 55, no. 2, pp. 250–262, 1989.

[44] D. W. Stephens and J. R. Krebs, *Foraging Theory*, Princeton University Press, Princeton, NJ, USA, 1986.

[45] J. Roughgarden, "Evolution of niche width," *The American Naturalist*, vol. 106, pp. 683–718, 1972.

[46] D. J. Futuyma and G. Moreno, "The evolution of ecological specialization," *Annual Review of Ecology and Systematics*, vol. 19, pp. 207–233, 1988.

[47] F. Palomares and T. M. Caro, "Interspecific killing among mammalian carnivores," *The American Naturalist*, vol. 153, no. 5, pp. 492–508, 1999.

[48] P. L. Bright, "Lessons from lean beasts: conservation biology of the mustelids," *Mammal Review*, vol. 30, no. 3-4, pp. 217–226, 2000.

[49] M. Papakosta, D. Bakaloudis, K. Kitikidou, C. Vlachos, and E. Chatzinikos, "Dietary overlap among seasons and habitats of red fox and stone marten in Central Greece," *European Journal of Scientific Research*, vol. 45, no. 1, pp. 122–127, 2010.

[50] L. Partridge and P. Green, "Intraspecific feeding specializations and population dynamics," in *Behavioural Ecology: Ecological Consequences of Adaptive Behaviour*, R. M. Sibly and R. H. Smith, Eds., pp. 207–226, Blackwell Scientific Publications, Oxford, UK, 1985.

[51] M. Begon, J. L. Harper, and C. R. Townsend, *Ecology: Individuals, Populations and Communities*, Blackwell Science, Oxford, UK, 3rd edition, 1996.

[52] F. di Castri, "Animal biogeography and ecological niche," in *Mediterranean Types Ecosystems*, F. di Castri and H. A. Mooney, Eds., pp. 279–283, Chapman & Hall, London, UK, 1973.

[53] M. Ben-David, R. W. Flynn, and D. M. Schell, "Annual and seasonal changes in diets of martens: evidence from stable isotope analysis," *Oecologia*, vol. 111, no. 2, pp. 280–291, 1997.

[54] D. I. Bolnick, R. Svanbäck, J. A. Fordyce et al., "The ecology of individuals: incidence and implications of individual specialization," *The American Naturalist*, vol. 161, no. 1, pp. 1–28, 2003.

[55] M. Posłuszny, M. Pilot, J. Goszczyński, and B. Gralak, "Diet of sympatric pine marten (*Martes martes*) and stone marten (*Martes foina*) identified by genotyping of DNA from faeces," *Annales Zoologici Fennici*, vol. 44, no. 4, pp. 269–284, 2007.

[56] A. Brangi, "Seasonal changes of trophic niche overlap in the stone marten (*Martes foina*) and the red fox (*Vulpes vulpes*) in a mountainous are of the northern Apennines (N Italy)," *Hystrix*, vol. 7, no. 1-2, pp. 113–118, 1995.

[57] P. J. Moors, "Sexual dimorphism in the body size of mustelids (Carnivora): the roles of food habits and breeding system," *Oikos*, vol. 34, pp. 147–158, 1980.

[58] J. A. Wiens, "Fat times, lean times and competition among predators," *Annual Review of Ecology, Evolution, and Systematics*, vol. 8, no. 10, pp. 348–349, 1993.

[59] A. Zalewski, "Geographical and seasonal variation in food habits and prey size of European pine martens," in *Martens and Fishers (Martes) in Human-Altered Environments: An International Perspective*, D. J. Harrison, A. K. Fuller, and G. Proulx, Eds., pp. 77–98, Springer, Berlin, Germany, 1994.

[60] S. Rosellini, I. Barja, and A. Pineiro, "The response of European pine marten (*Martes martes* L.) feeding to the changes of small mammal abundance," *Polish Journal of Ecology*, vol. 56, no. 3, pp. 497–503, 2008.

[61] E. Virgós, M. Llorente, and Y. Cortés, "Geographical variation in genet (*Genetta genetta* L.) diet: a literature review," *Mammal Review*, vol. 29, no. 2, pp. 119–128, 1999.

[62] C. Fischer, N. Ferrari, and J. M. Weber, "Exploitation of food resources by badgers (*Meles meles*) in the Swiss Jura Mountains," *Journal of Zoology*, vol. 266, no. 2, pp. 121–131, 2005.

4

Patterns of Diversity of the Rissoidae (Mollusca: Gastropoda) in the Atlantic and the Mediterranean Region

Sérgio P. Ávila,[1,2,3] Jeroen Goud,[4] and António M. de Frias Martins[1,2]

[1] *Departamento de Biologia, Universidade dos Açores, 9501-801 Ponta Delgada, Açores, Portugal*
[2] *CIBIO-Açores, Universidade dos Açores, 9501-801 Ponta Delgada, Açores, Portugal*
[3] *MPB-Marine PalaeoBiogeography Working Group of the University of the Azores, Rua da Mãe de Deus, 9501-801 Ponta Delgada, Açores, Portugal*
[4] *National Museum of Natural History, Invertebrates, Naturalis Darwinweg, Leiden, P.O. Box 9517, 2300 RA Leiden, The Netherlands*

Correspondence should be addressed to Sérgio P. Ávila, avila@uac.pt

Academic Editor: Cang Hui

The geographical distribution of the Rissoidae in the Atlantic Ocean and Mediterranean Sea was compiled and is up-to-date until July 2011. All species were classified according to their mode of larval development (planktotrophic and nonplanktotrophic), and bathymetrical zonation (shallow species—those living between the intertidal and 50 m depth, and deep species—those usually living below 50 m depth). 542 species of Rissoidae are presently reported to the Atlantic Ocean and the Mediterranean Sea, belonging to 33 genera. The Mediterranean Sea is the most diverse site, followed by Canary Islands, Caribbean, Portugal, and Cape Verde. The Mediterranean and Cape Verde Islands are the sites with higher numbers of endemic species, with predominance of *Alvania* spp. in the first site, and of *Alvania* and *Schwartziella* at Cape Verde. In spite of the large number of rissoids at Madeira archipelago, a large number of species are shared with Canaries, Selvagens, and the Azores, thus only about 8% are endemic to the Madeira archipelago. Most of the 542-rissoid species that live in the Atlantic and in the Mediterranean are shallow species (323), 110 are considered as deep species, and 23 species are reported in both shallow and deep waters. There is a predominance of nonplanktotrophs in islands, seamounts, and at high and medium latitudes. This pattern is particularly evident in the genera *Crisilla*, *Manzonia*, *Onoba*, *Porosalvania*, *Schwartziella*, and *Setia*. Planktotrophic species are more abundant in the eastern Atlantic and in the Mediterranean Sea. The results of the analysis of the probable directions of faunal flows support the patterns found by both the Parsimony Analysis of Endemicity and the geographical distribution. Four main source areas for rissoids emerge: Mediterranean, Caribbean, Canaries/Madeira archipelagos, and the Cape Verde archipelago. We must stress the high percentage of endemics that occurs in the isolated islands of Saint Helena, Tristan da Cunha, Cape Verde archipelago and also the Azores, thus reinforcing the legislative protective actions that the local governments have implemented in these islands during the recent years.

1. Introduction

Rapoport's latitudinal rule relates geographical distribution with latitude [1, 2]. This rule states that the range of the geographical distribution of species increases with latitude [3]. Several hypotheses were provided: the seasonal variability hypothesis [2, 4], the differential extinction hypothesis [3], the competition hypothesis [5–7], or the Milankovitch climate oscillations, which force larger distributional changes [8].

Although studies relating biological diversity with latitude usually use higher taxonomical categories, recent papers restricted to checklists of marine molluscs have been used to address this issue [9–12]. The papers of Roy et al. [9, 10], who used lists with 3,916 marine Caenogastropod species geographically distributed along the north and Central America shores between 10°S–83°N (west-Atlantic shores: 2,009 species; east-Pacific shores: 1,907 species), confirmed the latitudinal gradient pattern, with the number of species decreasing with latitude. However, they did not confirm the seasonal variability hypothesis, as the mean geographical range decreased from the tropics to higher latitudes (see [9], Figure 3). These authors concluded that the energy hypothesis [13, 14], that correlates the surface incident solar radiation and its correspondent average sea surface temperature (SST) with latitude and number of species, better explained this biogeographical pattern. This relationship is higher outside the tropical latitudes [9]. Notwithstanding solar radiation is a simple function of the latitude, SSTs are a complex function of climatic variables, oceanic currents, and other factors (e.g., local submarine topography, upwelling, estuarine systems with high discharge of both nutrients and sediments, etc.) [10, 15, 16].

The Rissoidae are a family of small-sized, marine to brackish-water gastropod molluscs. This very diverse family was taxonomically reviewed by Wenz [17], Coan [18], Nordsieck [19] and Ponder [20, 21]. Two subfamilies are presently recognized: Rissoinae and Rissoininae [22]. Given its species-abundance, easy preservation in fossilized form, and the fact that key-elements of the life cycle can be obtained from shell morphology, rissoids have a large potential for evolutionary studies [23].

Published information about the Atlantic and Mediterranean Rissoidae is vast and is scattered among a wide variety of journals but, with a few exceptions, these studies are typically geographically localized. Many species descriptions are usually based on shell morphology and on only a few specimens, most of them dead shells. At present, there is a lack of a background scenario of the geographical distribution for this family in the Atlantic and in the Mediterranean. No phylogeny has been established for this family.

To our knowledge, this is the first attempt to summarize present information about the geographic distributional pattern of this family in the Atlantic Ocean and the Mediterranean Sea, with the purpose of identifying the biotic similarities between areas.

2. Materials and Methods

2.1. Geographical Distribution. The geographical distribution of the Rissoidae in the Atlantic Ocean and Mediterranean Sea was compiled through an exhaustive search of the primary literature and is up to date until July 2011. The following sites and references were considered:

- ARC: Arctic: above 75° N: Warén [24, 25], Hansson [26],

- GRE: Greenland, western shores of Baffin Island, Baffin Bay, Davis Strait, and Labrador Sea: Bouchet and Warén [27], Hansson [26],

- ICE: Iceland: Warén [28, 29],

- SCA: Scandinavia: Norway Sea, Skagerrak and Kattegat, Baltic Sea and Faroe Islands: Fretter and Graham [30], Warén [24, 29], Hansson [26],

- BRI: British Isles: Smith [31], Fretter and Graham [30], Killeen and Light [32],

- POR: western Atlantic Iberian façade (from Cabo Vilán, western Galician shores, down to Cape São Vicente) and southern shores of Algarve, Portugal): Nobre [33, 34], Nobre and Braga [35], Macedo et al. [36], Rolán [37],

- MED: Mediterranean: Nordsieck [19], Aartsen and Fehr-de-Wal [38], Aartsen [39–47], Verduin [48–50], Aartsen and Verduin [23, 51], Palazzi [52], Aartsen et al. [53], Amati [54–56], Amati and Nofroni [57–59], Amati and Oliverio [56], Oliverio [60–62], Oliverio et al. [63], Aartsen and Linden [64], Linden and Wagner [65], Hoenselaar and Moolenbeek [66], Aartsen and Menkhorst [53], Giusti and Nofroni [67], Aartsen et al. [68], Amati et al [69], Hoenselaar and Hoenselaar [70], Nofroni and Pizzini [71], Oliverio et al. [72], Aartsen and Engl [73], Smriglio and Mariottini [74], Margelli [75], Bogi and Galil [76], Buzzurro and Landini [77], Peñas et al. [78], Oliver and Templado [79], CIESM, CLEMAM,

- AZO: Azores: Watson [80], Dautzenberg [81], Amati [82], Gofas [83], Oliverio et al. [72], Linden [84], Linden and van Aartsen [85], Ávila [86–88],

- LUS: Lusitanian group of seamounts (a chain of seamounts located between Portugal and Madeira): Gorringe, Josephine, Ampère, Seine: Ávila and Malaquias [89], Beck et al. [90], Gofas [91],

- MET: Meteor group of seamounts (located about 600 km south of the Azores): Great Meteor, Irving, Atlantis, Hyères, Plato, Tyro, Cruiser: Gofas [91],

- MAD: Madeira, Porto Santo and Desertas Islands, Nobre [92], van Aartsen [47], Palazzi [52], Verduin [93], Moolenbeek and Hoenselaar [94, 95], Segers et al. [96],

- SEL: Selvagens Islands: Verduin [93], Amati [97],

- CAN: Canary Islands: van Aartsen [47], Moolenbeek and Faber [98], Rolán [99], Verduin [93], Linden and Wagner [100], Moolenbeek and Hoenselaar [94, 95, 101], Segers [102], Hernández-Otero et al. [103],

- CAP: Cape Verde archipelago: Rolán [104], Moolenbeek and Rolán [105], Templado and Rolán [106], Rolán and Rubio [107], Rolán and Luque [108], Rolán [109], Rolán and Oliveira [110],

- STH: Saint Helena Island: Smith [111], MALACOLOG,

- TRS: Tristão da Cunha Island: Worsfold et al. [112], MALACOLOG,

- WAF: West African shores—Atlantic Morocco, from Straits of Gibraltar south, Western Sahara, and Mauritania, Cape Verde (Senegal): Verduin [49], Gofas and Warén [113], Moolenbeek and Piersma [114], Rolán and Fernandes [115], Gofas [116, 117],

- ANG: Angola: Rolán and Ryall [118], Rolán and Fernandes [115], Gofas [116],

- NSC: New Scotia biogeographical province—Atlantic shores of USA, between Newfoundland (50° N) and Cape Cod (42° N): MALACOLOG,

- VIR: Virginian biogeographical province *sensu* Engle and Summers [119]—Atlantic shores of USA, between Cape Cod (42° N) and Cape Hatteras, North Carolina (35° N): MALACOLOG,

- CRL: Carolinian biogeographical province–Atlantic shores of USA, between Cape Hatteras, North Carolina (35° N) and Cape Canaveral (28°30′ N): Rex et al. [120], MALACOLOG,

- TRO: Tropical biogeographical province (from now on generically designated as "Caribbean")—Atlantic shores of USA, south of Cape Canaveral (28°30′ N), including western and eastern shores of Florida, Gulf of Mexico (Louisiana and Texas shores, as well as Yucatan Peninsula, México), Bahamas, Caribbean Sea, south to Cabo Frio (Brazil) (23°S): Dall [121], Baker et al. [122], Faber and Moolenbeek [123], Jong and Coomans [124], Leal and Moore [125], Faber [126], Leal [127], Rolán [128], Espinosa and Ortea [129], Rosenberg et al. [130] MALACOLOG,

- BRA: Biogeographical province of Brazil (this includes the Paulista and Patagonic Provinces *sensu* Palacio [131])—from Cabo Frio (23°S) south to River Plate (35°S): MALACOLOG,

- SSA: southeast of South America—biogeographical province of Malvinas (*sensu* Palacio [131]) Atlantic shores from River Plate (35°S) south to Tierra del Fuego and Cape Horn, including Los Estados Island, Falkland Islands (Malvinas), Burdwood Bank and South Georgia Island: Ponder [132], Ponder and Worsfold [133],

- ANT: Antarctic—from 60°S south, including South Orkney Islands (Signy Island), South Shetland islands, Antarctic Peninsula and Weddell Sea: Ponder [132].

We have also consulted other bibliographical sources, with a wider systematical or geographical subject, such as Babio and Thiriot-Quiévreux [134], Aartsen [40], Fretter and Graham [30], Ponder [21], Verduin [135], Templado and Rolán [136], Hoenselaar and Moolenbeek [66], Moolenbeek and Hoenselaar [137], Moolenbeek and Faber [138–140], Sleurs [141–143], Hoenselaar [70, 144], Bouchet and Warén [27], Sleurs and Preece [145], Warén [146], Hoenselaar and Goud [147], Goud [148], Gofas et al. [149], Rolán [150], Gofas [91], and Garilli [151]. MALACOLOG, CIESM, CLEMAM, and WoRMS web databases were very useful and widely consulted.

2.2. Bathymetrical Zonation. The bathymetrical zonation considers shallow species (those living between the intertidal and 50 m depth) and deep species (those usually living below 50 m depth). The choice of the threshold at 50 m depth is related with the following reasons: (i) algal species to which Rissoidae are very often associated are rare below 50 m depth; (ii) direct sampling by scuba-diving is more frequent in waters less than 50 m depth; (iii) in waters deeper than 50 m depth, usually the samplings are obtained via indirect methodologies (grabs, most often).

The complete database was last updated in October 2011 and is available from the authors upon request.

2.3. Modes of Larval Development. All species were classified according to their mode of larval development and bathymetrical zonation. Rissoids lay ovigerous capsules in the substrate that originate larvae with different modes of larval development. The extension of the larval phase reflects on the capabilities of dispersal and this has important ecological and historical biogeographical implications, related with the geographical distribution of the species. Two types of larval development were considered: planktotrophic (with a free-swimming feeding stage) and nonplanktotrophic (either lecithotrophic or direct development, both without a free-swimming feeding stage) [21, 89, 152, 153]. As almost nothing is known about the life cycle of rissoids, the mode of larval development was determined indirectly, through the analysis of the protoconch. In the Rissoidae family, multi-spiral protoconchs, with a small nucleus (usually less than 200 μm) and with several whorls (typically more than 2), are associated with a planktotrophic mode of larval development. In most of the planktotrophic species, it is possible to discern between protoconch I and protoconch II. This is especially evident in species that possess a "sinusigera" larva, as is the case of *Alvania cancellata* [154]. Paucispiral protoconchs, with nucleus usually larger than 200 μm, typically more than 300 μm [87], and with about 1–1.5 whorls are related with a nonplanktotrophic mode of larval development [87, 88, 152, 155–157].

2.4. Biotic Similarities between Areas: Parsimony Analysis of Endemicity (PAE). Rosen and Smith [158] developed the Parsimony Analysis of Endemicity or PAE method, which, under a cladistic framework, classifies areas in accordance with their shared taxa. This method is particularly useful when the researchers have no phylogenetic information to incorporate into the analysis, therefore, providing a primary description of the distributional pattern of the taxa and the biotic similarities between areas [159]. The original data resumes to a matrix with the presence/absence of the selected species in the studied areas. Thus, data for PAE consists of Area x Taxa matrices and the cladograms that result from this analysis represent nested sets of areas and correspond to the most parsimonious solution [160, 161]. The original Area x Taxa matrix with the geographical distribution of the Rissoidae was split in two, the first containing the shallow (<50 m depth), and the second with the deep (>50 m) rissoid species. Both shallow and deep rissoid species' matrices were analysed as follows.

Four species considered as Lessepsian and thus reported to the Mediterranean were removed from the initial database: *Alvania dorbignyi* [162], *Rissoina bertholleti*, *Rissoina spirata* (Sowerby, 1825), and *Voorwindia tiberiana* [163]. Unique taxa (restricted endemics, autapomorphies) were also removed prior to the PAE analysis [158]. No cosmopolitan taxa (plesiomorphies) were found in either of the matrices (shallow and deep rissoids). Low diversity localities (e.g., Brazil, Virginian Province, Tristan da Cunha Island, and Antarctic) were kept in the analysis. One outgroup with an all-0 score was added to the first row of the data matrix, to allow topologies to be rooted [160, 164]. Tree reconstruction was based on the heuristic search algorithm in PAUP* version 4.0 beta 10 [165] including 1,000 random stepwise-addition sequence replicates with tree-bisection-reconnection (TBR) and MULTREES on, and Acctran optimization in effect, but restricting the number of optimal trees per replicate to 1. Consensus trees (50% majority rule) were generated when more than 1 parsimonious tree resulted from the analysis, using 200 random addition sequence replicates with TBR and MULTREES on, and Acctran optimization in effect. One thousand bootstrap replicates with 100 random additions per replicate, with TBR and MULTREES on, and decay indices ("branch support" of Bremer [166]), measured support for individual nodes.

2.5. Biotic Similarities between Areas: Probable Directions of Faunal Flows. The analysis of the historical relationships between the selected areas was complemented by using the following formulas (X_A and X_B) for each pair of areas (A and B) [167]:

$$X_A = \frac{(\text{number of species present in areas } A \text{ and } B)}{A},$$

$$X_B = \frac{(\text{number of species present in areas } A \text{ and } B)}{B}, \tag{1}$$

where A is the total number of rissoid species present in area A, and B is the total number of rissoid species present in area B. When a faunal flow happened in historical times, from a source area to the target area, we expect the target area to show a subset of the species present in the source area. So, different values of the two indices (X_A and X_B) are expectable, and the source area must have the smaller value [167].

3. Results

Rissoidae family comprises 47 valid genera, some with a worldwide distribution (*Manzonia, Rissoina, Zebina, Stosicia, Pusillina,* and *Alvania*), while others are geographically restricted. According to Ponder [21], the genera *Attenuata, Lamellirissoina, Lironoba, Lucidestea, Merelina, Parashiela, Striatestea,* and *Voorwindia* are restricted to the Indo-Pacific. However, Leal [127] reported a species of *Lironoba* from Brazil (Tropical Province). Some genera solely occur in the Indian Ocean, for example, *Fenella* (Madagascar and Red Sea), while the genera *Tomlinella* is restricted to Reunion Island and Mauritius and the recently described *Porosalvania* is restricted to the Meteor group of seamounts [91].

3.1. Geographical Distribution. Five hundred and forty-two species of Rissoidae are presently reported to the Atlantic Ocean and the Mediterranean Sea, belonging to 33 genera. Six genera are represented by a single species: *Amphirissoa cyclostomoides* Dautzenberg and Fischer, [168], a nonplanktotrophic deep species that occurs in the Caribbean, in the Azores and in the Meteor group of seamounts; *Benthonella tenella* (Jeffreys [169], *B. gaza* Dall [170], and *B. fischeri* Dall, 1889 are synonyms), a widespread planktotrophic deep

species; *Lironoba* sp., a nonplanktotrophic species restricted to the Atol das Rocas [127]; *Galeodinopsis tiberiana* (Coppi, 1876) (= *Alvania fariai* [115]), a shallow planktotrophic species reported to the Cape Verde archipelago, West Africa, and Angola; *Pontiturboella rufostrigata* [171], a species restricted to the Black Sea; *Voorwindia tiberiana* [163], a shallow planktotrophic *Lessepsian* species that occurs in the Mediterranean.

The Mediterranean Sea is the most diverse site, with 160 species of Rissoidae, followed by Canary Islands (89 species), Caribbean (77), Portugal (74), and Cape Verde (67). The lowest diversity sites are the Carolinian Province (18 species), Greenland (16), Arctic (13), Angola (11), New Scotia Biogeographical Province (10), Antarctic (8), Virginian Biogeographical Province, Tristan da Cunha Island, and Brazil (all with just 7 species, Table 1).

The genera *Alvania* (74 species), *Rissoa* (26), *Setia* (18), and *Pusillina* (11) are species-abundant in the Mediterranean and along the Portuguese shores. *Boreocingula* and *Frigidoalvania* (as the name indicates) are restricted to higher latitudes (Arctic, Greenland, Iceland, and Scandinavia). *Boreocingula* also occurs in the British Isles, and *Frigidoalvania* is reported to the Atlantic shores of North America (NSC and VIR). *Onoba* is a genus with high number of species at Iceland and Greenland (9 and 5, resp.), but the most diverse sites are in the South Atlantic: 22 species at southeast of South America and 6 species at Tristan da Cunha Island and Antarctic (Table 1).

Benthonellania, Folinia, Microstelma, Rissoina, and *Zebina* have a higher number of species in the Caribbean (Tropical Province). *Crisilla* and *Manzonia* are particularly species-diverse in the Macaronesian archipelagos, especially at Canary Islands, Selvagens and Madeira, Porto Santo, and Desertas Islands. *Manzonia* is a specious genus also at the Lusitanian group of seamounts. *Schwartziella* spp. is very abundant at Cape Verde archipelago (26 species) as well as in the Carolinian and Tropical Provinces, and at Saint Helena Islands (4, 9, and 5 species, resp.). *Porosalvania* is a newly described endemic genus to the Meteor group of seamounts where it radiated into a number of species (Table 1).

Stosicia and *Voorwindia* occur predominantly in the Indo-Pacific Region [21], with just two species reported to the Western Atlantic, *Stosicia aberrans* [172], and *Stosicia houbricki* [173] (= *Stosicia fernandezgarcesi* [129], all restricted to the Tropical Province; *Voorwindia tiberiana* [163] occurs in the Mediterranean, where this *Lessepsian* species is considered as alien (nonestablished, CIESM database). However, fossil species of *Stosicia* are known from the Lower Miocene of the Eastern Atlantic and the Mediterranean [173].

3.2. Endemic Species. The Mediterranean and Cape Verde Islands are the sites with higher numbers of endemic species (71 and 58, resp.), with predominance of *Alvania* (37) in the first site, and of *Schwartziella* (26) and *Alvania* (20) at Cape Verde. Caribbean also has a high number of endemisms (57 species), especially of the genera *Alvania* (19) and *Rissoina* (13). British Isles, Angola, New Scotia,

TABLE 1: Number of Rissoid species, by genus in the Atlantic Ocean and in the Mediterranean Sea. ARC: Arctic; GRE: Greenland; ICE: Iceland; SCA: Scandinavia; BRI: British Isles; POR: Portugal; MED: Mediterranean Sea; LUS: Lusitanian Sea; MET: Meteor seamounts; AZO: Azores; MAD: Madeira, Porto Santo and Desertas; SEL: Selvagens Islands; CAN: Canary Islands; CAP: Cape Verde; WAF: west-African coast; ANG: Angola; NSC: New Scotia biogeographic province; VIR: Virginian biogeographic province; CRL: Carolinian biogeographic province; TRO: Tropical biogeographic province (Caribbean); BRA: Brazil; STH: Santa Helena; TRS: Tristan da Cunha; SSA: southern South-America; ANT: Antarctic.

	ARC	GRE	ICE	SCA	BRI	POR	MED	LUS	MET	AZO	MAD	SEL	CAN	CAP	WAF	ANG	NSC	VIR	CRL	TRO	BRA	STH	TRS	SSA	ANT
Alvania	4	7	10	12	10	30	74	10	9	19	22	13	28	25	29	5	1	1	2	21	1			1	1
Amphirissoa									1	1										1					
Benthonella			1	1	1				1	1			1				1	1	1	1					
Benthonellania						1		1		1	2				3	1			4	5	2	1			
Boreocingula	2	2	1	1										1											
Botryphallus					1	1	1		1	1	1														
Cingula	1			1	1	1			1	1										2		9			
Crisilla				1	1	1	7	1		1	8	6	10	6	3										
Folinia																				2					
Frigidoalvania			1	1													3	2							
Galeodinopsis									2		1	1	1		1	1									
Gofasia						2		5	2																
Lironoba													1							1					
Manzonia				1	1	2	1	6		1	7	7	10		2					1					
Microstelma																				4		1			
Obtusella	3	2	2	2	2	1	2			2	1	1	1	2	1	1									
Onoba	3	4	9	3	2	3	7			1	2	1	5				4	2				2	6	22	7
Peringiella						1	2																		
Plagyostila						1	1								2										
Pontiturboella							1							1	2										
Porosalvania									8																
Powellisetia		1																					1	3	
Pseudosetia			1	2	1	3	2	1		1	3		2		1	1									
Pusillina				2	2	7	11	1	2	1	3	5	5	1	2	1	1	1						1	
Rissoa				3	4	12	26		1	2	7	5	15	2											
Rissoina						1	3				1		1	2	1	1			4	18	2	2		1	
Rudolphosetia						1	1						1												
Schwartziella									1					26	1				4	9	2				
Setia					1	6	18			5	3		6		1				1	1		5			
Simulamerelina																			1	3					
Stosicia																				2					
Voorwindia							1																		
Zebina							1				2		2	1	1				2	6					
Total of Rissoidae species	13	16	25	30	27	74	160	27	26	38	63	38	89	67	50	11	10	7	18	77	7	20	7	27	8

and Virginian Provinces do not have endemic species, and Brazil, Greenland, and Scandinavia only possess a single endemic species (Table 2). However, if these figures are viewed in percentages, Saint Helena Island, Cape Verde, and Tristan da Cunha are the sites with higher percentages of rissoid endemisms (90.0%, 86.6%, and 85.7%, resp.). Other sites with high percentage values are the Meteor group of seamounts (76.9%), the Caribbean (74.0%), southeastern shores of South America (66.7%), the Azores (44.7%), and the Mediterranean (44.4%). Antarctic (37.5%), the Lusitanian group of seamounts (37.0%), the West-African shores (24.0%), and Canary Islands (19.1%) also have a significant amount of endemic rissoids (Table 2).

In spite of the large number of rissoids at Selvagens (38 species), a large number of species are shared with Canaries (30) and Madeira (27) (Table 3), thus only 3 species are endemic to these islands (7.9%). Similar percentages of rissoid endemics occur at Greenland (6.3%) and Scandinavia (3.3%). Iceland has 12.0% of endemisms (Table 2).

3.3. Bathymetrical Zonation.
Most of the 542-rissoid species that live in the Atlantic and in the Mediterranean are shallow species (329). One hundred and forty-six are considered as deep species, living in waters with more than 50 m depth, and 23 species are reported to both shallow and deep waters. It was not possible to establish the bathymetrical zonation of 44 rissoid species.

Benthonella, Benthonellania, Frigidoalvania, Gofasia, Microstelma, and *Pseudosetia* typically are deep species, whereas *Botryphallus, Crisilla, Manzonia, Peringiella, Pusillina, Rissoa, Rissoina, Rudolphosetia, Schwartziella, Setia*, and *Zebina* are mostly constituted by shallow species. Some of these genera (e.g., *Rissoa* and *Rissoina*) are exclusively littoral. In the eastern-Atlantic shores and at latitudes higher than 55° N (Arctic, Greenland, Iceland, and Scandinavia), *Alvania* genus is mostly made of deep species; in all the other sites, usually this genus is predominantly dominated by shallow species (Table 4).

3.4. Modes of Larval Development.
It was possible to infer the mode of larval development of 450 out of the 542 rissoid species, with 375 nonplanktotrophic species, and 75 planktotrophic species (Table 5). There is a predominance of nonplanktotrophs in islands, seamounts, and at high and medium latitudes. This pattern is particularly evident in the genera *Crisilla, Manzonia, Onoba, Porosalvania, Schwartziella*, and *Setia*. Planktotrophic species are more abundant in the eastern Atlantic and in the Mediterranean Sea. The British Isles and Angola are the only sites with excess of planktotrophs in relation to nonplanktotrophic rissoids. *Rissoa* is a very diverse genus in the Mediterranean Sea and along the shores of Portugal, and most are planktotrophic species. In the Arctic, Greenland, southeastern South America, and Antarctic, all rissoid species are nonplanktotrophs (Table 5).

When the bathymetrical zonation of the rissoid species is analyzed in combination with the modes of larval development, some patterns emerge:

(a) most of the shallow nonplanktotrophic species occur in the Mediterranean sea, Cape Verde, and Canary islands, as well as Portugal, the Azores, Madeira archipelago, Selvagens, west African shores, Caribbean, and southeastern South America; *Alvania, Manzonia, Rissoa, Schwartziella*, and *Setia* are diverse genera in the north Atlantic archipelagos (Azores, Madeira, Selvagens, Canaries, and Cape Verde) (Table 6);

(b) shallow planktotrophic rissoid species are much more diverse along the European Atlantic shores, the west-African shores, the Mediterranean, and the Caribbean than in the Atlantic islands, with the exception of Canaries (Table 7);

(c) Scandinavia, British Isles, Portugal, Angola, and the Carolinian Province are the only sites with higher numbers of shallow planktotrophic species relative to the number of shallow non-planktotrophs (cf. Tables 6 and 7);

(d) deep nonplanktotrophic rissoid species are more diverse in the North Atlantic than in the South Atlantic; for instance, there are four such species in the Arctic and no species at all in the Antarctic: these species are also more diverse in the eastern Atlantic than in the western Atlantic shores (Table 8);

(e) deep planktotrophic rissoids are restricted to 4 genera, *Alvania, Benthonella, Benthonellania*, and *Obtusella* (Table 9);

(f) *Benthonella tenella* [169], the sole representative of this genus in the studied area, is the rissoid species with wider geographical range in the Atlantic; other species with large geographical ranges are *Obtusella intersecta* [174] and *Alvania cimicoides* [175]; all of them are deep planktotrophic species, although *Obtusella intersecta* may also occur in the littoral.

3.5. Biotic Similarities between Areas: Parsimony Analysis of Endemicity.
We used PAE separately on the shallow and on the deep rissoid species. After removing all the endemic species (no cosmopolitan species were found), 115 shallow species and 41 deep species of rissoids were analysed with the PAE methodology, using PAUP*.

PAE of the shallow Atlantic and Mediterranean rissoids produced a single most parsimonious tree ($L = 180$, Ci = 0.6389, Ri = 0.7005) with three main groups. The first one strongly clusters Portugal, the Mediterranean, British Isles, and Scandinavia, with bootstrap values higher than 91%. A second group subdivides in two: the first subgroup, the Macaronesian archipelagos of Madeira, Canary Islands, Selvagens, and the Azores clusters; the second subgroup has West-African coast, Angola, and Cape Verde Islands. In a third group, western Atlantic sites are clustered: Caribbean and Carolinian Province cluster to Brazil at 65% bootstrap value. Saint Helena Island weakly clusters to the previous sites (bootstrap value of only 51%). New Scotia and Virginian Provinces cluster in an independent group (66%), as well as Southern South America and Antarctic (95%) (Figure 1).

TABLE 2: Number of endemic Rissoidae, other abbreviations as in Table 1.

	ARC	GRE	ICE	SCA	BRI	POR	MED	LUS	MET	AZO	MAD	SEL	CAN	CAP	WAF	ANG	NSC	VIR	CRL	TRO	BRA	STH	TRS	SSA	ANT
Alvania			2			1	37	1	7	10	3	2	7	20	5				1	19				1	
Amphirissoa																									
Benthomella																									
Benthonellania															1				1	2	1				
Boreocingula	1	1																							
Botryphallus										1				1											
Cingula	1																			2					
Crisilla							3	1			1		1	6	1							9			
Folinia																				2					
Frigidoalvania																									
Galeodinopsis																									
Gofasia								3	2																
Lironoba																				1					
Manzonia								5		1	2	1	4		1					1					
Microstelma																				4		1			
Obtusella							1			1			1	1											
Onoba			1			2	6			1												2	6	14	2
Peringiella							1																		
Plagyostila															1										
Pontiturboella																									
Porosalvania									8																
Powellisetia				1																				2	1
Pseudosetia																									
Pusillina							3		1		1			1										1	
Rissoa							11						3	1								2			
Rissoina									1					1					1	13					
Schwartziella									1					26	1					5		4			
Setia							9		1	3	1		1		1										
Simulamerelina																				2					
Stosicia																				2					
Voorwindia																									
Zebina														1	1					4					
Total of endemic Rissoidae	2	1	3	1	0	3	71	10	20	17	8	3	17	58	12	0	0	0	3	57	1	18	6	18	3
Total of Rissoidae species	13	16	25	30	27	74	160	27	26	38	63	38	89	67	50	11	10	7	18	77	7	20	7	27	8
% endemics	15.4	6.3	12.0	3.3	0.0	4.0	44.4	37.0	76.9	44.7	12.7	7.9	19.1	86.6	24.0	0.0	0.0	0.0	16.7	74.0	14.3	90.0	85.7	66.7	37.5

TABLE 3: Number of shared Rissoidae species, other abbreviations as in Table 1.

	ARC	GRE	ICE	SCA	BRI	POR	MED	LUS	MET	AZO	MAD	SEL	CAN	CAP	WAF	ANG	NSC	VIR	CRL	TRO	BRA	STH	TRS	SSA	ANT
ARC	13																								
GRE	10	16																							
ICE	8	13	25																						
SCA	7	9	16	30																					
BRI	2	4	11	22	27																				
POR	0	1	7	18	22	74																			
MED	0	0	6	17	21	62	160																		
LUS	0	0	2	5	6	11	9	27																	
MET	0	0	1	1	1	1	1	2	26																
AZO	0	0	3	5	6	8	7	5	5	38															
MAD	0	0	4	5	8	17	17	7	1	12	63														
SEL	0	0	0	0	1	11	11	4	0	3	27	38													
CAN	1	2	7	14	17	32	36	7	2	9	42	30	89												
CAP	1	0	2	3	4	5	5	2	0	3	4	1	4	67											
WAF	0	0	3	7	9	18	22	5	0	4	12	6	17	7	50										
ANG	0	0	1	2	2	3	3	0	0	2	2	0	2	3	11	11									
NSC	4	5	6	2	2	1	1	1	1	1	1	0	2	0	0	0	10								
VIR	2	3	4	2	2	1	1	1	1	1	1	0	2	0	0	0	7	7							
CRL	0	0	1	1	1	1	1	1	1	1	2	0	2	0	1	0	1	1	18						
TRO	0	0	1	1	1	2	2	1	2	2	3	0	3	0	0	0	1	1	15	77					
BRA	0	0	0	0	0	0	0	0	0	0	0	0	0	0	1	0	0	0	3	5	7				
STH	0	0	0	0	0	0	0	0	0	0	1	0	0	0	0	0	0	0	1	1	1	20			
TRS	0	0	0	0	0	0	0	0	0	0	0	0	0	0	0	0	0	0	0	0	0	0	7		
SSA	0	0	0	0	0	0	0	0	0	0	0	0	0	0	0	0	0	0	0	0	0	0	0	27	
ANT	0	0	0	0	0	0	0	0	0	0	0	0	0	0	0	0	0	0	0	0	0	0	0	4	8

TABLE 4: Bathymetric zonation of the Rissoidae. Lit—littoral species (usually living at depths less than 50 m depth); deep—deep species (usually living at depths higher than 50 m depth). Other abbreviations as in Table 1.

(a)

	ARC		GRE		ICE		SCA		BRI		POR		MED		LUS		MET		AZO		MAD		SEL		CAN		CAP	
	lit	deep	lit	deep	lit	deep	lit	deep	lit	deep	lit	deep	lit	deep	lit	deep	lit	deep	lit	deep	lit	deep	lit	deep	lit	deep	lit	deep
Alvania		2	2	4	2	7	2	8	5	5	17	13	50	20		9		6	11	8	13	9	10	2	17	11	17	8
Amphirissoa																1				1								
Benthonella						1				1				1		1		1	1	1		1						
Benthonellania																		1		1		2						
Boreocingula	1	1				1		1		1																		
Botryphalus							1		1		1		1						1		1		1		1			
Cingula		1					1		1		1								1		1							
Crisilla							1		1		1		6					1			7		6		9	1	6	
Folinia																												
Frigidoalvania		1																										
Galeodinopsis																									1		1	
Gofasia												2				2		5				1		1		1		
Lironoba																												
Manzonia							1		2		2		1					5	1		6		7	1	10			
Microstelma																												
Obtusella														1						1					1		1	
Onoba	1			2		3	1	1	2		2	1	3	3					1		1	1	1		3	1		
Peringiella									1		1		2									1	1					
Plagyostila									1		1		1												1			
Pontiturboella																												
Porosalvania																8												
Powellisetia																												
Pseudosetia								2		1		3		2		1		1		1		1		1		2		1
Pusillina							1		5		9	1	9			2		1	1	1	1		2		2	2		1
Rissoa							3		5		12		23						2		7		5		14			
Rissoina											1		2			1							1		1			
Rudolphosetia											1		1								1		1					
Schwartziella																											20	6
Setia											6		17			1			5		3		1		6		6	
Simulamerelina																												
Stosicia																												
Voorwindia													1															
Zebina																					2				2			
Total number of Rissoidae	2	5	2	6	2	12	11	13	17	8	51	20	118	28	0	26	0	21	24	13	43	16	33	5	66	19	50	16

(b)

	WAF lit	WAF deep	ANG lit	ANG deep	TRS lit	TRS deep	STH lit	STH deep	NSC lit	NSC deep	VIR lit	VIR deep	CRL lit	CRL deep	TRO lit	TRO deep	BRA lit	BRA deep	SSA lit	SSA deep	ANT lit	ANT deep
Alvania	18	11	6											1	10	3						
Amphirissoa																1				1		
Benthonella										1				1		1						
Benthonellania		3		1				1				1		3		4		2				
Boreocingula																						
Botryphallus																						
Cingula							4									2						
Crisilla	1	2	1																			
Folinia															1							
Frigidoalvania									1	1	1	1										
Galeodinopsis	1									1	1											
Gofasia																						
Lironoba																						
Manzonia	1	1													1							
Microstelma							1									3						
Obtusella																						
Onoba					2	4	1		1	1		1							14	4	5	
Peringiella																						
Plagyostila	2																					
Pontiturboella																						
Porosalvania		1																				
Powellisetia		1				1													3			
Pseudosetia																					1	
Pusillina	1											1							1			
Rissoa	1						2															
Rissoina	1		1										4		13							
Rudolphosetia																						
Schwartziella	1						1						4		8		2					
Setia	1															1						
Simulamerelina													1		3							
Stosicia															2							
Voorwindia																						
Zebina	1												2		4							
Total number of Rissoidae	29	19	8	1	2	5	9	1	2	4	2	4	11	5	42	15	2	2	18	5	6	0

TABLE 5: Mode of larval development of the Rissoidae: np: nonplanktotrophic species; p: planktotrophic species, other abbreviations as in Table 1.

Region		Alvania	Amphirissoa	Benthonella	Benthonellania	Borocingula	Botryphallus	Cingula	Crisilla	Folinia	Frigidoalvania	Galeodinopsis	Gofasia	Lironoba	Manzonia	Microstelma	Obtusella	Onoba	Peringiella	Plagyostila	Pontituroboella	Porosalvania	Powellisetia	Pseudosetia	Pusillina	Rissoa	Rissoina	Rudolphosetia	Schwartziella	Setia	Simulamerelina	Stosicia	Voorwindia	Zebina	Total number of rissoids
ARC	np	4				2		1			3							3																	13
	p																																		0
GRE	np	7				2					2							5																	16
	p																																		0
ICE	np	9		1		1					1							8																	20
	p	1															1	1						1											4
SCA	np	7		1		1		1	1		1						1	1							1										15
	p	5													1		1	1						2	2	2									14
BRI	np	3		1		1		1	1								1	1																	9
	p	7					1								1									1	2	3	1			1					17
POR	np	17					1		1				2		1			2	1					3	2	3				6					39
	p	13													1										2	9	1	1							35
MED	np	51		1			1		2									1	1					2	5	10	1			17					93
	p	17							3						1		2	4	1	1					5	12	1	1					1	1	48
LUS	np	5							1				5		6			1	1	1															20
	p	5		1	1																														7
MET	np	9	1		1								2									8		1			1		1						25
	p																																		1
AZO	np	17	1	1	1		1	1	1						1		1	1						1		2	1			5					33
	p	2															1	1							1										5
MAD	np	20		1	1		1	1	3				1		7			2						1	2	6		1		3				1	50
	p	2			1		1	1									1								1	1								1	8
SEL	np	12					1		2				1		7			1						1		4		1		1				1	32
	p	1																								1									2
CAN	np	23					1		3				1		10			4						2	2	8		1		4				1	60
	p	5		1				2	2								1	1							3	6	1							1	21

TABLE 5: Continued.

		Alvania	Amphirissoa	Benthonella	Benthonellania	Boreocingula	Botryphallus	Cingula	Crisilla	Folinia	Frigidoalvania	Galeodinospis	Gofasia	Lironoba	Manzonia	Microstelma	Obtusella	Onoba	Peringiella	Plagyostila	Pontiturboella	Porosalvania	Powellisetia	Pseudosetia	Pusillina	Rissoa	Rissoina	Rudolphosetia	Schwartziella	Setia	Simulamerelina	Stosicia	Voorwindia	Zebina	Total number of rissoids
CAP	np	20					1		6																1	1	1		26					1	57
CAP	p	5																									1								10
WAF	np	18			2				2			1			2		2			1				1	1	1	1		1	1				1	29
WAF	p	11			1				1			1					1			2					2	1									21
ANG	np	2																																	2
ANG	p	3			1				1			1					1								1		1								9
NSC	np				1						3	1						4																	9
NSC	p			1																															1
VIR	np	1									2							2							1										6
VIR	p			1																															1
CRL	np	1																									1		1		1			1	5
CRL	p																										2		3			1		1	8
TRO	np	7	1																								7		4		3	1		3	27
TRO	p	1		1						1				1													7		3					2	16
BRA	np				1																														1
BRA	p	1																									1		2						4
STH	np							8																					1						9
STH	p							1										1																	2
TRS	np																	1																	1
TRS	p																																		0
SSA	np																	18					2		1										21
SSA	p																																		0
ANT	np																	6					1												7
ANT	p																																		0

TABLE 6: Number of littoral Rissoidae with nonplanktotrophic mode of larval development, other abbreviations as in Table 1.

	ARC	GRE	ICE	SCA	BRI	POR	MED	AZO	MAD	SEL	CAN	CAP	WAF	ANG	NSC	VIR	CRL	TRO	BRA	STH	TRS	SSA	ANT
Alvaria						8	36	10	12	9	14	15	10	2				5					
Amphirissoa																							
Benthomella																							
Benthomellania																							
Boreocingula	2	1	1	1	1																		
Botryphallus				1	1	1	1	1	1	1	1	1											
Cingula						1		1	1											3			
Crisilla							2	1	3	2	2	6											
Folinia																							
Frigidoalvania															1	1							
Gofasia																							
Lironoba																							
Manzonia						1			6	7	10		1										
Microstelma																							
Obtusella																							
Onoba	1	1	1	1	1	2	3	1	1	1	3				1	1						13	5
Peringiella																							
Plagyostila																							
Powellisetia																						2	1
Pseudosetia																							
Pusillina				1	1	1	4															1	
Rissoa						3	10	2	6	4	8	1											
Rissoina												1					1	5					
Rudolphosetia						1	1		1	1	1												
Schwartziella												20					1	4					
Setia					1	6	17	5	3	1	4		1										
Simulamerelina													1				1	3					
Stosicia																		1					
Voorwindia																							
Zebina									1	1	1		1				1	2					
Total number of Rissoidae	3	2	2	4	5	24	74	21	35	27	44	44	14	2	2	2	4	20	0	3	0	16	6

TABLE 7: Number of littoral Rissoidae with planktotrophic mode of larval development, other abbreviations as in Table 1.

	ARC	GRE	ICE	SCA	BRI	POR	MED	LUS	MET	AZO	MAD	SEL	CAN	CAP	WAF	ANG	NSC	VIR	CRL	TRO	BRA	STH	TRS	SSA	ANT
Alvania				2	5	9	12	3		1	1	1	3	2	8	3									
Amphirissoa																									
Benthonella																									
Benthonellania																									
Boreocingula																									
Botryphallus																									
Cingula																						1			
Crisilla				1	1	1	3						2		1	1									
Folinia																				1					
Frigidoalvania																									
Galeodinopsis														1	1	1									
Gofasia																									
Lironoba																									
Manzonia				1	1	1	1																		
Microstelma																									
Obtusella														1											
Onoba																									
Peringiella						1	1																		
Plagyostila						1	1							1	2										
Pontiturboella																									
Porosalvania																									
Powellisetia																									
Pseudosetia																									
Pusillina				1	1	4	5				1		2		1										
Rissoa				2	3	9	12				1	1	6		1										
Rissoina						1	1	1					1	1	1	1			2	5					
Rudolphosetia																									
Schwartziella																			3	3					
Setia																					2	1			
Simulamerelina																									
Stosicia																				1					
Voorwindia							1																		
Zebina							1				1		1						1	2					
Total number of Rissoidae	0	0	0	7	11	27	38	4	0	1	4	2	15	6	15	6	0	0	6	12	2	2	0	0	0

TABLE 8: Number of deep Rissoidae with nonplanktotrophic mode of larval development, other abbreviations as in Table 1.

	ARC	GRE	ICE	SCA	BRI	POR	MED	LUS	MET	AZO	MAD	SEL	CAN	CAP	WAF	ANG	NSC	VIR	CRL	TRO	BRA	STH	TRS	SSA	ANT
Alvania	2	4	6	5	3	9	14	4	9	7	8	3	9	5	8				1	1					
Amphirissoa									1	1										1					
Benthomella																									
Benthonellaria								1	1	1	1				2							1			
Boreocingula																									
Botryphallus																									
Cingula	1																								
Crisilla								1					1		2										
Folinia																									
Frigidoalvania	1																1	1							
Galeodinopsis																									
Gofasia						2		5	2		1	1													
Lironoba												1	1												
Manzonia								5																	
Microstelma															1										
Obtusella										1															
Onoba		2	3	1			1				1		1				1	1					1	1	
Peringiella																									
Plagyostila									8																
Pontiturboella																									
Porosalvania																									
Powellisetia																									
Pseudosetia			1	2	1	3	2	1	1	1	1		2		1										
Pusillina						1	1	1	2		1		2	1	1		1	1							
Rissoa																									
Rissoina									1																
Rudolphosetia																									
Schwartziella														6											
Setia																									
Simulamerelina																									
Stosicia																									
Voorwindia																									
Zebina														1											
Total number of Rissoidae	4	6	10	8	4	15	18	18	25	11	13	5	16	13	15	0	3	3	1	2	0	1	1	1	0

TABLE 9: Number of deep Rissoidae with planktotrophic mode of larval development, other abbreviations as in Table 1.

	ARC	GRE	ICE	SCA	BRI	POR	MED	LUS	MET	AZO	MAD	SEL	CAN	CAP	WAF	ANG	NSC	VIR	CRL	TRO	BRA	STH	TRS	SSA	ANT
Alvania			1	3	2	4	5	2		1	1		2	3	3										
Amphirissoa																									
Benthonella			1	1	1	1	1	1	1	1	1		1				1	1	1	1					
Benthonellania											1				1	1									
Boreocingula																									
Botryphallus																									
Cingula																									
Crisilla																									
Folinia																									
Frigidoalvania																									
Galeodinopsis																									
Gofasia																									
Lironoba																									
Manzonia																									
Microstelma																									
Obtusella							1																		
Onoba																									
Peringiella																									
Plagyostila																									
Pontiturboella																									
Porosalvania																									
Powelliseia																									
Pseudosetia																									
Pusillina																									
Rissoa																									
Rissoina																									
Rudolphosetia																									
Schwartziella																									
Setia																									
Simulamerelina																									
Stosicia																									
Voorwindia																									
Zebina																									
Total number of Rissoidae	0	0	2	4	3	5	7	3	1	2	3	0	3	3	4	1	1	1	1	1	0	0	0	0	0

The consensus tree (L = 82, Ci = 0.5000, Ri = 0.5816) that results from the PAUP* analysis of the deep rissoids is given in Figure 2. Bootstrap values are higher than 50% only for three groups: Portugal-Mediterranean (78%), Caribbean-Carolinian Province (75%), and New Scotia Province-Virginian Province. Some other sites also cluster, but at values lower than 50% (Figure 2).

3.6. Biotic Similarities between Areas: Probable Directions of Faunal Flows. The results of the analysis of the probable directions of faunal flows (using the X_A and X_B indices; see Section 2) are summarized in Figures 3–8, and support the patterns found by both the PAE analysis and the geographical distribution. Four main source areas for rissoids emerge: Mediterranean, Caribbean, Canaries/Madeira archipelagos, and the Cape Verde archipelago. In the western Atlantic, a rissoid movement originating in the Caribbean seems to have developed southwards to Brazil (Figure 3) and northwards to the Carolinian Province (Figure 4). A southward movement of rissoids, from the Arctic down to the Virginean Province, is envisaged in Figure 4, with a faunal break zone, between the Carolinian and the Virginian Provinces (Figure 4), as well as between Brazil and Southern South America (Figure 5).

In the eastern Atlantic, the patterns are more complicated (Figures 6 and 7). It seems clear that the Mediterranean is the source area for a northwards movement to Portugal, British Isles, and Scandinavia (Figure 7). Scandinavia seems to be the source area for both Iceland and Arctic (64 and 54%, resp.) and Iceland probably played an important role as a source for both Greenland and Arctic. The Mediterranean is weakly related with the West-African shores (44%, Figure 6). The relationships between the Azores and both Madeira and Canaries are weak (32 and 24%, resp.), and Canaries seem to be the main source of the rissoid fauna of Madeira (67%) and Selvagens (79%). Cape Verde archipelago is isolated from all sites (Figure 8), its highest relationship being with Angola (27%).

4. Discussion

4.1. Geographical Distribution. It is beyond the scope of this paper to discuss in detail all hypotheses related with Rapoport's latitudinal rule (e.g., the seasonal variability hypothesis [2, 4], the differential extinction hypothesis [3], the competition hypothesis [5–7], or the Milankovitch climate oscillations [8]), but one of the corollaries of the seasonal variability hypothesis is that, at low latitudes, the expected bathymetrical range of a given species, in average, should be lower than at high latitudes. Stevens [3] verified this pattern for the Pacific fishes, and a similar pattern is shown by the rissoids of the Norteastern Atlantic (Figure 9(a)).

Roy et al. [9, 10], who used lists with 3,916 marine Caenogastropod species geographically distributed along the north and Central America shores between 10°S–83°N, confirmed the latitudinal gradient pattern, with the number of species decreasing with latitude. A similar analysis using our Rissoidae database (littoral species only) conforms to the general pattern of latitudinal diversity gradient described in

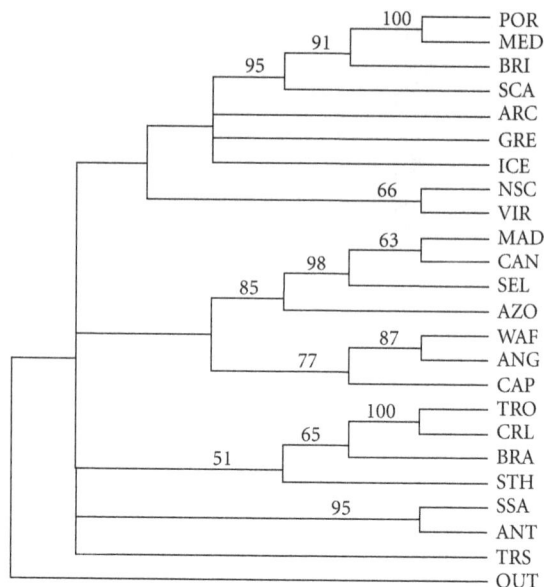

FIGURE 1: Consensus tree with bootstrap values for the shallow Rissoidae species.

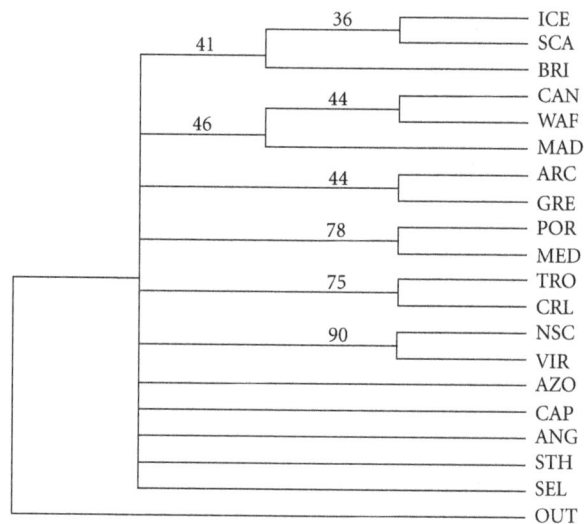

FIGURE 2: Consensus tree with bootstrap values for the deep Rissoidae species, abbreviations as in Table 1.

[9, 15, 16, 176, 177] and shows an evident decline of the number of rissoid species with latitude (Figures 9(b)–9(d)). Important asymmetries in the geographical distribution of the mollusc species were found in [12], when they compared the north and south hemispheres of the east-Pacific coast of America. In the northern hemisphere, there is correspondence between the diversity latitudinal gradient and SSTs, but in the southern hemisphere, in particular from 40 to 60°S, the number of species increases with latitude, even though SSTs decrease monotonically with this variable. This pattern is also evident with the shallow Rissoidae along the west-Atlantic coasts of South America (Figures 9(d)) and the explanation is dependent on the coastal area (comprising

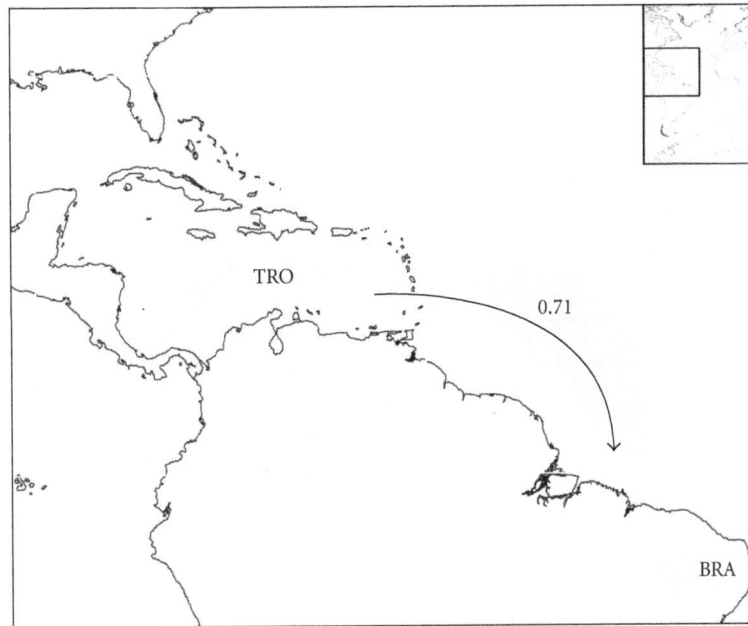

FIGURE 3: Probable colonization patterns of rissoid fauna in the central west-Atlantic. The arrows represent the probable main flux direction of faunas, and the associated numbers represent, for each pair of areas, the higher of the two similarity index values computed as described in the methods, abbreviations as in Table 1.

depths less than 200 m) which, according to Valdovinos et al. [12], is a factor that better explains biodiversity than SSTs. Thus, the increase of the number of shallow rissoids with latitude along the southern South-America shores (Figure 9(d)) is due to the area effect of the Magellan fjords, which played an important role as refugia during glacial periods, locally enhancing the speciation evolutionary processes [12, 178].

It is noteworthy to emphasize that the Mediterranean area has more species than expected for similar latitudes (31–43°N) (Figure 9(b)). This is certainly due to the high sampling effort for this region, but we think that other reasons are also behind this fact (see below). A similar trend was also reported in several other taxonomic groups (Hydromedusae, Siphonophora, Chaetognatha, Appendicularia, Salpida, Cephalopoda, Euphausiacea, Decapoda, and Pisces) [15, 179], reinforcing the Mediterranean as an area of high marine biodiversity. This is even more interesting if we think that the Mediterranean area was repopulated just 5.33 Ma ago, when the "Messinian Salinity Crisis" ended [180–182]. This dramatic event occurred between 5.96–5.33 Ma and provoked an almost complete annihilation of the Mediterranean marine fauna and flora [183, 184]. The desiccation of the Mediterranean Sea happened because of the closure of the Rifian corridor, a marine pathway in the northwest of Morocco, which connected the Mediterranean with the Atlantic [185–187]. This caused an impressive drop in sea level, exceeding 1,500 m, and thick evaporitic series deposited in the Mediterranean basin [188]. The reopening of the connection between the Atlantic and the Mediterranean happened 5.33 Ma ago and, although there are different hypothesis under discussion such as tectonic movements

of the crust [189], the most plausible explanation for the reflooding of the Mediterranean Sea is the retrogressive erosion in the Gibraltar strait [190]. The marine molluscs that recolonized the Mediterranean basins were the remnants of an impoverished Miocene Lusitano-Atlantic fauna [184], and contemporaneous of the "Messinian Salinity Crisis," lacking several Tethyan relics [191]. This malacofauna was severely impacted by climatic changes—the decrease of the SSTs, increase of the ice volume at Antarctic to more than 50% than the present one [192], and the lowering of the mean sea level in about 40 m [184]. The Pliocene and Pleistocene glacial cycles heavily affected the eastern Atlantic shores, but the "buffer" zone provided by the Mediterranean and acted as a refugia zone, especially in the southern shores [193], and this may be the reason for the high number of rissoid species that this area possesses nowadays.

By contrast, the low number of rissoids on the Virginian Province (only 7 species) is probably related with the predominance of sandy bottoms on the littoral of this biogeographical Province, and with the multiple lagunar and estuarine systems, which are inhospitable to the benthic algae where many species of these micromolluscs live [194]. Van Reine et al. [195] provide a similar explanation to explain the poverty of the west-African tropical algal flora. This happens because suitable hard surfaces for seaweed attachment are scarce, with large areas of sand and mangrove, the wave-exposure is high (very few sheltered sites occur) and also because the inshore salinity is often reduced. This has a profound impact on the epibenthonic malacofauna that lives associated with these algae, and the scarcity of algae by the reasons previously discussed is the most likely explanation for the low number of the west-African rissoids.

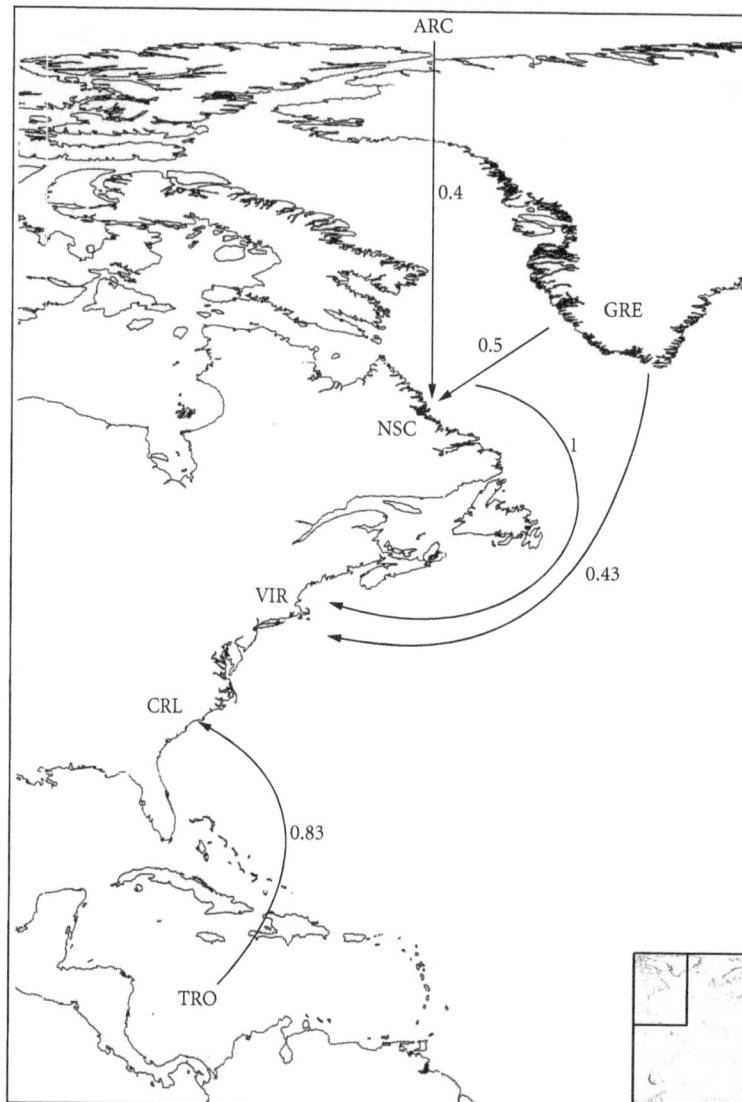

FIGURE 4: Probable colonization patterns of rissoid fauna in the Northwest Atlantic. The arrows represent the probable main flux direction of faunas, and the associated numbers represent, for each pair of areas, the higher of the two similarity index values computed as described in the methods, abbreviations as in Table 1.

4.2. Endemic Species. By definition, "a species can be endemic to an area for two different reasons: (a) because it has originated in that place and never dispersed, or (b) because it now survives in only a part of its former wider range" [196]. We do not know any endemic rissoid to the Azores, Madeira, or Canaries that is documented in the fossil record as formerly having a broader geographic distribution [197, 198]. So, they are autochthonous descendents of immigrants, rather than geographic relics.

In some areas, a few genera went through a speciation process that led to a high number of both species and endemics, for example, *Alvania, Crisilla, Onoba, Pusillina, Rissoa,* and *Setia* at the Mediterranean; *Benthonellania* and *Rissoina* at the Caribbean; *Manzonia* and *Crisilla* at the Madeira, Selvagens, and Canaries archipelagos; *Crisilla* and *Schwartziella* at the Cape Verde archipelago; *Onoba* at Iceland; and *Cingula* at the geographically isolated Saint Helena Island.

Mironov [199] proposed the concept of "centers of marine fauna redistribution" as a dynamic concept that accommodates an evolutionary perspective for a geographical area, integrating two usually opposite concepts: the center of origin, and the center of accumulation. He emphasized that a center of redistribution should be regarded as a biogeographic unit with three developmental stages, succeeding each other, in time: first, it is a stage of accumulation of species, it then evolves to a stage of speciation and, the last phase, is the dispersal stage. Thus, such concept is the "consecutive stage of development of an integrated and complicated event," uniting the opposing accumulation and dispersal concepts in a given area, which is designated as a "center of redistribution" [196]. According to these authors, in the past, during the Cretaceous and the Palaeogene, the Mediterranean part of the Tethys Sea acted as a center of origin, but it lost the role of a centre of speciation and dispersal during the

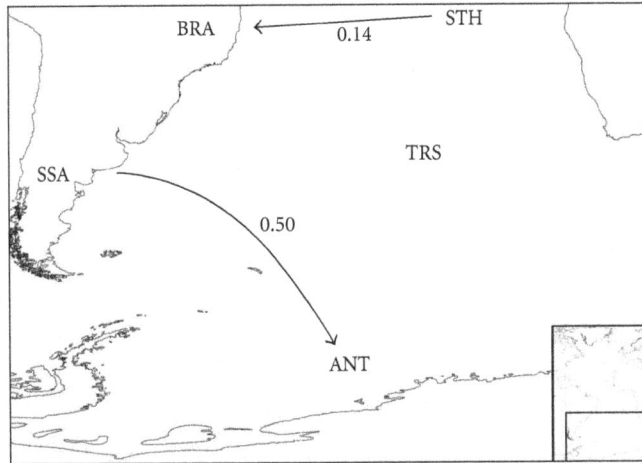

FIGURE 5: Probable colonization patterns of rissoid fauna in the South Atlantic. The arrows represent the probable main flux direction of faunas, and the associated numbers represent, for each pair of areas, the higher of the two similarity index values computed as described in the methods, abbreviations as in Table 1.

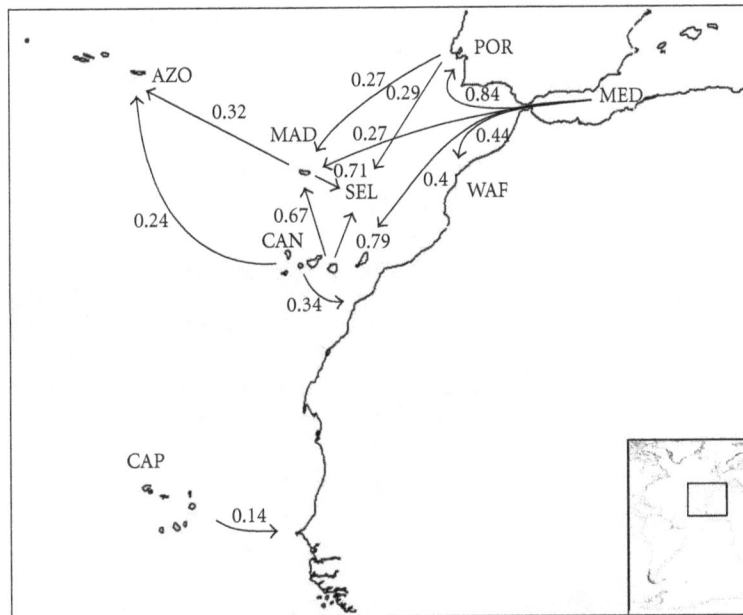

FIGURE 6: Probable colonization patterns of rissoid fauna in the Macaronesian islands, Northeast-Atlantic, and Mediterranean. The arrows represent the probable main flux direction of faunas, and the associated numbers represent, for each pair of areas, the higher of the two similarity index values computed as described in the methods, abbreviations as in Table 1.

Miocene and Pliocene. However, our data show that the high number of rissoids (160 species), as well as the high number of endemics (71) that are reported to the Mediterranean area, must be related to the first two stages proposed by Mironov [199]—accumulation and speciation—which are certainly related to the protective role of the Mediterranean as a refugee during glacial episodes. In the present, the other center of speciation in the Atlantic is the Caribbean area, which is described by Krylova [196] as a center of redistribution, and considered by Briggs [200] as a centre of origin.

We must stress the high percentage of endemics that occur in the isolated islands of Saint Helena, Tristan da Cunha, and at Cape Verde archipelago (more than 85% of endemics), and also at the Azores (44.7%) thus reinforcing the legislative protective actions that the local governments have implemented in these islands during the recent years. The Cape Verde islands probably received the first rissoids from West-African shores, from where it distances nowadays just about 500 km but must have undergone a long period of isolation, which explains such a high number of endemics.

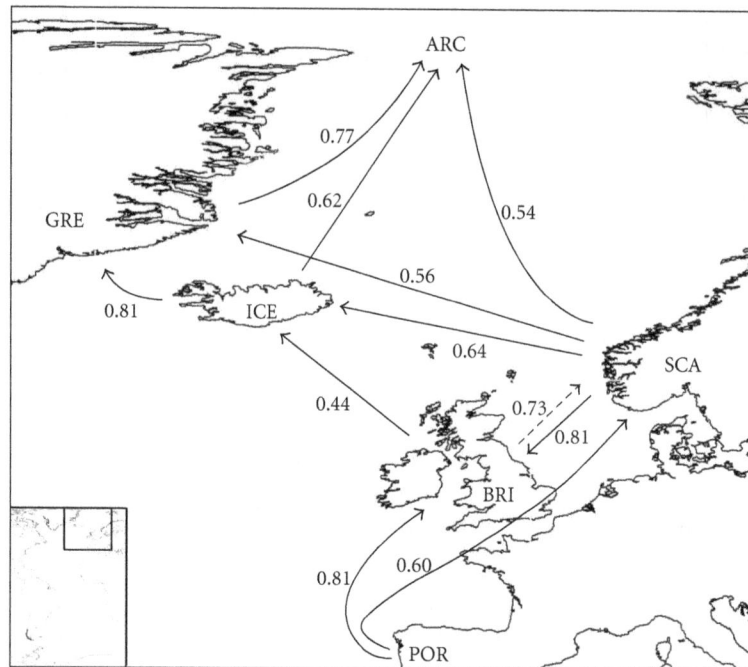

FIGURE 7: Probable colonization patterns of rissoid fauna in the Northeast Atlantic. The arrows represent the probable main flux direction of faunas, and the associated numbers represent, for each pair of areas, the higher of the two similarity index values computed as described in the methods, abbreviations as in Table 1.

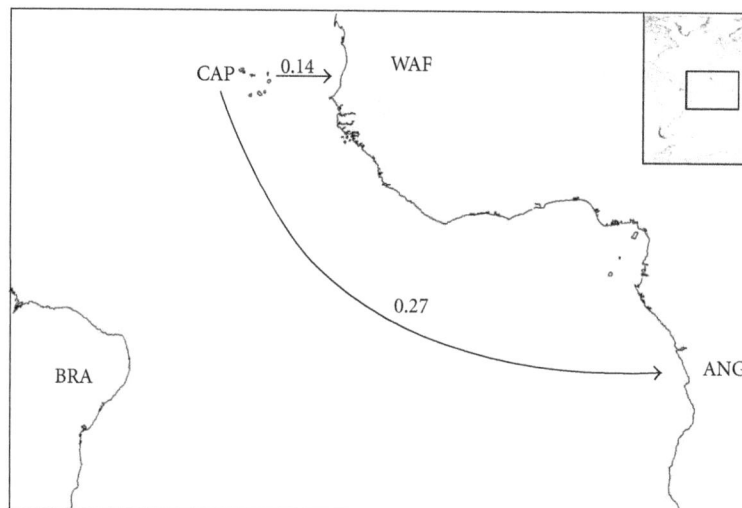

FIGURE 8: Probable colonization patterns of rissoid fauna in the Central East-Atlantic. The arrows represent the probable main flux direction of faunas, and the associated numbers represent, for each pair of areas, the higher of the two similarity index values computed as described in the methods, abbreviations as in Table 1.

Also, our results (both PAE and X_A and X_B indices) point to a faunal breakdown zone between Cape Verde archipelago and the nearest areas, with very weak relationships with the Canary Islands (7%) and West-African shores (14%), and a little bit higher similarities with Angola (27%).

4.3. Bathymetrical Zonation and Modes of Larval Development. It is a well-known fact that biotic communities in high latitudes are usually rich in nonplanktotrophic species [157,

201]. This is Thorson's rule "pelagic development reveals a clear biological polarity: from low towards high latitude pelagic development disappears progressively and becomes replaced by direct development, demersal development, and viviparity" [202, 203]. Thorson's original formulation related also pelagic development with depth, saying that the number of species with such a type of development would gradually diminish from the shallow shelf downwards to the abyssal depths, until its complete disappearance [202] a concept that did not hold [204, 205].

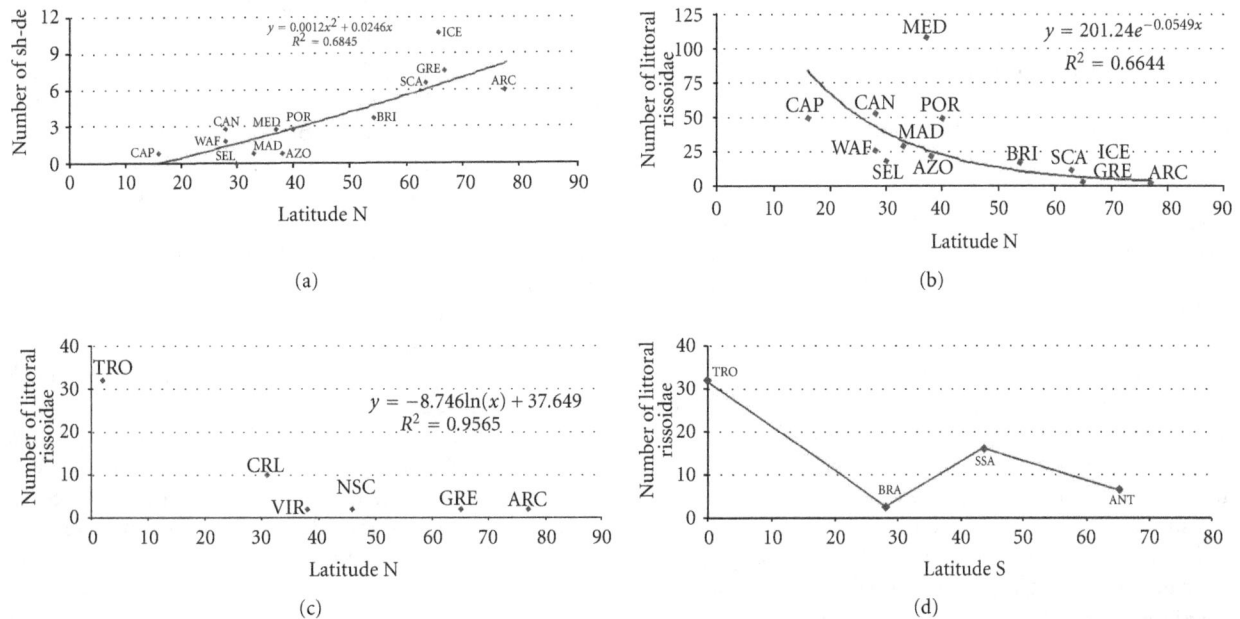

FIGURE 9: Relation between (a) number of rissoids with large bathymetrical range (#sh-de: shallow-deep) and latitude in the eastern Atlantic; (b) number of littoral rissoids and latitude in the eastern Atlantic; (c) number of littoral rissoids and latitude in the western Atlantic; (d) number of littoral rissoids and latitude in the western Atlantic.

It is interesting that the few planktotrophic deep rissoid species are indeed those with higher density and with wider geographical ranges (e.g., *Obtusella intersecta* and *Benthonella tenella* [27]. Rex and Warén [206] provided a plausible explanation for this. These authors noted that the higher predominance of planktotrophic carnivore species among the Meso and Neogastropoda with increasing depths, a phenomenon detected by Grahame and Branch [207], may be due to the enhancement of this mode of development at bathyal depths, because of the advantages that the larval dispersal confers in an environment where resources are unevenly distributed.

The first fossil record of the Rissoidae family is from the lower Jurassic of the Tethys Sea [21, 208] and the oldest records in Europe are from the Toarcian of Italy [209] and from the Bajocian of England [210]. Some modern rissoid genera (e.g., *Rissoa* and *Alvania*) are known since the Early Miocene (Eggenburgian) of the Central Paratethys [211]. The protoconch morphologies of Middle Miocene (Badenian) rissoids belonging to modern genera (*Rissoa*, *Manzonia*, and *Alvania*) are similar to their extant Mediterranean and NE Atlantic congeners, and, interestingly, all of these species possess a planktotrophic mode of larval development [211].

The ancestral of the Rissoidae presumably had a planktotrophic mode of development [208]. Jablonski [212] stated that species with a planktotrophic larval development and long-distance dispersal usually have wider geographical ranges, longer geological ranges, and smaller speciation and extinction rates than nonplanktotrophic relatives. He detected changes in the modes of larval development of Cretacian molluscs, from planktotrophic to nonplanktotrophic, but the reverse was not found. In the Mediterranean, sibling

species were found, almost identical in their teleoconchs and differing mostly in their protoconchs: one multispiral, denoting a planktotrophic larval development, the other one paucispiral, denoting a nonplanktotrophic larval development [213]. This unidirectional change in the mode of larval development may also explain the high percentage of nonplanktotrophic rissoid species that are present in the Atlantic archipelagos/islands (Tables 6 and 7). Moore [214] at Rockall, an isolated and inhabited islet located in the north Atlantic (57°N, 13°W), first noted this phenomenon and stated that a pelagic phase may be an advantage for dispersal, but it may exclude species from certain habitats, namely, from oceanic islands. Moreover, larvae of nonplanktotrophic species are protected from the surrounding environment during the initial phase of their development, by the walls of the egg. In addition, they do not need external food supply, as vitelum provides them all energy they need to account for the completion of the metamorphosis [207]. Last but not the least, when the juveniles emerge from the egg, they are inserted in their natural habitat, not incurring into the risks that adversely affect the larvae of planktotrophic species, which may settle into nonsuitable substrates [204]. Thus, it is not such a surprise that at high latitudes, where the abiotic conditions are more extreme for shallow species (low salinity, low temperatures during a large period of the year) [215] and resources for larval stages are available only during short periods, there exists such a predominance of nonplanktotrophic rissoid species.

In the absence of a phylogenetic analyses for this family, one can only speculate that during geological times, the ancestral(s) (either planktotrophic or nonplanktotrophic) dispersed, reached an oceanic island by natural means (see [198, 216] for a review), and was able to establish there a viable population. Once in the island, there are evolutive advantages in diminishing the dispersal capabilities [217, 218]. If the ancestral was a planktotrophic species, these selective forces may induce the change to a nonplanktotrophic mode of development. In doing so, this automatically reduces the chances of the larvae to be lost into the open ocean. This may be achieved by two ways: (i) reduction of the period of dispersion, or (ii) by larval behaviour related with the selection of favourable marine currents that keep the larvae nearby the substrates that are appropriated to the adult benthonic life [219]. In the first case, the change of larval development produces a new species, with identical teleoconch (or very similar) to the ancestral, but with a different protoconch. This geologically "instantaneous" speciation event, without intermediary forms, is common in both the fossil record and among the extant caenogastropods, of which several pairs of sibling species are reported [47, 213, 220–224]. Speciation may also occur from ancestral planktotrophic species by means of the disruption of a formerly wider geographical range (e.g., due to the glacial-interglacial cycles). Gofas [116] reported four pairs of such rissoid species from West African shores, all of them with planktotrophic development and closely related to European congenerics: *Alvania africana/A. beani, Alvania marioi/A. cancellata, Alvania gofasi/A. zetlandica*, and *Crisilla transitoria/C. semistriata*.

As it is not possible to invert the loss of the planktotrophic phase, in a relatively short interval, the planktotrophic ancestral may originate one or several species, by adaptive radiation, each species occupying a different niche. This promotes the increase of the number of nonplanktotrophic species, usually with a restricted range of dispersion (Strahtmann, 1986). Several examples are known from oceanic islands that elucidate the above-mentioned mechanisms: at Madeira, Selvagens, and Canary Islands, the rissoid genera *Alvania, Crisilla, Manzonia*, and *Rissoa* produced a high number of species, whereas at Cape Verde, the most specious genera were *Alvania, Crisilla*, and *Schwartziella*. In the Azores, *Alvania* and *Setia* are also good examples (Table 1). Similar speciation events also happened in other families of gastropods (e.g., in Cape Verde, Conidae, with about 45 species/subspecies, and *Euthria*, with 20 species) [225, 226].

Currently, we are unable to choose between the following hypotheses:

(1) colonization by an ancestral (either planktotrophic or nonplanktotrophic), followed by speciation with adaptive radiation;

(2) several independent colonizations, spaced in timed, by an ancestral that originates a different species, without adaptive radiation;

(3) both hypotheses above described.

4.4. Biotic Similarities between Areas: Parsimony Analysis of Endemicity and Probable Directions of Faunal Flows. Although many authors postulate a stepping-stone dispersal through the chain of seamounts located between Portugal and Madeira archipelago (Gorringe, Josèphine, Ampère, and Seine), especially during the sea level low stands associated with the Pleistocene glacial periods [88, 167, 198], our analysis shows that the rissoid relationships between these areas are weak (Figure 6). This methodology does not provide a clear migratory route from the Mediterranean, Portuguese shores, or West-African shores towards the Macaronesian archipelagos ("*sensu strictum*": Azores, Madeira, Selvages, and Canaries), all computed pairs of areas show low relationships (POR-MAD = 27%; POR-SEL = 29%; MED-MAD = 27%; MED-CAN = 40%; MAD-WAF = 44%; CAN-WAF = 34%, Figure 6). Interestingly, Canaries act as a probable source of rissoid fauna, instead of recipient, when compared to all areas, with the sole exception of the Mediterranean.

Another interesting feature is the isolation of the Cape Verde archipelago, which is very weakly related with the "*sensu strictum*" Macaronesian archipelagos (Azores, Madeira, Selvagens, and Canaries archipelagos, Figure 8). This reinforces the idea that, at least for the marine species, Cape Verde should not be included in the Macaronesian area, as already suggested by a number of authors [167, 227, 228].

Acknowledgments

The authors are thankful for Anders Warén, John J. van Aartsen, Emilio Rolán, Willy Segers, Enzo Campani, Constantine Mifsud, and Colin Redfern for their contribution to the mode of development and bathymetry of several Rissoidae species. They also would like to thank an anonymous reviewer and the editor for comments that improved this manuscript. S. P. Ávila was supported by Grant SFRH/BPD/22913/2005 (FCT, Fundação para a Ciência e Tecnologia) of the Portuguese Government, and Synthesys Grants ABC-PD81 and ABC-PD93 (Access to Belgian Collections: European Community, Access to Research Infrastructure action of the Improving Human Research Potential Programme).

References

[1] E. H. Rapoport, *Areography: Geographical Strategies of Species*, Pergamon Press, New York, NY, USA, 1982.

[2] G. C. Stevens, "The latitudinal gradient in geographical range: how so many species coexist in the tropics," *American Naturalist*, vol. 133, no. 2, pp. 240–256, 1989.

[3] G. C. Stevens, "Extending Rapoport's rule to Pacific marine fishes," *Journal of Biogeography*, vol. 23, no. 2, pp. 149–154, 1996.

[4] H. L. Sanders, "Marine benthic diversity: a comparative study," *American Naturalist*, vol. 102, pp. 243–282, 1968.

[5] M. L. Rosenzweig, "On continental steady states of species diversity," in *Ecology and Evolution of Communities*, M. L. Cody and J. M. Diamond, Eds., pp. 121–140, Belknap Press of Harvard University Press, Cambridge, Mass, USA, 1975.

[6] E. R. Pianka, "Latitudinal gradients in species diversity," *Trends in Ecology and Evolution*, vol. 8, pp. 223–224, 1989.

[7] J. H. Brown, *Macroecology*, University of Chicago Press, Chicago, Ill, USA, 1995.

[8] M. Dynesius and R. Jansson, "Evolutionary consequences of changes in species' geographical distributions driven by Milankovitch climate oscillations," *Proceedings of the National Academy of Sciences of the United States of America*, vol. 97, no. 16, pp. 9115–9120, 2000.

[9] K. Roy, D. Jablonski, J. W. Valentine, and G. Rosenberg, "Marine latitudinal diversity gradients: tests of causal hypotheses," *Proceedings of the National Academy of Sciences of the United States of America*, vol. 95, no. 7, pp. 3699–3702, 1998.

[10] K. Roy, D. Jablonski, and J. W. Valentine, "Dissecting latitudinal diversity gradients: functional groups and clades of marine bivalves," *Proceedings of the Royal Society B*, vol. 267, no. 1440, pp. 293–299, 2000.

[11] S. R. Floeter and A. Soares-Gomes, "Biogeographic and species richness patterns of Gastropoda on the southwestern Atlantic," *Revista Brasileira de Biologia*, vol. 59, no. 4, pp. 567–575, 1999.

[12] C. Valdovinos, S. A. Navarrete, and P. A. Marquet, "Mollusk species diversity in the Southeastern Pacific: why are there more species towards the pole?" *Ecography*, vol. 26, no. 2, pp. 139–144, 2003.

[13] D. H. Wright, "Species-energy theory: an extension of species-area theory," *Oikos*, vol. 41, no. 3, pp. 496–506, 1983.

[14] D. H. Wright, D. J. Currie, and B. A. Maurer, "Energy supply and patterns of species richness on local and regional scales," in *Species Diversity in Ecological Communities*, R. E. Ricklefs and D. Schluter, Eds., pp. 66–74, University of Chicago Press, 1993.

[15] E. Macpherson, "Large-scale species-richness gradients in the Atlantic Ocean," *Proceedings of the Royal Society B*, vol. 269, no. 1501, pp. 1715–1720, 2002.

[16] A. Astorga, M. Fernández, E. E. Boschi, and N. Lagos, "Two oceans, two taxa and one mode of development: latitudinal diversity patterns of South American crabs and test for possible causal processes," *Ecology Letters*, vol. 6, no. 5, pp. 420–427, 2003.

[17] Wenz, "Gastropoda. teil 1, allgemeiner teil und prosobranchia," in *Handbuch der Paläozoologie*, O. H. Schindewolf, Ed., vol. 6, pp. 231–1, Gebrüer Bornträger, Berlin, Germany, 1944.

[18] E. Coan, "A proposed revision of the rissoacean families rissoidae, rissoinidae and cingulopsidae," *Veliger*, vol. 6, no. 3, pp. 164–171, 1964.

[19] F. Nordsieck, *Die Europäischen Meeresschnecken (Opistho-Branchia mit Pyramidellidae; Rissoacea)*, Vom Eismeer bis Kapverden, Mittelmeer und Schwarzes Meer. Gustav Fischer, Stuttgart, Germany, 1972.

[20] W. F. Ponder, "The classification of the Rissoidae and Orbitestellidae with descriptions of some new taxa," *Transactions of the Royal Society of New Zealand, Zoology*, vol. 9, no. 17, pp. 193–224, 1967.

[21] W. F. Ponder, "A Review of the Genera of the Rissoidae (Mollusca: Mesogastropoda: Rissoacea)," *Records of the Australian Museum*, vol. 4, pp. 1–221, 1985.

[22] W. F. Ponder and R. G. De Keyzer, "Family rissoidae," in *Mollusca: The Southern Synthesis. Fauna of Australia*, P. L. Beesley, G. J. B. Ross, and A. Wells, Eds., vol. 5, pp. 749–751, 1998.

[23] J. J. van Aartsen and A. Verduin, "European marine Mollusca: notes on less well-known species V. *Cingula (Setia) macilenta* (Monterosato, 1880) and Rissoa concinnata Jeffreys, 1883," *Basteria*, vol. 46, pp. 127–128, 1982.

[24] A. Warén, "Revision of the Rissoidae of the Norwegian North Atlantic Expedition 1876–1878," *Sarsia*, vol. 53, pp. 1–14, 1973.

[25] A. Warén, "Revision of the Arctic-Atlantic Rissoidae (Gastropoda, ProsoBranchia)," *Zoologica Scripta*, vol. 3, pp. 121–135, 1974.

[26] H. G. Hansson, *NEAT (North East Atlantic Taxa): Scandinavia marine Mollusca Check-List*, 1998, http://www.tmbl.gu.se.

[27] P. Bouchet and A. Warén, "Revision of the Northeast Atlantic bathyal and abyssal Mesogastropoda," *Bolletino Malacologico*, vol. 3, supplement, pp. 579–840, 1993.

[28] A. Warén, "New and little known mollusca from Iceland," *Sarsia*, vol. 74, pp. 1–28, 1989.

[29] A. Warén, "New and little known Mollusca from Iceland and Scandinavia—part 3," *Sarsia*, vol. 81, no. 3, pp. 197–245, 1996.

[30] V. Fretter and A. Graham, "The prosoBranch molluscs of Britain and Denmark—part 6," *Journal of Molluscan Studies*, supplement 6, pp. 153–241, 1978.

[31] S. M. Smith, "*Rissoa violacea* Desmarest, Rissoa lilacina Récluz, *Rissoa rufilabrum* Alder and *Rissoa porifera* Lovén and their distribution in British and Irish waters," *Journal of Conchology*, vol. 27, pp. 235–248, 1970.

[32] I. J. Killeen and J. M. Light, "Observations on *Onoba semicostata* and *O. aculeus* around British and northern French coasts," *Journal of Conchology*, vol. 36, no. 2, pp. 7–12, 1998.

[33] A. Nobre, *Moluscos Marinhos de Portugal*, vol. 1, Imprensa Portuguesa, Porto, Portugal, 1931.

[34] A. Nobre, *Moluscos Marinhos de Portugal*, vol. 2, Imprensa Portuguesa, Porto, Portugal, 1936.

[35] A. Nobre and J. M. Braga, *Notas Sobre a Fauna das Ilhas Berlengas e Farilhões*, Coimbra Editora, Coimbra, Portugal, 1942.

[36] M. C. C. Macedo, M. I. C. Macedo, and J. P. Borges, *Conchas Marinhas de Portugal*, Editorial Verbo, Lisbon, Portugal, 1999.

[37] E. Rolán, "The genus *Onoba* (Mollusca, Caenogastropoda, Rissoidae) from NW Spain, with the description of two new species," *Zoosymposia*, vol. 1, pp. 233–245, 2008.

[38] J. J. van Aartsen and M. C. Fehr-de-Wal, "Some remarks about *Alvania deliciosa* (Jeffreys, 1884)," *Basteria*, vol. 37, pp. 71–76, 1973.

[39] J. J. van Aartsen, "*Alvania vermaasi* nov. spec., a new species of gastropod from the Gulf of Algeciras (Spain)," *Basteria*, vol. 39, pp. 91–96, 1975.

[40] J. J. van Aartsen, "European marine Mollusca: notes on less well-known species. 1. *Alvania* (Alcidiella) *spinosa* Monterosato, 1890," *Basteria*, vol. 40, pp. 127–132, 1976.

[41] J. J. van Aartsen, "Synoptic tables of Mediterranean and European conchology (Gen. *Alvania*) (tav XVIII)," *La Conchiglia*, vol. 14, no. 158-159, pp. 4–5, 1982.

[42] J. J. van Aartsen, "Synoptic tables of Mediterranean and European conchology (Gen. *Alvania*) (tav XIX)," *La Conchiglia*, vol. 14, no. 160-161, pp. 16–17, 1982.

[43] J. J. van Aartsen, "Synoptic tables of Mediterranean and European conchology. Gen. *Alvania* (Subgen. *Alvinia* & *Galeodina*) (tav XX)," *La Conchiglia*, vol. 14, no. 162-163, pp. 8–9, 1982.

[44] J. J. van Aartsen, "Synoptic tables of Mediterranean and European conchology. Gen. *Alvania* (Subgen. *Arsenia* & *Alvaniella*) (tav XXI)," *La Conchiglia*, vol. 14, no. 164-165, pp. 4–6, 1982.

[45] J. J. van Aartsen, "Synoptic tables of Med. and Europ. conchology. Genere *Alvania* (sottogenere Alcidiella). tab. XXII," *La Conchiglia*, vol. 15, no. 166-167, pp. 8–9, 1983.

[46] J. J. van Aartsen, "Sinoptic tables of Mediterranean and European conchology. Genus *Alvania* (subgenus *Actonia*, *Thapsiella* and *Moniziella*). Genus *Manzonia*. (tab. XXIII)," *La Conchiglia*, vol. 15, no. 168-169, pp. 4–5, 1983.

[47] J. J. van Aartsen, "*Manzonia overdiepi*, a new marine gastropod (Rissoidae) from Canary and Madeira is," *La Conchiglia*, vol. 15, no. 168-169, pp. 6–7, 1983.

[48] A. Verduin, "On the systematics of recent *Rissoa* of the subgenus Turboella Gray, 1847, from the Mediterranean and European Atlantic coasts," *Basteria*, vol. 40, pp. 21–73, 1976.

[49] A. Verduin, "*Rissoa (Turgidina) testudae* subg. nov., sp. nov., a marine gastropod from the Straits of Gibraltar," *Basteria*, vol. 43, pp. 47–50, 1979.

[50] A. Verduin, "*Alvania cimex* (L.) s. l. (Gastropoda, Proso-Branchia), an aggregate species," *Basteria*, vol. 50, pp. 25–32, 1986.

[51] J. J. van Aartsen and A. Verduin, "On the conchological identification of *Cingula (Setia) fusca* (Philippi, 1841), C. (S.) *turriculata* (Monterosato, 1884), and C. (S.) inflata (Monterosato, 1884), marine gastropods from the Mediterranean," *Basteria*, vol. 42, pp. 27–47, 1978.

[52] S. Palazzi, "Taxonomic notes on the Rissoidae and related families. VI. Description of two new species of *Pisinna* Monterosato, 1878," *Notiziario CISMA*, vol. 4, no. 1-2, pp. 11–15, 1982.

[53] J. J. van Aartsen, H. P. M. G. Menkhorst, and E. Gittenberger, "The marine Mollusca of the Bay of Algeciras, Spain, with general notes on Mitrella, Marginellidae and Turridae," *Basteria*, no. 2, supplement, pp. 1–135, 1984.

[54] B. Amati, "*Alvania gagliniae* sp. n. (Gastropoda; Proso-Branchia)," *Notiziário CISMA*, vol. 6, no. 1-2, pp. 35–41, 1984.

[55] B. Amati, "Il genere *Obtusella* Cossmann, 1921, nei mari Europei (Gastropoda; ProsoBranchia)," *Notiziário CISMA*, vol. 7-8, pp. 57–63, 1986.

[56] B. Amati and M. Oliverio, "*Alvania (Alvaniella) hallgassi* sp. n. (Gastropoda; ProsoBranchia)," *Notiziario CISMA*, vol. 2, no. 7, pp. 28–34, 1985.

[57] B. Amati and I. Nofroni, "*Alvania settepassii* sp. n. (Gastropoda: ProsoBranchia)," *Notiziario CISMA*, vol. 2, no. 7, pp. 19–27, 1985.

[58] B. Amati and I. Nofroni, "*Alvania datchaensis* sp. n. (Gastropoda; ProsoBranchia)," *Notiziario CISMA*, vol. 10, pp. 46–63, 1987.

[59] B. Amati and I. Nofroni, "Designazione del lectotipo di "Setia" gianninii F. Nordsieck, 1974 e descrizione di *Onoba dimassai* nuova specie (ProsoBranchia: Rissoidae)," *Notiziario CISMA*, vol. 12, no. 13-14, pp. 30–37, 1991.

[60] M. Oliverio, "*Alvania Amatii* n, sp. (Gastropoda: Proso-Branchia)," *Notiziário CISMA*, vol. 7-8, pp. 29–34, 1986.

[61] M. Oliverio, "A new prosoBranch from the Mediterranean sea, *Alvania dianensis* n. sp. (Mollusca; Gastropoda)," *Bulletin Zoölogisch Museum*, vol. 11, no. 13, pp. 117–120, 1988.

[62] M. Oliverio, "Sull'identità di *Alvania fractospira* Oberling, 1970 (ProsoBranchia, Rissooidea)," *Notiziário CISMA*, vol. 14, pp. 33–36, 1993.

[63] M. Oliverio, B. Amati, and I. Nofroni, "Proposta di adeguamento sistematico dei Rissoidaea (sensu Ponder) del mar Mediterraneo—parte I: famiglia Rissoidae Gray, 1847 (Gastropoda: ProsoBranchia)," *Notiziário CISMA*, vol. 7-8, pp. 35–52, 1986.

[64] J. J. van Aartsen and J. van der Linden, "*Alvania gothica* a new species from the Mediterranean," *La Conchiglia*, vol. 18, no. 202-203, pp. 14–15, 1986.

[65] J. van der Linden and W. M. Wagner, "*Cingula antipolitana* spec. nov., a new marine gastropod species from southern France (ProsoBranchia, Rissoacea)," *Basteria*, vol. 51, pp. 59–61, 1987.

[66] H. J. Hoenselaar and R. G. Moolenbeek, "Two new species of *Onoba* from southern Spain (Gastropoda: Rissoidae)," *Basteria*, vol. 51, pp. 17–20, 1987.

[67] F. Giusti and I. Nofroni, "*Alvania dipacoi* new species from the Tuscan Archipelago," *La Conchiglia*, vol. 21, no. 242–245, pp. 54–56, 1989.

[68] J. J. van Aartsen, C. Bogi, and F. Giusti, "Remarks on the genus *Benthonella* (Rissoidae) in Europe, and the description of *Laeviphitus* (nov.gen.) *Verduini* (nov.spec.) (Epitonidae)," *La Conchiglia*, vol. 246–249, pp. 19–22, 1989.

[69] B. Amati, I. Nofroni, and M. Oliverio, "New species and rediscoveries within the *Alvania*-group from 1980 for the Mediterranean Sea (ProsoBranchia Truncatelloidea)," *La Conchiglia*, vol. 253–255, pp. 47–48, 1990.

[70] H. J. Hoenselaar and J. Hoenselaar, "A new *Setia* species from southern Spain," *Basteria*, vol. 55, pp. 173–175, 1991.

[71] I. Nofroni and M. Pizzini, "New data of the group *Alvania rudis* (Philippi, 1844) and description of *Alvania clarae*, nova species (ProsoBranchia: Rissoidae)," *La Conchiglia*, vol. 260, pp. 48–51, 1991.

[72] M. Oliverio, B. Amati, and I. Nofroni, "Revision of the *Alvania testae* group of species (Gastropoda, ProsoBranchia, Truncatelloidea = Rissooidea)," *Lavori S.I.M*, vol. 24, pp. 249–259, 1992.

[73] J. J. van Aartsen and W. Engl, "*Cingula anselmoi* n. sp., a new European Rissoid," *La Conchiglia*, vol. 290, pp. 21–22, 1999.

[74] C. Smriglio and P. Mariottini, "*Onoba Oliverioi* n. sp. (Proso-Branchia, Rissoidae), a new gastropod from the Mediterranean," *Iberus*, vol. 18, no. 1, pp. 15–19, 2000.

[75] A. Margelli, "Further remarks on *Alvania elisae* Margelli," *La Conchiglia*, vol. 33, no. 301, pp. 24–27, 2001.

[76] C. Bogi and B. S. Galil, "*Setia levantina* n. sp., una nuova specie di Rissoidae Dalle coste israeliane," *Bollettino Malacologico*, vol. 43, no. 9–12, pp. 171–173, 2007.

[77] G. Buzzurro and F. Landini, "Descrizione di una nuova specie di Rissoidae (Gastropoda: ProsoBranchia) per le coste laziali (Mar Tirreno)," *Bollettino Malacologico*, vol. 42, no. 1–4, pp. 24–26, 2007.

[78] A. Peñas, E. Rolán, and M. Ballesteros, "Segunda adición a la fauna malacológica del litoral de Garraf (NE de la Península Ibérica)," *Iberus*, vol. 26, no. 2, pp. 15–42, 2008.

[79] J. D. Oliver and J. Templado, "Dos nuevas especies del género *Alvania* (Caenogastropoda, Rissoidae)," *Iberus*, vol. 27, no. 1, pp. 57–66, 2009.

[80] R. B. Watson, Report on the Scaphopoda and Gasteropoda collected by H.M.S. "Challenger" during the years 1873–1876. Reports on the Scientific Results of the "Challenger" Expedition 1873–1876. Zoology, Vol. XV, part XLII, 756 pp., LIII pls, 1886.

[81] P. Dautzenberg, "Contribution à la faune malacologique des Iles Açores. Résultats des draGages effectués par le yacht l'Hirondelle pendant sa campagne scientifique de 1887. Révision des mollusques marins des Açores," *Résultats des Campagnes Scientifiques du Prince de Monaco*, vol. 1, p. 112, 1889.

[82] B. Amati, "*Manzonia (Alvinia) Sleursi* sp. n. (Gastropoda, ProsoBranchia)," *Notiziário CISMA*, vol. 10, pp. 25–30, 1987.

[83] S. Gofas, "Two new species of *Alvania* (Rissoidae) from the Azores," *Publicações Ocasionais da Sociedade Portuguesa de Malacologia*, vol. 14, pp. 39–42, 1989.

[84] J. van der Linden, "*Alvania obsoleta* spec. nov. from the Azores (Gastropoda, ProsoBranchia: Rissoidae)," *Basteria*, vol. 57, no. 1–3, pp. 79–82, 1993.

[85] J. van der Linden and J. J. van Aartsen, "*Alvania abstersa* nom. nov., a new name for *A. obsoleta* van der Linden, 1993, non *A. obsoleta* (S. V. Wood, 1848) (Gastropoda ProsoBranchia: Rissoidae)," *Basteria*, vol. 58, p. 2, 1994.

[86] S. P. Ávila, "Shallow-water marine molluscs of the Azores: biogeographical relationships," *Arquipélago. Life and Marine Sciences*, supplement 2, pp. 99–131, 2000.

[87] S. P. Ávila, "The shallow-water Rissoidae (Mollusca, Gastropoda) of the Azores and some aspects of their ecology," *Iberus*, vol. 18, no. 2, pp. 51–76, 2000.

[88] S. P. Ávila, *Processos e padrões de dispersão e colonização nos rissoidae (Mollusca: Gastropoda) dos Açores*, Ph.D. thesis, Universidade dos Açores, Ponta Delgada, Portugal, 2005.

[89] S. P. Ávila and M. A. E. Malaquias, "Biogeographical relationships of the molluscan fauna of the Ormonde Seamount (Gorringe Bank, Northeast Atlantic Ocean)," *Journal of Molluscan Studies*, vol. 69, no. 2, pp. 145–150, 2003.

[90] T. Beck, T. Metzger, and A. Freiwald, Biodiversity inventorial atlas of macrobenthic seamount animals. Eu-ESF project OASIS, 126 pp. Oceanic seamounts: an integrated study; EVK2-CT-2002-00073, 2006, http://www1.uni-hamburg.de/OASIS/Pages/publications/BIAS.pdf.

[91] S. Gofas, "Rissoidae (Mollusca: Gastropoda) from northeast Atlantic seamounts," *Journal of Natural History*, vol. 41, no. 13-16, pp. 779–885, 2007.

[92] A. Nobre, "Moluscos testáceos marinhos do arquipélago da Madeira," in *Memórias e Estudos do Museu Zoológico da Universidade de Coimbra, Série I*, vol. 98, p. 101, Coimbra Editora, Coimbra, Portugal, 1937.

[93] A. Verduin, "On the taxonomy of some Rissoacean species from Europe. Madeira and the Canary Islands (Gastropoda, ProsoBranchia)," *Basteria*, vol. 52, pp. 9–35, 1988.

[94] R. G. Moolenbeek and H. J. Hoenselaar, "The genus *Alvania* on the Canary Islands and Madeira (Mollusca: Gastropaoda)—part 1," *Bulletin Zoölogisch Museum*, vol. 11, no. 27, pp. 215–228, 1989.

[95] R. G. Moolenbeek and H. J. Hoenselaar, "The genus *Alvania* on the Canary Islands and Madeira (Mollusca: Gastropoda)—part 2 [final part]," *Bulletin Zoölogisch Museum*, vol. 16, no. 8, pp. 53–64, 1998.

[96] W. Segers, F. Swinnen, and R. de Prins, *Marine Molluscs of Madeira*, Snoeck Publishers, Heule, Belgium, 2009.

[97] B. Amati, "On a new species of Manzonia from Selvagens islands, (Gastropoda, ProsoBranchia, Rissoidae)," *Publicações Ocasionais da Sociedade Portuguesa de Malacologia*, vol. 16, pp. 9–12, 1992.

[98] R. G. Moolenbeek and M. J. Faber, "A new micromollusc from the Canary Islands (Mollusca, Gastropoda: Rissoacea)," *Basteria*, vol. 50, pp. 177–180, 1986.

[99] E. Rolán, "Aportaciones al estudio de los Risoaceos de las Islas Canarias—I. Description de tres especies nuevas," *Publicações Ocasionais da Sociedade Portuguesa de Malacologia*, vol. 8, pp. 1–4, 1987.

[100] J. van der Linden and W. M. wagner, "*Alvania multiquadrata* spec. Nov. from the Canary Islands (Gastropoda Proso-Branchia: Rissoidae)," *Basteria*, vol. 53, pp. 35–37, 1989.

[101] R. G. Moolenbeek and H. J. Hoenselaar, "New additions to the *Manzonia* fauna of the Canary Islands (Gastropoda: Rissoidae)," *Publicações Ocasionais da Sociedade Portuguesa de Malacologia*, vol. 16, pp. 13–16, 1992.

[102] W. Segers, "*Alvania grancanariensis* new species from the Canary Islands (Gastropoda: PrososBranchia)," *Gloria Maris*, vol. 37, no. 5-6, pp. 82–87, 1999.

[103] J. M. Hernández-Otero, F. García-Talavera, and M. Hernández-García, "División apogastropoda," in *Lista de Especies Marinas de Canarias (Algas, Hongos, Plantas y Animales)*, L. Moro, J. L. Martín, M. J. Garrido, and I. Izquierdo, Eds., pp. 83–91, Consejería de Política Territorial y Medio Ambiente del Cobierno de Canarias, 2003.

[104] E. Rolán, "El genero *Manzonia* Brusina, 1870 en el archipielago de Cabo Verde," *Publicações Ocasionais da Sociedade Portuguesa de Malacologia*, vol. 9, pp. 27–36, 1987.

[105] R. G. Moolenbeek and E. Rolán, "New species of Rissoidae from the Cape Verde Islands (Mollusca: Gastropoda)—part 1," *Bulletin Zoölogisch Museum*, vol. 11, no. 14, pp. 121–126, 1988.

[106] J. Templado and E. Rolán, "Las especies del género *Crisilla* y afines (Gastropoda: ProsoBranchia: Rissoidae) en el archipiélago de Cabo Verde," *Iberus*, vol. 11, no. 2, pp. 1–25, 1993.

[107] E. Rolán and F. Rubio, "New information on the malacological fauna (Mollusca: Gastropoda) of the Cape Verde Archipelago, with the description of five new species," *Apex*, vol. 14, no. 1, pp. 1–10, 1999.

[108] E. Rolán and Á. A. Luque, "The subfamily Rissoininae (Mollusca: Gastropoda: Rissoidae) in the Cape Verde Archipelago (West Africa)," *Iberus*, vol. 18, no. 1, pp. 21–94, 2000.

[109] E. Rolán, *Malacological Fauna From the Cape Verde Archipelago—part 1 Polyplacophora and Gastropoda*, Conchbooks, 2005.

[110] E. Rolán and Á. de Oliveira, "A new species of *Rissoa* (ProsoBranchia, Rissoidae) from Cape Verde Archipelago," *Gloria Maris*, vol. 47, no. 4, pp. 73–77, 2008.

[111] E. A. Smith, "Report on the marine molluscan fauna of the island of St. Helena," *Proceedings of the Zoological Society of Londoon Part 2*, pp. 247–317, 1890.

[112] T. M. Worsfold, G. Avern, and W. F. Ponder, "Shallow water rissoiform gastropods from Tristan da Cunha, South Atlantic Ocean, with records of species from Gough Island," *Zoologica Scripta*, vol. 22, no. 2, pp. 153–166, 1993.

[113] S. Gofas and A. Warén, "Taxonomie de quelques especes du genre *Alvania* (Mollusca, Gastropoda) des côtes Iberiques et Marocaines," *Bolletino Malacologico*, vol. 18, no. 1–4, pp. 1–16, 1982.

[114] R. G. Moolenbeek and T. Piersma, "A new *Setia* species from Mauritania (Gastropoda: Rissoidae)," *Gloria Maris*, vol. 29, no. 2, pp. 31–33, 1990.

[115] E. Rolán and F. Fernandes, "Tres nuevas especies del genero *Manzonia* (Mollusca, Gastropoda) para la costa occidental de Africa," *Publicações Ocasionais da Sociedade Portuguesa de Malacologia*, vol. 15, pp. 63–68, 1990.

[116] S. Gofas, "The West African Rissoidae (Gastropoda: Rissooidea) and their similarities to some European species," *Nautilus*, vol. 113, no. 3, pp. 78–101, 1999.

[117] S. Gofas, "A new *Manzonia* (Gastropoda: Rissoidae) from Nothwestern Morocco," *Iberus*, vol. 28, no. 1, pp. 91–96, 2010.

[118] E. Rolán and P. Ryall, "Checklist of the Angolan marine molluscs," *Reseñas Malacológicas*, vol. 10, pp. 1–132, 1985.

[119] V. D. Engle and J. K. Summers, "Biogeography of benthic macroinvertebrates in estuaries along the Gulf of Mexico and western Atlantic coasts," *Hydrobiologia*, vol. 436, pp. 17–33, 2000.

[120] M. A. Rex, M. C. Watts, R. J. Etter, and S. O'neill, "Character variation in a complex of rissoid gastropods from the upper Continental slope of the western North Atlantic," *Malacologia*, vol. 29, no. 2, pp. 325–339, 1988.

[121] W. H. Dall, "Reports on the results of dredging, under the supervision of Alexander Agassiz, in the Gulf of Mexico, and in the Caribbean Sea, 1877–1879, by the United States Coast Survey Steamer "Blake"," *Bulletin of the Museum of Comparative Zoology*, vol. 9, pp. 33–144, 1881.

[122] F. Baker, G. D. Hanna, and A. M. Strong, "Some rissoid Mollusca from the Gulf of California," *Proceedings of the California Academy of Sciences*, vol. 19, no. 4, pp. 23–40, 1930.

[123] M. J. Faber and R. G. Moolenbeek, "On the doubtful records of *Alvania* platycephala, *Alvania* pagodula and *Alvania* didyma, with the description of two new rissoid species (Mollusca; Gastropoda: Rissoidae)," *Beaufortia*, vol. 37, no. 4, pp. 67–71, 1987.

[124] K. M. de Jong and H. E. Coomans, *Marine Gastropods from Curaçao, Aruba and Bonaire*, E. J.Brill, Leiden, The Netherlands, 1988.

[125] J. H. Leal and D. R. Moore, "*Rissoina indiscreta*, a new rissoid species from the tropical southwestern Atlantic with Indo-West Pacific affinities (Mollusca, Gastropoda, Rissooidea)," *Bulletin of Marine Science*, vol. 45, no. 1, pp. 139–147, 1989.

[126] M. J. Faber, "Studies on West Indian marine molluscs—19. On the identity of *Turbo Bryereus* Montagu , 1803, with the description of a new species of *Rissoina* (Gastropoda ProsoBranchia: Rissoidae)," *Basteria*, vol. 54, pp. 115–121, 1990.

[127] J. H. Leal, *Marine ProsoBranch Gastropods from Oceanic Islands Off Brazil, Species Composition and Biogeography*, Universal Book Services/Dr. W. Backhuys, Oegstgeest, The Netherlands, 1991.

[128] E. Rolán, "A new species of *Zebina* (Gastropoda: Rissoidae: Rissoininae) from Yucatán (Mexico)," *Apex*, vol. 13, no. 4, pp. 177–179, 1998.

[129] J. Espinosa and J. Ortea, "Descripción de cuatro nuevas especies de la familia Rissoinidae (Mollusca: Gastropoda)," *Avicennia*, vol. 15, pp. 141–149, 2002.

[130] G. Rosenberg, F. Moretzsohn, and E. F. García, "Gastropoda (Mollusca) of the Gulf of Mexico," in *Gulf of Mexico—Origins, Waters, and Biota Biodiversity*, D. L. Felder and D. K. Camp, Eds., pp. 579–699, Texas A&M Press, College Station, Tex, USA, 2009.

[131] F. J. Palacio, "Revisión zoogeográfica marina del sur del Brasil," *Boletim do Instituto Oceanográfico de São Paulo*, vol. 31, no. 1, pp. 69–92, 1980.

[132] W. F. Ponder, "Rissoaform Gastropods from the Antarctic and sub-Antarctic. The Eatoniellidae, Rissoidae, Barleeidae, Cingulopsidae, Orbitestellidae and Rissoellidae (Mollusca: Gastropoda) of Signy Island, South Orkney Islands, with a review of the Antarctic and sub-Antarctic (excluding southern South America and the New Zealand sub-Antarctic islands) species," *British Antarctic Survey Scientific Reports*, vol. 108, pp. 1–96, 1983.

[133] W. F. Ponder and T. M. Worsfold, "A review of the Rissoiform Gastropods of Southwestern South America (Mollusca, Gastropoda)," *Contributions in Science, Natural History Museum of Los Angeles County*, vol. 445, pp. 1–63, 1994.

[134] C. R. Babio and C. Thiriot-Quiévreux, "Gastéropodes de la région de Roscoff. Étude particulière de la protoconque," *Cahiers de Biologie Marine*, vol. 15, pp. 531–549, 1974.

[135] A. Verduin, "On the systematics of some recent Rissoa (Gastropoda, ProsoBranchia)," *Basteria*, vol. 50, pp. 13–24, 1986.

[136] J. Templado and E. Rolán, "El genero *Onoba* H. & A. Adams, 1854 (Gastropoda: Rissoidae) en las costas Europeas—1," *Iberus*, vol. 6, pp. 117–124, 1986.

[137] R. G. Moolenbeek and H. J. Hoenselaar, "On the identity of *Onoba moreleti* Dautzenberg, 1889 (Gastropoda: Rissoidae), with the description of *Onoba josae* n. sp.," *Basteria*, vol. 51, pp. 153–157, 1987.

[138] R. G. Moolenbeek and M. J. Faber, "The Macaronesian species of the genus *Manzonia* (Gastropoda: Rissoidae)—part I," *De Kreukel*, vol. 1, pp. 1–16, 1987.

[139] R. G. Moolenbeek and M. J. Faber, "The Macaronesian species of the genus *Manzonia* (Gastropoda: Rissoidae)—part II," *De Kreukel*, vol. 2-3, pp. 23–31, 1987.

[140] R. G. Moolenbeek and M. J. Faber, "The Macaronesian species of the genus *Manzonia* (Gastropoda: Rissoidae)—part III," *De Kreukel*, vol. 10, pp. 166–179, 1987.

[141] W. J. M. Sleurs, "Mollusca Gastropoda: four new rissoinine species (Rissoininae) from deep water in New Caledonian region," in *Résultats des Campagnes MUSORSTOM*, A. Crosnier and P. Bouchet, Eds., vol. 7 of *Mémoires du Muséum National d'Histoire Naturelle, series A*, pp. 163–178, 1991.

[142] W. J. M. Sleurs, "A revision of the Recent species of *Rissoina* (*Moerchiella*), *R.* (*Apataxia*), *R.* (*Alinzebina*) and *R.* (*Pachyrissoina*) (Gastropoda: Rissoidae)," *Bulletin de l'Institut Royal des Sciences Naturelles de Belgique*, vol. 63, pp. 71–135, 1993.

[143] W. J. M. Sleurs, "Two new *Rissoina* (s.s.) sister species from the Western Pacific," *Molluscan Research*, vol. 15, pp. 13–19, 1994.

[144] H. J. Hoenselaar and J. Hoenselaar, "Conchological differences between *Crisilla marioni* (Fasulo & Gaglini, 1987) and *Crisilla cristallinula* (Manzoni, 1868) (Gastropoda Prosobranchia: Rissoidae)," *Basteria*, vol. 58, pp. 195–197, 1994.

[145] W. J. M. Sleurs and R. C. Preece, "The Rissoininae (Gastropoda: Rissoidae) of the Pitcairn Islands, with the description of two new species," *Journal of Conchology*, vol. 35, pp. 67–82, 1994.

[146] A. Warén, "Ecology and systematics of the north European species of *Rissoa* and *Pusillina* (ProsoBranchia: Rissoidae)," *Journal of the Marine Biological Association of the United Kingdom*, vol. 76, no. 4, pp. 1013–1059, 1996.

[147] H. J. Hoenselaar and J. Goud, "The Rissoidae of the CANCAP expeditions—I: the genus *Alvania* Risso, 1826 (Gastropoda ProsoBranchia)," *Basteria*, vol. 62, pp. 69–115, 1998.

[148] J. Goud, "*Setia lidyae* Verduin, 1988, a junior synonym of *Alvania iunoniae* Palazzi, 1988, with additional data on the distribution of some *Setia* species described by Verduin (Gastropoda ProsoBranchia, Rissoidae)," *Basteria*, vol. 63, pp. 69–71, 1999.

[149] S. Gofas, J. le Renard, and P. Bouchet, "Mollusca," in *European Register of Marine Species. A Check-List of the Marine Species in Europe and a Bibliography of Guides to their Identification. Collection Patrimoines Naturels*, M. J. Costello, C. S. Emblow, and R. White, Eds., vol. 50, pp. 180–213, 2001.

[150] E. Rolán, "A new species of *Alvania* (Mollusca, Rissoidae) from Annobón (Gulf of Guinea, West Africa)," *Iberus*, vol. 19, no. 1, pp. 49–52, 2001.

[151] V. Garilli, "On some neogene to recent species related to *Galeodina* Monterosato, 1884, *Galeodinopsis* Sacco, 1895, and *Massotia* Bucquoy, Dautzenberg, and Dollfus, 1884 (Caenogastropoda: Rissoidae) with the description of two new *Alvania* species from the Mediterranean Pleistocene," *Nautilus*, vol. 122, no. 1, pp. 19–51, 2008.

[152] D. Jablonski and R. A. Lutz, "Molluscan larval shell morphology: ecological and paleontological applications," in *Skeletal Growth of Aquatic Organisms*, D. C. Rhoads and R. A. Lutz, Eds., pp. 323–377, Plenum, New York, NY, USA, 1980.

[153] C. Thiriot-Quiévreux, "Protoconches et coquilles larvaires de mollusques rissoidés mediterranées," *Annales de la Institue Océanographique*, vol. 56, pp. 65–76, 1980.

[154] E. M. Da Costa, *Historia Naturalis Testaceorum Britanniae*, Millan, White, Elmsley & Robson, London, UK, 1778.

[155] T. Shuto, "Larval ecology of prosoBranch gastropods and its bearing on biogeography and paleontology," *Letahia*, vol. 7, no. 3, pp. 239–256, 1974.

[156] R. S. Scheltema, "On the relation between dispersal of pelagic larvae and the evolution of marine prosoBranch gastropods," in *Marine Organisms: Genetics, Ecology, and Evolution*, B. Battaglia and J. A. Beardmore, Eds., NATO Conference Series. Series 4: Marine Sciences, pp. 303–322, Plenum Press, New York, NY, USA, 1978.

[157] D. Jablonski and R. A. Lutz, "Larval ecology of marine benthic invertebrates: paleobiological implications," *Biological Reviews*, vol. 58, no. 1, pp. 21–89, 1983.

[158] B. R. Rosen and A. B. Smith, "Tectonics from fossils? Analysis of reef-coral and sea-urchin distributions from late Cretaceous to Recent, using a new method," *Geological Society Special Publication*, vol. 37, no. 1, pp. 275–306, 1988.

[159] I. J. Garzón-Orduña, D. R. Miranda-Esquivel, and M. Donato, "Parsimony analysis of endemicity describes but does not explain: An illustrated critique," *Journal of Biogeography*, vol. 35, no. 5, pp. 903–913, 2008.

[160] J. J. Morrone, "On the identification of areas of endemism," *Systematic Biology*, vol. 43, no. 3, pp. 438–441, 1994.

[161] J. J. Morrone and J. V. Crisci, "Historical biogeography: introduction to methods," *Annual Review of Ecology and Systematics*, vol. 26, pp. 373–401, 1995.

[162] J. V. Audouin, "Explication sommaire des planches de Mollusques de l'Egypte et de la Syrie publiés par J.C. Savigny," in *Savigny J.C. Description de l'Egypte ou recueil des observations et des recherches qui ont été faites en Egypte pendant l'expédition de l'armée française, publié par les ordres de Sa Majesté l'empereur Napoléon le grand. Histoire Naturelle, Animaux Invertébrés*, vol. 1, pp. 7–56, Imprimerie Nationale, Paris, France, 1826.

[163] R. Issel, Malacologia del Mar Rosso, ricerche zoologiche e paleontologiche. Biblioteca Malacologica, Pisa, pp. XI + 387; pl. 1–5, 1869.

[164] J. E. Watrous and Q. D. Wheeler, "The out-group comparison method of character analysis," *Systematic Zoology*, vol. 30, pp. 1–11, 1981.

[165] D. L. Swofford, *PAUP*: Phylogenetic Analysis Using Parsimony (* and other methods), version 4, Sinauer, Sunderland, Mass, USA, 2002.

[166] K. Bremer, "Branch support and tree stability," *Cladistics*, vol. 10, no. 3, pp. 295–304, 1994.

[167] V. C. Almada, R. F. Oliveira, E. J. Gonçalves, A. J. Almeida, R. S. Santos, and P. Wirtz, "Patterns of diversity of the northeastern Atlantic blenniidfish fauna (Pisces: Blennnidae)," *Global Ecology and Biogeography*, vol. 10, no. 4, pp. 411–422, 2001.

[168] P. Dautzenberg and H. Fischer, "Dragages effectués par l'Hirondelle et par la Princesse Alice 1888–1896. Gastropodes et Pélécypodes," *Mémoires de la Société Zoologique de France*, vol. 10, pp. 139–234, 1897.

[169] J. G. Jeffreys, *British Conchology*, vol. 5, 1869.

[170] W. H. Dall, "Reports on the results of dredging, under the supervision of Alexander Agassiz, in the Gulf of Mexico (1877-78) and in the Carribean Sea (1879-80), by the U. S. Coast Survey steamer "Blake", Lieut.-Commander C. D. Sigsbee, U. S. N. and Commander J. R. Bartlett, U. S. N. commanding. XXIX. Report on the Mollusca. Part II: Gastropoda and Scaphopoda," *Bulletin of the Museum of Comparative Zoology*, vol. 18, pp. 1–492, 1889.

[171] P. Hesse, "Mollusken von varna und umgebung," *Nachrichtsblatt der Deutschen Malakozoologischen Gesellschaft*, vol. 48, no. 4, pp. 143–158, 1916.

[172] C. B. Adams, "Descriptions of supposed new species of marine shells, which inhabit Jamaica," *Contributions to Conchology*, vol. 7, pp. 109–123, 1850.

[173] W. J. M. Sleurs, "A revision of the recent species of the genus Stosicia (Gastropoda: Rissoidae)," *Academiae Analecta*, vol. 1, pp. 117–158, 1996.

[174] S. V. Wood, A monograph of the Crag Mollusca with descriptions of shells from the Upper Tertiaries of the British Isles. 2. Bivalves, London: printed for the Palaeontographical society, Part 2: 217–342, pl. 21–31, 1875.

[175] E. Forbes, "Report on the Mollusca and Radiata of the Aegean sea, and on their distribution, considered as bearing on geology," *Reports of the British Association for the Advancement of Science*, pp. 130–193, 1844.

[176] B. Santelices and P. A. Marquet, "Seaweeds, latitudinal diversity patterns, and Rapoport's Rule," *Diversity and Distributions*, vol. 4, no. 2, pp. 71–75, 1998.

[177] J. W. Valentine, K. Roy, and D. Jablonski, "Carnivore/noncarnivore ratios in northeastern Pacific marine gastropods," *Marine Ecology Progress Series*, vol. 228, pp. 153–163, 2002.

[178] J. A. Crame, "An evolutionary framework for the polar regions," *Journal of Biogeography*, vol. 24, no. 1, pp. 1–9, 1997.

[179] N. Myers, R. A. Mittermeler, C. G. Mittermeler, G. A. B. Da Fonseca, and J. Kent, "Biodiversity hotspots for conservation priorities," *Nature*, vol. 403, no. 6772, pp. 853–858, 2000.

[180] K. J. Hsü, W. B. F. Ryan, and M. B. Cita, "Late miocene desiccation of the mediterranean," *Nature*, vol. 242, no. 5395, pp. 240–244, 1973.

[181] W. Krijgsman, F. J. Hilgen, I. Raffi, F. J. Sierro, and D. S. Wilson, "Chronology, causes and progression of the Messinian salinity crisis," *Nature*, vol. 400, no. 6745, pp. 652–655, 1999.

[182] W. Krijgsman, M. M. Blanc-Valleron, R. Flecker et al., "The onset of the Messinian salinity crisis in the Eastern Mediterranean (Pissouri Basin, Cyprus)," *Earth and Planetary Science Letters*, vol. 194, no. 3-4, pp. 299–310, 2002.

[183] K. J. Hsü, L. Montadert, D. Bernoulli et al., "History of the Mediterranean Salinity crisis," *Initial Report of the Deep Sea Drilling Project*, vol. 42, pp. 1053–1078, 1978.

[184] S. Raffi and R. Marasti, "The Mediterranean bioprovince from the Pliocene to the Recent: observations and hypothesis based on the evolution of the taxonomic diversity of molluscs," in *Proceedings of the 1st International Meeting on Palaeontology, Essential of Historical Geology*, E. M. Gallitelli, Ed., pp. 151–177, Venice, Italy, 1982.

[185] W. Krijgsman, C. G. Langereis, W. J. Zachariasse et al., "Late Neogene evolution of the Taza-Gercif Basin (Rifian Corridor, Morocco) and implications for the Messininan salinity crisis," *Marine Geology*, vol. 153, pp. 147–160, 1999.

[186] W. Krijgsman, M. Garcés, J. Agustí, I. Raffi, C. Taberner, and W. J. Zachariasse, "The "Tortonian salinity crisis" of the eastern Betics (Spain)," *Earth and Planetary Science Letters*, vol. 181, no. 4, pp. 497–511, 2000.

[187] R. Barbieri and G. G. Ori, "Neogene palaeoenvironmental evolution in the Atlantic side of the Rifian Corridor (Morocco)," *Palaeogeography, Palaeoclimatology, Palaeoecology*, vol. 163, no. 1-2, pp. 1–31, 2000.

[188] S. V. Popov, I. G. Shcherba, L. B. Ilyina et al., "Late Miocene to Pliocene palaeogeography of the Paratethys and its relation to the Mediterranean," *Palaeogeography, Palaeoclimatology, Palaeoecology*, vol. 238, no. 1–4, pp. 91–106, 2006.

[189] S. Duggen, K. Hoernie, P. Van den Bogaard, L. Rüpke, and J. P. Morgan, "Deep roots of the Messinian salinity crisis," *Nature*, vol. 422, no. 6932, pp. 602–606, 2003.

[190] N. Loget and J. Van Den Driessche, "On the origin of the Strait of Gibraltar," *Sedimentary Geology*, vol. 188-189, pp. 341–356, 2006.

[191] M. Harzhauser, W. E. Piller, and F. F. Steininger, "Circum-Mediterranean Oligo—miocene biogeographic evolution—the gastropods' point of view," *Palaeogeography, Palaeoclimatology, Palaeoecology*, vol. 183, no. 1-2, pp. 103–133, 2002.

[192] N. J. Shackleton and J. P. Kennett, "Late Cenozoic oxygen and carbon isotope changes at DSDP site 284: implications for glacial history of the Northern Hemisphere and Antarctica," *Initial Reports of the Deep Sea Drilling Project*, vol. 47, pp. 433–445, 1975.

[193] J. Thiede, "A Glacial Mediterranean," *Nature*, vol. 276, no. 5689, pp. 680–683, 1978.

[194] C. van den Hoek, "Phytogeographic provinces along the coasts of the northern Atlantic Ocean," *Phycologia*, vol. 14, no. 4, pp. 317–330, 1975.

[195] W. F. P. van Reine, D. M. John, G. W. Lawson, and L. B. T. Kostermans, Eds., *Seaweeds of the Northwestern Coast of Africa and Adjacent Islands*, Foundation for the Promotion of Scientific Research in Africa, Brussels, Belgium, 2006.

[196] A. N. Mironov and E. M. Krylova, "Origin of the fauna of the Meteor Seamounts, north-eastern Atlantic," in *Biogeography of the North Atlantic Seamounts*, A. N. Mironov, A. V. Gebruk, and A. J. Southward, Eds., pp. 22–57, 2006.

[197] S. P. Ávila, R. Amen, J. M. N. Azevedo, M. Cachão, and F. García-Talavera, "Checklist of the Pleistocene marine molluscs of Prainha and Lagoinhas (Santa Maria Island, Azores)," *Açoreana*, vol. 9, no. 4, pp. 343–370, 2002.

[198] S. P. Ávila, C. M. da Silva, R. Schiebel, F. Cecca, T. Backeljau, and A. M. de Frias Martins, "How did they get here? Palaeobiogeography of the Pleistocene marine molluscs of the Azores," *Bulletin of the Geological Society of France*, vol. 180, pp. 295–307, 2009.

[199] A. N. Mironov, "Centers of marine fauna redistribution," *Entomological Review*, vol. 86, supplement 1, pp. S32–S44, 2006.

[200] J. C. Briggs, "Proximate sources of marine biodiversity," *Journal of Biogeography*, vol. 33, no. 1, pp. 1–10, 2006.

[201] V. Fretter and A. Graham, "British ProsoBranch Molluscs, their functional anatomy and ecology," *Ray Society*, vol. 164, pp. 1–820, 1994.

[202] G. Thorson, "Reproduction and larval ecology of marine bottom invertebrates," *Biological Review*, vol. 25, pp. 1–45, 1950.

[203] G. Thorson, "The distribution of benthic marine mollusca along the N.E. Atlantic shelf from Gibraltar to Mursmank," in *Proceedings 1st European Malacological Congress*, pp. 5–25, 1962.

[204] S. A. Mileikovsky, "Types of larval development in marine bottom invertebrates, their distribution and ecological significance: a re-evaluation," *Marine Biology*, vol. 10, pp. 193–213, 1971.

[205] M. A. Rex, R. J. Etter, and C. T. Stuart, "Large-scale patterns of species diversity in the deep-sea benthos," in *Marine Biodiversity. Patterns and Processes*, R. F. G. Ormond, J. D. Gage,

and M. V. Angel, Eds., pp. 94–121, Cambridge University Press, Cambridge, UK, 1997.

[206] M. A. Rex and A. Warén, "Planktotrophic development in deep-sea prosoBranch snails from the western North Atlantic," *Deep Sea Research Part A, Oceanographic Research Papers*, vol. 29, no. 2, pp. 171–184, 1982.

[207] J. Grahame and G. M. Branch, "Reproductive patterns of marine invertebrates," *Oceanogrraphy and Marine Biology Annual Review*, vol. 23, pp. 373–398, 1985.

[208] W. F. Ponder, "The truncatelloidean (rissoacean) radiation—a preliminary phylogeny," in *ProsoBranch Phylogeny*, W. F. Ponder, Ed., supplement 4, Malacological Review, pp. 129–166, 1988.

[209] M. A. Conti, S. Monari, and M. Oliverio, "Early rissoid gastropods from the Jurassic of Italy: the meaning of first appearences," *Scripta Geologica*, vol. 2, pp. 67–74, 1993.

[210] W. H. Hudeston, "A monograph of the inferior Oolite Gastropoda, 1(5, 6)," *Monographs of the Palaeontographic Society*, vol. 45, pp. 225–272, 1891, vol. 46, pp. 273–324, 1892.

[211] T. Kowalke and M. Harzhauser, "Early ontogeny and palaeoecology of the Mid-Miocene rissoid gastropods of the Central Paratethys," *Acta Palaeontologica Polonica*, vol. 49, no. 1, pp. 111–134, 2004.

[212] D. Jablonski, "Larval ecology and macroevolution in marine invertebrates.," *Bulletin of Marine Science*, vol. 39, no. 2, pp. 565–587, 1986.

[213] M. Oliverio, "Developmental vs genetic variation in two Mediterranean rissoid gastropod complexes," *Journal of Molluscan Studies*, vol. 60, no. 4, pp. 461–465, 1994.

[214] P. G. Moore, "Additions to the littoral fauna of Rockall, with a description of *Areolaimus penelope* sp. nov. (Nematoda: Axonolaimidae)," *Journal of the Marine Biological Association of the United Kingdom*, vol. 57, pp. 191–200, 1977.

[215] E. Poulin, A. T. Palma, and J. P. Féral, "Evolutionary versus ecological success in Antarctic benthic invertebrates," *Trends in Ecology and Evolution*, vol. 17, no. 5, pp. 218–222, 2002.

[216] S. P. Ávila, "Oceanic islands, rafting, geographical range and bathymetry: a neglected relationship?" in *Proceedings of the 5th International Symposium on the Fauna and Flora of Atlantic Islands*, T. J. Hayden, D. A. Murray, and J. P. O'connor, Eds., vol. 9, pp. 22–39, Irish Biogeographical Society, Dublin, Ireland, 2006.

[217] M. R. Bhaud and J.-C. Duchêne, "Biologie larvaire et stratégie de reproduction des Annélides Polychètes en province subantartique," in *Actes du Colloque sur la Recherche Française dans les Terres Australes*, pp. 145–152, C.N.F.R.A., Paris, France, 1988.

[218] J. C. Duchêne, "Adelphophagie et biologie larvaire chez Boccardia polybranchia (Carazzi) (Annélide Polychète Spionidae) en province subantartique," *Vie et Milieu*, vol. 39, no. 3-4, pp. 143–152, 1992.

[219] M. Bhaud and J. C. Duchêne, "Change from planktonic to benthic development: is life cycle evolution an adaptive answer to the constraints of dispersal?" *Oceanologica Acta*, vol. 19, no. 3-4, pp. 335–346, 1996.

[220] R. Colognola, P. Masturzo, G. F. Russo, M. Scardi, D. Vinci, and E. Fresi, "Biometric and genetic analysis of the marine rissoid Rissoa auriscalpium (Gastropoda, ProsoBranchia) and its ecological implications," *Marine Ecology*, vol. 7, no. 3, pp. 265–285, 1986.

[221] M. Oliverio and L. Tringali, "Two sibling species of Nassariinae in the Mediterranean sea (ProsoBranchia: Muricidae: Nasariinae)," *Bolletino Malacologico*, vol. 28, no. 5–12, pp. 157–160, 1992.

[222] M. Oliverio, "Larval development and allozyme variation in the East Atlantic *Columbella* (Gastropoda: ProsoBranchia: Columbellidae)," *Scientia Marina*, vol. 59, no. 1, pp. 77–86, 1995.

[223] M. Oliverio, "Biogeographical patterns in developmental strategies of gastropods from Mediterranean Posidonia beds," *Bolletino Malacologico*, vol. 32, no. 1–4, pp. 79–88, 1996.

[224] C. Gili and J. Martinell, "Phylogeny, speciation and species turnover. The case of the Mediterranean gastropods of genus *Cyclope* Risso, 1826," *Lethaia*, vol. 33, no. 3, pp. 236–250, 2000.

[225] A. Monteiro and E. Rolán, "Study of three samples of *Euthria* (Mollusca: Buccinidae) from the Cape Verde archipelago with the description of two new species," *Gloria Maris*, vol. 44, no. 5, pp. 90–103, 2005.

[226] E. Rolán and A. Monteiro, "New information on the genus *Euthria* (Mollusca: Buccinidae) from the Cape Verde archipelago, with the description of three new species," *Gloria Maris*, vol. 46, no. 1-2, pp. 1–22, 2007.

[227] A. Vanderpoorten, F. J. Rumsey, and M. A. Carine, "Does Macaronesia exist? Conflicting signal in the bryophyte and pteridophyte floras," *American Journal of Botany*, vol. 94, no. 4, pp. 625–639, 2007.

[228] P. Wirtz, "Three shrimps, five nudibranchs, and two tunicates new for the marine fauna of Madeira," *Boletim do Museu Municipal do Funchal*, vol. 46, no. 257, pp. 167–172, 1994.

5

Quantitative Analysis of Driving Factors of Grassland Degradation: A Case Study in Xilin River Basin, Inner Mongolia

Yichun Xie[1] and Zongyao Sha[2]

[1] Department of Geography and Geology, Eastern Michigan University, Ypsilanti, MI 48197, USA
[2] International School of Software, Wuhan University, Wuhan 430079, China

Correspondence should be addressed to Zongyao Sha, zongyaosha@yahoo.com.cn

Academic Editors: J. Dodson and B. Tóthmérész

Current literature suggests that grassland degradation occurs in areas with poor soil conditions or noticeable environmental changes and is often a result of overgrazing or human disturbances. However, these views are questioned in our analyses. Based on the analysis of satellite vegetation maps from 1984, 1998, and 2004 for the Xilin River Basin, Inner Mongolia, China, and binary logistic regression (BLR) analysis, we observe the following: (1) grassland degradation is positively correlated with the growth density of climax communities; (2) our findings do not support a common notion that a decrease of biological productivity is a direct indicator of grassland degradation; (3) a causal relationship between grazing intensity and grassland degradation was not found; (4) degradation severity increased steadily towards roads but showed different trends near human settlements. This study found complex relationships between vegetation degradation and various microhabitat conditions, for example, elevation, slope, aspect, and proximity to water.

1. Introduction

Natural grasslands and savannas occupy nearly half of the terrestrial globe [1, 2] and provide important services to modern societies. However, grasslands are sensitive to changing edaphic conditions, management regimes, and climate and weather variables [3, 4]. With growing human populations and intensifying development, degradation of natural grasslands has been observed in many regions of the world and is a serious concern [5]. Therefore, understanding the factors driving grassland degradation is increasingly critical to the conservation and, in some cases, the restoration of these fragile ecosystems [6, 7]. Studies on the driving factors of grassland degradation can provide information for understanding vegetation deterioration pathways and, thus, maintain ecosystem functioning and services. Heras et al. [8] identified soil quality, revegetation treatments, and climatic conditions as main driving forces in a Mediterranean dry environment. There is a need for adopting proactive grassland conservation measures and for forecasting vegetation responses to future environmental changes [9]. It is essential for policy makers to understand how vegetation

responses to environmental and social changes. However, due to our limited understanding of these socioecological systems when identifying potential drivers and their possible vegetation responses, policy initiatives aimed at sustainability in vegetation ecology may fail [10].

Grassland degradation may be a complex collection of dynamic processes (e.g., desertification, salinisation, soil compaction, soil water-logging, wind erosion, water erosion, etc.) [11, 12]. Evaluation of grassland health involves assessing a large number of ecological attributes with a set of well-defined indicators, which are usually difficult or costly to measure [13]. One of these indicators is the "state" of grassland health [14]. A state usually includes one or more different biological (including soil) communities that occur on a particular ecological site and have three attributes (soil/site stability, hydrologic function, and biotic integrity) [13]. For instance, a state may include different plant communities that are connected by community pathways [15, 16]. Changes between states are referred to as "transitions." Unlike community pathways, these "threshold" transitions are not reversible by simply altering the intensity or direction of factors that produced the changes [13]. Different patterns of vegetation transitions may reflect different stages of ecosystem stability [7, 17]. Based on the studies of the relations between ecosystem structure, function, degradation, restoration, and transition, Cortina et al. [18] confirmed that those aspects of an ecosystem were related to each other. A plant community transition is often regarded as an indicator of grassland degradation. Few studies, however, have attempted to examine vegetation transitions with the purpose of land restoration [19].

It is of practical interest to underpin causal relations between vegetation degradation (transition) and their driving factors. Several studies developed plant functional-type-based models to explore physical and biological mechanisms between vegetation transitions and environments [20, 21]. Others used empirical models to investigate the causes of vegetation degradation [22–24]. For example, Zhao et al. [25] presented a composite index of VWR (vegetation water ratio), combining land surface water index and enhanced vegetation index, to facilitate an identification of vegetation transitions by simply comparing the values of VWR at different stages. It is noted that natural processes rarely cause vegetation transitions, which are often induced by human disturbances. In many cases, natural processes are intensified by social factors, leading to vegetation degradation. Hence social sciences should be integrated into these plant functional models or empirical models in order to construct more effective models to study grassland degradation [24]. There is an urgent need to build causal diagrams of human-nature interactions through interdisciplinary collaboration [26].

Based on literature reviews, five groups of factors are identified to induce grassland degradation to noticeable degrees. The first group is the biophysical variables, including those reflecting global climate change [27]. The second group includes the botanic (or biotic) variables. Among these variables are plant cover and plant productivity. The Normalized Difference Vegetation Index (hereafter NDVI) provides a measure of the greenness of vegetation. NDVI or its derived forms mayindicate the productivity of vegetation [25]. The third group of variables deals with the impacts of livestock and wildlife. Increased grazing intensity is the most significant driving force of grassland degradation [28]. The fourth group of variables describes socioeconomic development and human interferences on grassland [29]. Land degradation in the dry-lands is mainly expressed by a reduction of biomass productivity, and it is also a manifestation of unsustainable development often associated with poverty [30]. Finally local habitat conditions, such as, water accessibility, elevation, slope, and slope aspect, may affect grassland degradation.

It is worth pointing out that evaluating the driving forces of change and projecting future changes requires a commitment to methodological pluralism and critical interpretation of social and environmental data [31]. For instance, studies of causal relationships between grassland transitions and their driving factors were confronted with several challenges. First, frequent or continuous vegetative time series data are needed to detect vegetation transition patterns [25]. Second, socioeconomic data synchronous with the data of vegetation transitions rarely exist. Third, due to

the limitation of data availability, studies on grassland degradation through vegetation transitions have been mostly done on a plot basis [32]. Plot-based studies rely heavily on field surveys, which are too time-consuming and costly to conduct for large areas, although they are effective for obtaining accurate vegetative data. However, new techniques for vegetation data collection and mapping raise the possibility of quantifying grassland properties remotely [33, 34]. Satellite platforms, in particular, offer an effective means of collecting contemporary data over vast areas and in short periods of times [35, 36]. The approaches based on remotely sensed data have been increasingly applied in vegetation transition studies [25, 37].

Here we describe a systematic approach for applying a Binary Logistic Regression (BLR) model to explore grassland degradation and its driving factors in the Xilin River Basin of Inner Mongolia, China—representative of the world's largest contiguous terrestrial biome, the Eurasian steppe [38, 39]. Section 2 describes the data and analysis methodology. The results of our analysis are provided in Section 3. Finally, our findings and conclusions regarding vegetation degradation are discussed in Section 4.

2. Study Area, Data, and Methods

2.1. Study Area. The Xilingol River Basin, situated between 43°26′ and 44°29′ North and 115°32′ and 117°12′ East, is representative of the vast steppe of northern China [40]. More than 90% of the land in the region is covered by grassland [5]. From southeast to northeast, the altitude gradually decreases from 1608 to 902 meters above sea level (Figure 1). The basin's total area is about 10,000 square kilometers (km²) and has an average annual temperature range from 1 to 2°C [41]. Annual mean precipitation is around 300 mm, 60–80% of which occurs between June and August, coinciding with the highest temperatures (May to September) [5, 42, 43]. The whole region is divided into 27 administrative units, including a Xilinhot urban area, 4 pastures, and 22 villages. Herd husbandry provides the main income for local farmers. Over the last several decades, the human population has grown and overgrazing is now an important concern for local governments and ecologists due to its vegetation degradation. Previous studies and onsite surveys showed that continuous overgrazing imposes a severe threat to the sustainability of this grassland.

The study region includes 11 vegetation communities (as indicated by climax species) [44]. Two climax vegetation communities, that is, *Stipa grandis* (SG) and *Leymus chinensis* (LC), are widely distributed over the study area and consist of the local reference states [38, 45]. It has been confirmed through field survey that SG and LC communities were partially replaced by degraded vegetation communities within the past two decades, most of which are *Cleistogenes squarrosa* (CS) and *Artemisia frigida* (AF) [39]. For example, on the Xier Plain, located at the middle to the upper reach of the Xilin River, AF and CS almost dominate the area, where the primary climax vegetation was supposed to be SG and LC, indicating a significant vegetation transition. The transition of either SG or LC communities to any other

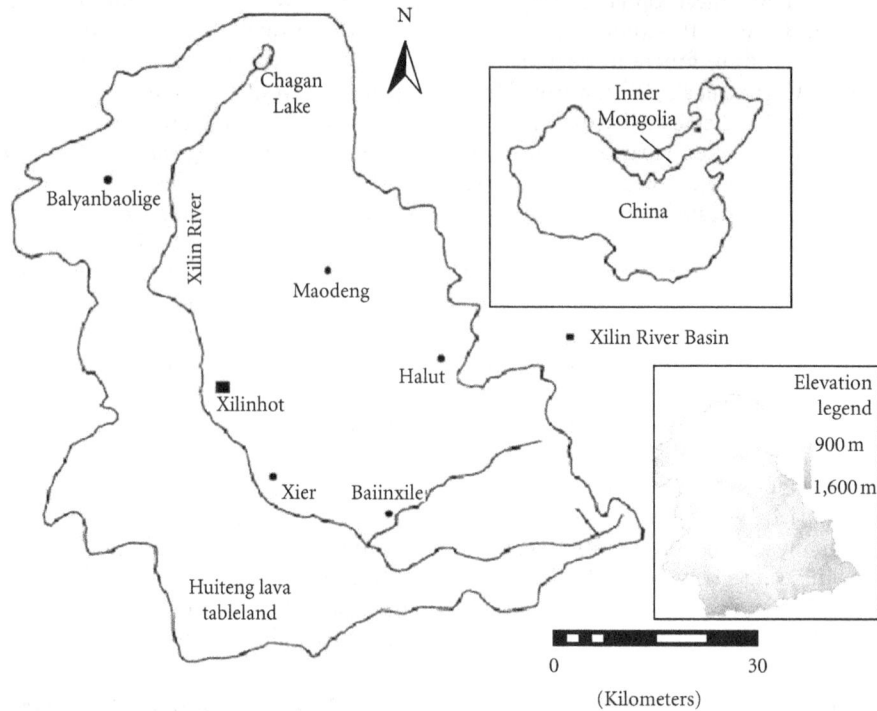

FIGURE 1: The research region.

vegetation community is defined as grassland degradation in this study.

2.2. The Data Sources and Analysis Methods. Four binary logistic regression models were constructed and the dependent variables of grassland degradation were described as transitions of LC and AG to any other vegetation communities within two periods, 1985–1998 and 1998–2004 (Table 1(a)). Two sets of nine independent variables from two years (1985 and 1998) were obtained (Table 1(b)). The first two independent variables indicating the biotic conditions are the density of the base vegetation (DV) and the normalized difference vegetation index (NDVI). The third independent variable is the average grazing intensity (AGI), reflecting the impact of livestock. Two other independent variables, the distance to road (DR) networks and the distance to settlement (DS) centers, are intended to reflect human disturbances. Finally the local habitat conditions are described by four variables, elevation (ALT), slope (SLP), slope orientation (ORI), and the distance to water (river) body (DW).

The four dependent variables of grassland degradation and two independent variables (DV and NDVI) are obtained from Landsat ETM+ and TM image processing. Landsat ETM+ images were used to classify the vegetation communities in 2004 [44], while Landsat TM images were processed to create the vegetation cover maps in 1985 and 1998 [46]. Therefore, three vegetation cover maps at an almost identical date in different years (1985, 1998, and 2004) were produced and the spatial distributions of SG and LC were mapped (Figures 2(a) and 2(b)). The dependent variables

are generated through the change detection analysis. They are binary, 1 indicating transitions of LC or AG to any other degraded vegetation communities in two periods, 1985–1998 and 1998–2004, and 0 denoting no transitions (Figure 3).

The density of base vegetation (DV) was computed on the basis of the spatial distribution of SG and LC (of 1985 for succession between 1985 and 1998, and of 1998 between 1998 and 2004). A 7×7 density kernel (Figure 4) is first applied to resample SG and LC (denoted by DK used in (1)) over the study region, which ensures a continuous interpolation of the density of the vegetation communities. This kernel places more weight on the central pixel but assigns less weight to adjacent pixels according to Tobler's first law of geography "things that are closer are more alike" [47, 48].

The density function is defined as

$$\mathrm{DV}_{ij} = \sum_{i'=i-3}^{i+3} \sum_{j'=j-3}^{j+3} \left(P_{i'j'} \times \mathrm{DK}_{i'j'} \right), \qquad (1)$$

where DV_{ij} is the *DV* value for the current pixel (i, j) in the density image vegetation communities (SG or LC), i' and j' are the relative coordinate locations of the pixel, $P_{i'j'}$ is a binary DV value (1 when the pixel is covered by the base vegetation type (SG or LC) and 0 when covered by other types of vegetation communities) of the pixel located at i' and j', and $\mathrm{DK}_{i'j'}$ is the kernel density value. The relative value of DV_{ij}, which has a minimum value of 0 and maximum 90, takes into consideration the nearby vegetation communities. When it equals the maximum value of 90, it indicates that all pixels within the sampling window are SG (or LC). On the contrary, the minimum value of 0 indicates that all pixels

TABLE 1: Candidate variables and the succession cases.

(a) Variables used to fit the probability of vegetation successions by BLR

Variable abbr.	Description	Minimum	Maximum	Mean	Std. dev
†NDVI$_{85}$	Normalized difference vegetation index in 1985	−0.22	0.59	0.19	0.07
†NDVI$_{98}$	Normalized difference vegetation index in 1998	−0.21	0.66	0.16	0.06
†DV$_{(SG\_85)}$	Density of vegetation LC in 1985	0.00	90.00	52.91	29.63
†DV$_{(SG\_98)}$	Density of vegetation LC in 1998	0.00	90.00	66.01	20.62
†DV$_{(LC\_85)}$	Density of vegetation SG in 1985	0.00	90.00	12.62	21.59
†DV$_{(LC\_98)}$	Density of vegetation SG in 1998	0.00	80.00	5.84	11.14
‡DS (km)	Distance to village settlement center	0.00	8.10	3.53	0.59
‡DR$_{85}$ (km)	Density of road in 1985	0.00	2.10	1.40	1.22
‡DR$_{98}$ (km)	Density of road in 1998	0.00	2.05	1.17	1.18
‡DW (km)	Distance to water (river) body	0.00	2.55	0.87	0.73
♀SLP (degree)	Slope in degree	0.00	66.00	4.79	6.30
♀ORI (degree)	Orientation from North in degree	0.00	180.00	76.46	59.00
♀ALT (m)	Altitude in meters	902.00	1608.00	1160.81	73.91
♯AGI$_{(85-98)}$ (sheep/km^2)	Average grazing intensity during 1985~1998	25.95	92.57	74.32	23.69
♯AGI$_{(98-04)}$ (sheep/km^2)	Average grazing intensity during 1998~2004	79.28	170.12	120.69	18.88

(b) Dependent variable (vegetation succession) description

Case no.	Succession Type	Period	Description
1	SG to CS/AF	1985~1998	Vegetation succession from SG to CS/AF during 1985~1998
2	SG to CS/AF	1998~2004	Vegetation succession from SG to CS/AF during 1998~2004
3	LC to CS/AF	1985~1998	Vegetation succession from LC to CS/AF during 1985~1998
4	LC to CS/AF	1998~2004	Vegetation succession from LC to CS/AF during 1998~2004

Data extracted from †: Landsat imagery and the ground truthing is referred to Xie et al. [46] and Sha et al. [44]; ♀: digital elevation model; ‡: road network map, annual economic statistics of the local governments and Landsat imagery; ♯: annual economic statistics of the local governments.

within the window are filled with other vegetation types. Generally, the larger the value is, the greater the chance is that the vegetation community is SG or LC. Therefore, the maps produced by (1) are referred to as the base vegetation cover density. Since there are the two periods (1985~1998 and 1998~2004) and two vegetation communities considered, 4 density distribution maps (DV$_{SG-1985}$, DV$_{SG-1998}$, DV$_{LC-1985}$, and DV$_{LC-1998}$) can be produced, which are shown in Figures 2(c), 2(d), 2(e), and 2(f), respectively.

NDVI is an indicator of vegetative greenness. The use of NDVI in studying vegetation changes has a long history in large area ecological research [25]. NDVI is calculated using a near infrared and red spectral band from Landsat TM data for 1985 and 1998, respectively:

$$NDVI = \frac{NIR - RED}{NIR + RED}. \qquad (2)$$

The average grazing intensity (AGI) is computed based on an extensive field survey conducted in 2004. The statistical data of grazing intensity at the village level on a yearly basis (from 1985 to 2004) were systematically collected from the local village committees. There are five recorded animal species, that is, sheep, horse, buffalo, camel, and donkey.

Because different animals have varied impacts on grass consumption and vegetation damage, each of the animals is converted to a standard unit, the sheep unit, to compute the grazing intensity. Based on the survey with the local herdsmen, an average grazing intensity (AGI) is calculated on the basis of the following equation:

$$AGI = \sum_{i=1}^{5} C_i \times N_i, \qquad (3)$$

where C_i is the coefficient for animal i, and N_i is the total number of the animal. To determine each coefficient of C_i, a questionnaire survey was done and the answers from the local farmers were synthesized to compute the coefficient. Specifically, one horse is equivalent to 6 sheep and thus is multiplied by 6 to transform into the sheep unit. From the same survey, cattle, camel, and donkey are multiplied by 5, 7, and 3, respectively. As a result, the average AGI during the period of 1985 to 1998 is noted as AGI$_{85-98}$. Similarly, AGI$_{98-04}$ is the averaged value of AGIs from 1998 to 2004. As shown in Table 1, the average grazing intensity during 1998~2004 almost doubles that during 1985~1998.

The road network is mainly derived from the topographic data and local road maps, with quality checking using

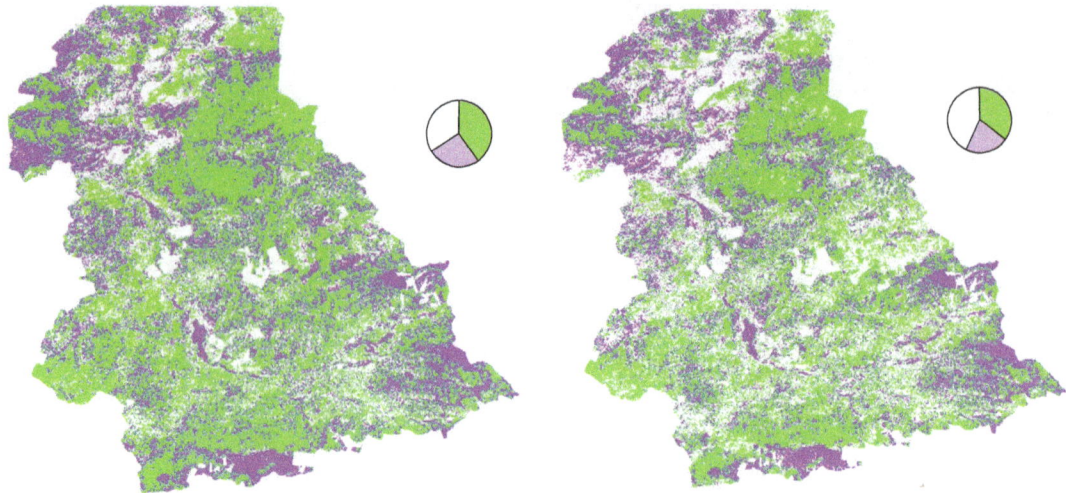

(a) SG and LC distribution in 1985

(b) SG and LC distribution in 1998

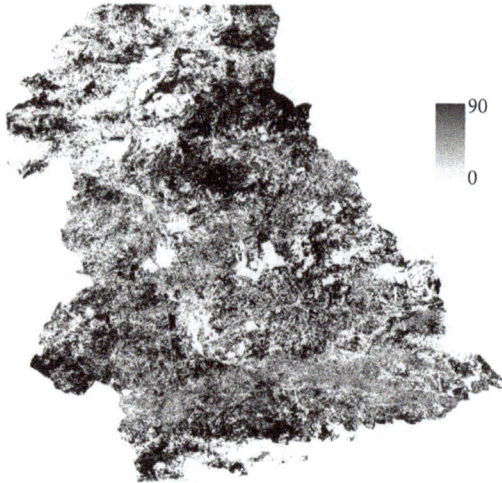

(c) SG density in 1985 (from (a))

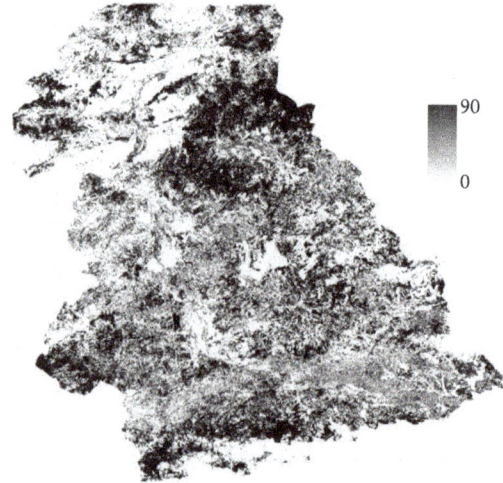

(d) SG density in 1998 (from (b))

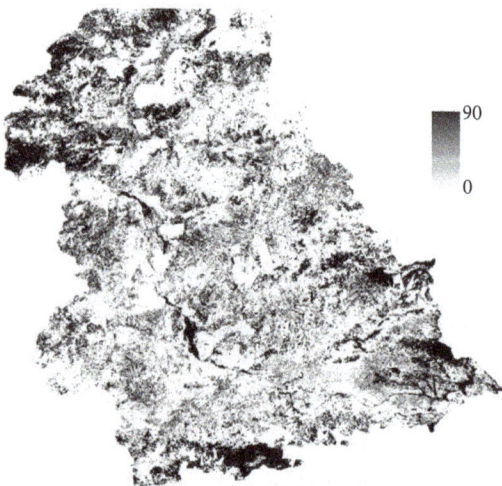

(e) SG density in 1985 (from (a))

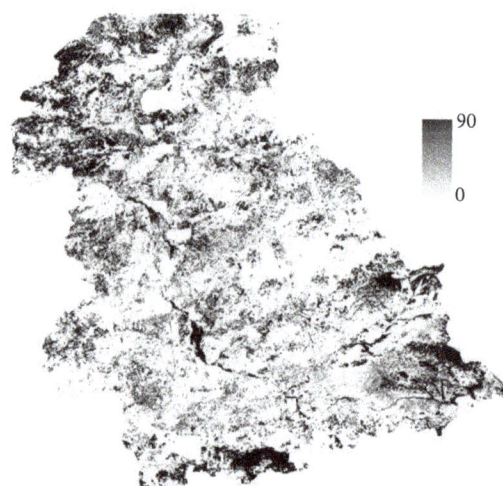

(f) SG density in 1998 (from (b))

FIGURE 2: The base vegetation community maps and the density of vegetation communities.

(a) SG transition during 1985~1998

(b) LC transition during 1985~1998

(c) SG transition during 1998~2004

(d) LC transition during 1998~2004

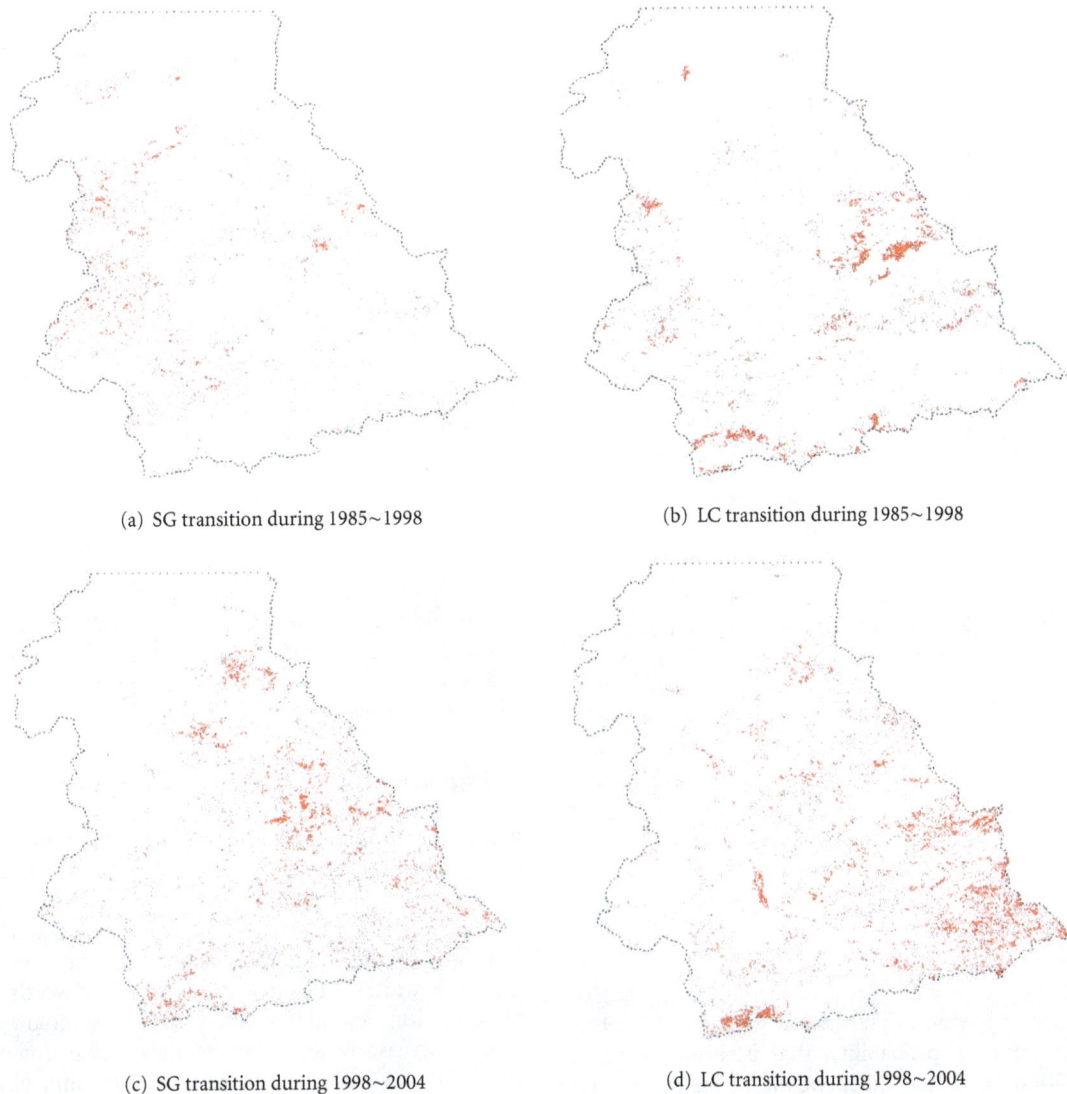

FIGURE 3: Spatial distributions of vegetation successions of SG and LC. Red: occurrence of vegetation transitions; white: nonoccurrence of vegetation transitions.

the TM and ETM+ images. The road network was also checked during the field survey in 2004. Settlement centers and water bodies (rivers) were derived from a similar approach as the road network. The ERDAS analysis function "SEARCH" was applied to create the three maps for DR, DS, and DW, respectively, with three vector layers showing the road network, water bodies (river), and village settlement centers.

The remaining independent variables reflecting local habitat conditions were computed from a digital elevation model (DEM) at a scale of $1:25,000$. The DEM was resampled to create a new DEM at the spatial resolution of $30 \times 30\,m$ so that the new DEM had the same spatial resolution as the vegetation community maps. This resampled DEM was then used to generate three topographical maps of elevation (ALT), slope (SLP), and slope orientation (ORI). Considering the sensitivity of vegetation growth in certain sunshine directions, the original aspect value (OLD_{ORI})

derived from DEM is further orientated with the following equation:

$$NEW_{ORI} = 360 - OLD_{ORI}, \quad \text{if } OLD_{ORI} > 180. \quad (4)$$

This transformation (4) indicates that the incident sun direction was an important factor for green grass growing and thus provides a better indicator to impact vegetation transition than the original aspect.

2.3. Binary Logistic Regression Model. Binary Logistic Regression (BLR) is a type of predictive modeling that tries to predict the binary probability of an outcome, for example, the occurrence or nonoccurrence of an event. In this study, BLR model was applied to fit the relationships between the dependent variable, the occurrence of vegetation degradation at any location (pixel i), and the independent variables listed in Table 1. A total of 5,000 samples labeled with one

0	0	0	1	0	0	0
0	1	2	3	2	1	0
0	2	5	6	5	2	0
1	3	6	10	6	3	1
0	2	5	6	5	2	0
0	1	2	3	2	1	0
0	0	0	1	0	0	0

FIGURE 4: Filter kernel for density mapping.

of two climax vegetation communities (SG or LC) were randomly chosen independently for each model. For example, let us take 5000 pixels labeled as "SG" in the year of 1985 as shown in Figure 2(a) to analyze SG degradation during 1985~1998. Four fifth out of all the selected samples were randomly selected to build a BLR model while the rest were kept for validation. The probability of having vegetation degradation (or SG transition to another vegetation type) at any location (pixel) was estimated by

$$\ln\left[\frac{p(y = 1 \mid X)}{1 - p(y = 1 \mid X)}\right] = \beta_0 + \sum_{i=1}^{n}\beta_i\chi_i, \qquad (5)$$

where p means the probability of occurrence (y), that is, the probability of having SG transition during 1985–1998. $1 - p$ is the opposite probability, that is, nonoccurrence of SG transition. x_i is the independent variables. β_0 and β_i $(i = 1, 2, \ldots, n)$ are the estimated parameters, and $p(y = 1 \mid X)$ is the probability that y takes the value 1, given the vector of independent variables X. The quantity $p(y = 1 \mid X)/(1 - p(y = 1 \mid X))$ is referred to as the odds, whereas $\ln[p(y = 1 \mid X)/(1 - p(y = 1 \mid X))]$ is called the logit.

The same approach has been taken for the SG transition during 1998–2004, as well as the LC transitions during 1985–1998 and 1998–2004. Therefore, four BLR models of vegetation degradation were obtained. In addition, two more tries of building these BLR models were tested to validate the applicability of these models by introducing two additional sets of 5000 randomly selected samples from each base vegetation map and for each model, respectively. Similar results in terms of the coefficients $(\beta_i, i = 0, 1, 2, \ldots, n)$ in the models and the prediction accuracies of the models were obtained. Little variation (within 5.0%) was noticed among all of the three trials in terms of the value of each of the coefficients and the overall prediction accuracy.

The unstandardized logit coefficient, that is, the β in (5) measures the absolute contribution of a variable in determining the probability that a particular vegetation degradation occurred. However, this information may be misleading when the unit adopted is not consistent from variable to variable as a result of disparities in units and scales of measurement. Thus, prior to performing a logistic regression, we standardized the independent variables with zero mean and the unit standard deviation, using the formula:

$$x_i' = \frac{x_i - \bar{x}_i}{\sigma_x}, \qquad (6)$$

where x_i' is the standardized value of a variable, x_i the value of the original variable, $\bar{x}_i$ the mean, and σ_x the standard deviation of the original variable.

After substituting x_i with x_i' in (5), the result of the regression can be expressed in terms of conditional probability at any spatial location (pixel i) to be predicted:

$$p_i = \frac{\exp\left(\beta_0 + \sum_{i=1}^{n}\beta_i\chi_i\right)}{\left(1 + \exp\left(\beta_0 + \sum_{i=1}^{n}\beta_i\chi_i\right)\right)}. \qquad (7)$$

The probability p_i of vegetation degradation at a location i in the study region could be calculated according to (7). It is a straightforward computation based on (7) to make a probability map of vegetation degradation over a given region when the required variables for the model are available.

3. The Results

Regression models were run using the SPSS forward likelihood ratio (LR) method (see http://www.ats.ucla.edu/stat/spss/examples/alda/default.htm). The relative contributions of the independent variables to vegetation degradation are expressed by the "odds ratio", $\exp(\beta)$. An odds ratio greater than 1 indicates a positive effect. In other words, the odds of vegetation degradation increase by 1 standard unit with a unit increase in an independent variable. An odds ratio smaller than 1 indicates a negative relationship, which means that an increase in the independent variable decreases the odds of vegetation degradation, whereas an odds ratio of 1 indicates that the odds of vegetation degradation is neutral to an increase in the independent variable.

3.1. Spatial and Temporal Patterns of Vegetation Transitions. Two dominant vegetation communities, *Stipa grandis* (SG) and *Leymus chinensis* (LC), are widely distributed in the study area as reflected on the base vegetation community maps shown in Figures 2(a) and 2(b). Statistical analyses of these two maps revealed that SG or LC amounted to more than 50% of the total area. Patches labeled with LC occupied 25.9% in 1985 and 22.0% in 1998, while SG displays even wider distributions, 39.9% in 1985 and 34.9% in 1998. From 1985 to 1998, the areas covered by LC and SG all decreased by 3.9% and 5.0%, respectively. Four density maps (Figures 2(c), 2(d), 2(e), and 2(f)) are derived from Figures 2(a) and 2(b). The density maps of LC and SG showed a slightly decrease in 1998 (Figures 2(d) and 2(f)) when compared with the density maps in 1985 (Figures 2(c) and 2(e)), which demonstrated a certain degree of vegetation degradation from either SG or LC to other vegetation communities over this study period.

TABLE 2: Results of BLR analysis for SG succession.

Variable	β	S.E. (β)	Sig.	Exp (β)	95.0% C.I. for Exp(β)	
					Lower	Upper
(a) During 1985~1998						
$DV_{SG\text{-}85}$	.022	.003	<.001	1.017	1.017	1.028
$AGI_{8\text{-}98}$	−.016	.004	<.001	.976	.976	.993
DR_{85}	−.115	.012	<.001	.891	.832	.945
ALT	.122	.021	<.001	1.130	1.116	1.214
DS	−0.109	0.23	<.001	0.890	0.877	0.913
*SLP						
*ORI						
*NDVI$_{85}$						
*DW						
Intercept	−1.912	.706	<.001	.148		
(b) During 1998~2004						
$DV_{SG\text{-}98}$	.040	.003	<.001	1.041	1.036	1.046
$AGI_{(9\text{-}04)}$	.109	.002	.031	1.115	1.000	1.208
DR_{98}	−.161	.021	<.001	.851	.800	.889
ALT	.191	.019	<.001	1.210	1.112	1.277
*SLP						
*ORI						
*NDVI$_{98}$						
*DW						
*DS						
Intercept	−14.411	.853	<.001	.000		
Hosmer and Lemeshow Goodness-of-Fit Test: Chi-square = 1.802, Pr > Chi-square = .213						

n = 4000.
Maximum likelihood estimate of the parameter. S.E. (β): estimated standard error of the parameter estimate; Wald χ^2: Wald chi-squared statistic; Sig.: P value of the Wald chi-squared statistic; Exp(β): odd ratio.
*variables excluded by the logistic regression model after the run.
C.I.: confidence intervals.
The cut value is 500.

Based on the vegetation cover recorded for three dates (1985, 1998, and 2004), actual vegetation transition maps were produced (Figure 3). Visual interpretation of the spatial distribution of the vegetation transitions showed that transitions were not evenly distributed over the study area and that significant transitions occurred at the middle or southern part of the region. Though this distribution might be correlated with the distribution of the base vegetation communities (SG and LC), other factors need to be further examined to understand how they affect the vegetation degradation process.

3.2. Driven Factors of Vegetation Successions. On the basis of the significance levels of the model coefficients and Goodness-of-Fit tests (Pr > Chi-square is greater than 0.05 for all cases) (Tables 2 and 3), the BLR models performed well explaining the probability of vegetation transition, and their fitted models showed moderate predicting accuracy (over 75.0% for all models, Table 4). One of the biggest contributors to the vegetation transition is the density of base vegetation communities (DV$_{SG}$ and DV$_{LC}$). DV$_{SG}$ and DV$_{LC}$

in both periods (1985–1998 and 1998–2004) showed significantly positive correlations with the occurrence of vegetation degradation (indicated by the value of Exp(β), and the odds ratio in Tables 2 and 3). This finding is somewhat opposite to the findings about grassland degradation conducted by others, which suggest that degradation often happens in rangeland of poor health [49].

Another factor picked up by the BLR models is the grazing intensity. It is commonly reported in current literature that the increase of grazing intensity leads to intensified grassland degradation [50, 51] and decreased ANPP and species richness [52]. Although three out of four BLR models reached a similar finding, the SG transition during the period of 1985~1998 showed a negative relationship with the increase of grazing intensity. In other words, it could be interpreted that SG showed less degradation during 1985–1998 when the grazing intensity increased (Table 2(a), Exp(β) = 0.976). Our analysis points out that there exist complex relationships between grazing intensity and grassland degradation. As can be seen from the statistical result of the data (Table 1), the grazing intensity over the study region during

TABLE 3: Results of BLR analysis for LC succession.

Variable	β	S.E. (β)	Sig.	Exp (β)	95.0% C.I. for Exp(β)	
					Lower	Upper
(a) LC succession: during 1985~1998						
DV_{LC-85}	0.040	0.003	<0.001	1.041	1.035	1.046
$AGI_{(85-98)}$	0.013	0.005	0.005	1.013	1.004	1.022
DR_{85}	−0.123	0.021	0.007	0.884	0.801	0.923
ALT	0.006	0.001	<0.001	1.006	1.005	1.008
DS	−0.201	0.027	<0.001	0.818	0.779	0.900
*SLP						
*ORI						
*$NDVI_{85}$						
*DW						
Intercept	−13.830	1.015	<0.001	0.000		
Hosmer and Lemeshow Goodness-of-Fit Test: Chi-square = 9.712, Pr > Chi-square = 0.286						
(b) LC succession: during 1998~2004						
DV_{LC-85}	0.047	0.002	<0.001	1.048	1.043	1.052
$AGI_{(98-04)}$	0.088	.019	<0.001	1.092	1.090	1.095
DR_{98}	−0.098	0.011	<0.001	0.907	0.872	0.974
ALT	0.018	0.001	<0.001	1.018	1.008	1.010
*SLP						
*ORI						
*$NDVI_{98}$						
*DW						
*DS						
Intercept	−15.420	0.783	<0.001	0.000		
Hosmer and Lemeshow Goodness-of-Fit Test: Chi-square = 11.867, Pr > Chi-square = 0.157						

n = 4000.

Maximum likelihood estimate of the parameter. S.E. (β): estimated standard error of the parameter estimate; Wald χ^2: Wald chi-squared statistic; Sig.: P value of the Wald chi-squared statistic; Exp (β): odd ratio.

*variables excluded by the logistic regression model after the run.

C.I.: confidence intervals.

The cut value is 500.

1998~2004 (averaged 120.7 sheep unit/km^2) nearly doubled that during 1985~1998 (averaged 74.3 sheep unit/km^2). SG was more resilient to degradation when the grazing level was low while LC was more vulnerable to the intensification of grazing. The probability of the LC degradation increased by 1.12 and 1.10 over both periods (1985~1998 and 1998~2004), respectively (Table 3). Our finding also suggests that the average grazing intensity (AGI) at current level could have exceeded the carrying capability of the grassland ecosystems, as the degradation is a common phenomenon over the entire study area at present.

The contribution that the road network makes to the vegetation degradation can be seen from Tables 2 and 3. The density of the road network had a significant impact to the degradation of both SG and LC over the periods of 1985~1998 and 1998~2004. On average, an increase of the distance to the nearby road by 1 distance unit (i.e., 1.4 km in 1985~1998 and 1.2 km in 1998~2004, see Table 1) decreases the odds of SG degradation by a factor of 0.89 and 0.80 during the two periods, respectively. Similarly, an increase of

the distance to the nearby road network by 1 distance unit decreases the odds of SG degradation by a factor of 0.80 and 0.87, respectively.

The variable of DS is an indicator that describes possible impact of human disturbance on vegetation degradation. DS was negatively related to the probability of both SG and LC degradations during the period of 1985~1998 (Tables 2 and 3). An increase of DS by 1 distance unit (i.e., 3.5 km) decreased the odds of SG and LC degradation by 11% and 22% during this period, respectively. However, during the other period of 1998~2004, DS was excluded by BLR models, indicating that the distance to DS was no longer a significant factor for predicting the probability of vegetation degradation. It was reported that human activities had a complicated impact on vegetation dynamics, depending on spatial and temporal scales at which assessments were conducted [53]. In addition, it was confirmed through our field survey that many settlement centers appeared during 1998~2004 due to rapid socioeconomic growth in the study region. Many of these newly developed settlement centers

TABLE 4: Classification test of BLR models.

Observed		Predicted					
		Fitting cases			Validation cases		
		Succession		Percentage correct	Succession		Percentage correct
		Yes	No		Yes	No	
(a) SG succession: during 1985~1998							
Succession	Yes	1812	187	90.6	411	57	87.8
	No	434	1567	78.3	188	344	64.7
Overall percentage				84.5			75.5
(b) LC succession: during 1985~1998							
Succession	Yes	1405	160	89.8	245	43	85.1
	No	534	1901	78.1	174	538	75.6
Overall percentage				82.7			78.3
(c) LC succession: during 1998~2004							
Succession	Yes	1201	134	90.0	199	50	80.0
	No	456	2209	82.9	174	577	76.8
Overall percentage				85.2			77.6
(d) SG succession: during 1998~2004							
Succession	Yes	789	144	84.6	142	33	81.1
	No	506	2561	83.5	187	642	77.8
Overall percentage				83.8			78.4

The cut value is 500.

were not primarily dependent on grazing. Therefore, the impact of DSs on grassland degradation was significantly reduced during the second study period.

The impacts of the topographical factors on grassland degradation showed complex patterns. Elevation was the sole factor that significantly correlated with the occurrence of vegetation degradation based on the BLR models. High degradation probabilities of both SG and LC took place largely in the areas with high elevations. In the middle and upper reaches of Xilin river (south-east part) where the elevations are higher than other places, vegetation degradations were widely noticed (Figure 3). However, the other topographical factors of aspect and slope were excluded by all BLR models. Our null hypothesis that the topographical factors, aspect and slope, might affect the degradation occurrence was rejected by the result.

Validation of the above models is given in Table 4. For the selected sites used to build the models, the overall accuracies for SG and LC degradation are 84.5% and 82.7% during 1985~1998, and 85.2% and 83.8% during 1998~2004, respectively. For the validation sites, the overall accuracies for SG and LC degradation are a little lower, which is 75.5% and 78.3% during 1985~1998, and 77.6% and 78.4% during 1998~2004. The results demonstrate that the fit models can generally be used to predict vegetation degradation over the study area with a moderate prediction accuracy (over 75%).

Based on the BLR models, four maps showing the probabilities of the vegetation degradation (i.e., SG and LC degradations during 1985~1998 and 1998~2004) are produced.

The pixel value in the maps shows the modeled probability of vegetation degradation over the study area (Figure 5). When compared with the observed incidences of vegetation degradation (Figure 3), the probability maps have assigned relatively high probability values to the locations where the vegetation degradations were actually observed, indicating that the fit models were useful for predicting vegetation degradation. With an increasingly improved access to latest natural and socioeconomic data, it is feasible to update these degradation maps for the purpose of prediction and for best practices of grassland (ecological) management.

4. The Discussion and Conclusion

Grassland degradation has been regarded as an important indicator of grassland ecosystem health. However, there are limited researches dealing with the relationship between vegetation degradation and its driving factors (both natural and socioeconomic) that may influence the degradation pathways. Researches on grassland degradation have a twofold practical application, identifying possible driven forces that cause and accelerate the degradation processes, and providing scientific data for making informed decisions in order to change the degradation pathways in the direction of keeping grassland ecology sustainable.

4.1. How Vegetation Degradation Responds to Various Driving Factors. There are four important findings regarding the responses of vegetation degradation to various driving factors.

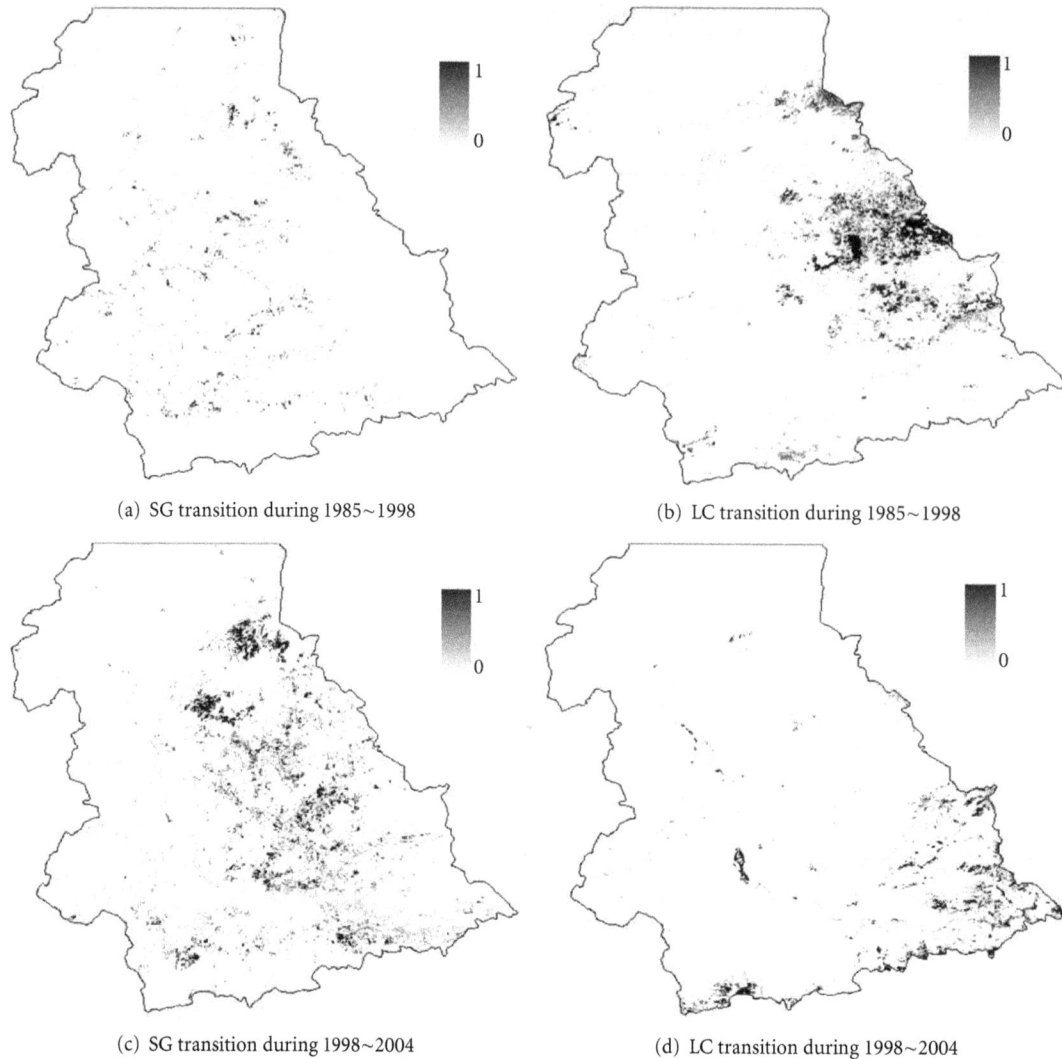

(a) SG transition during 1985~1998

(b) LC transition during 1985~1998

(c) SG transition during 1998~2004

(d) LC transition during 1998~2004

FIGURE 5: Probability mapping of vegetation transitions.

First, grassland degradation has shown a strong positive correlation with high density or productivity of climax communities. This finding might be contradictory to natural ecological processes but is a tragic outcome of human greed of seeking maximum profit. Under the eager of catching up with economic booms and enjoying material wealth, herdsmen made their best bets in most productive grassland in order to have maximum investment returns. Second, the decrease of biological productivity (NDVI in our case study) cannot be simply regarded as a direct indicator of grassland degradation. NDVI showed no significant relationship with vegetation degradation in this study. Similar studies on vegetation degradation did show that the indexes (NDVI and other forms) derived from remotely sensed imagery exhibited significant spatial and temporal correlations with vegetation degradation [54]. The different finding from our research was largely attributed to the fact that heavy (or excessive) grazing and subsequent degradation occurred in climax plant communities with high productivity.

Third, a causal relationship between grazing intensity and grassland degradation only holds when grazing intensity

reaches a certain threshold (or a balance point). This threshold also varies with different plant communities. Under this threshold of grazing pressure (e.g., SG during 1985~1998), SG was resilient to degradation. However, due to the lack of yearly data and longitudinal data (before 1985), we were not able to identify the threshold values of grazing intensities for SG or LC degradation. It will be an important piece of future research to establish grazing intensity threshold values for various plant communities so that grassland strategic conservation plans could be made on the basis of scientific data about grazing intensity or grassland bearing capacity.

Fourth, complicated relationships have been found between grassland degradation and direct human disturbances. The development of road network was a leading cause of grassland degradation, which was consistent with our field observations. Degradation severity increased steadily towards road network. However the impact of human settlements on grassland degradation showed a clear temporal change. Earlier settlements served primarily for animal husbandry and were closely linked grazing activities and thus had significant impact on grassland degradation. Newly village centers and towns have played more roles as economic or urban centers and their direct impacts on grazing have been reduced.

4.2. Policy Implications for Preventing Vegetation Degradation in the Study Region. Overgrazing occurred widely in the study area. With this in mind, there have been areas of fenced grazing, which were excluded in our case study. It is believed that such practices could be expanded over larger areas as an effective measure to protect grassland vegetation and prevent degradation [55]. Another phenomenon is the fast development of the unplanned road network. As evidenced in our field investigation, wherever a road was extended, the grassland was destroyed adjacent to the road. On the other hand, a road, if well planned, could not only bring more convenience to the local population but also be ecologically beneficial because roads extended to remote areas allow the rapid movement of animals to balance grazing intensity over a large region. This may reduce overgrazing in one area while other areas remain ungrazed. Unfortunately, because of immediate economical incentives, the development of roads was unevenly distributed, making vegetation degradation even worse since these were concentrated near town or residential centers, or current husbandry centers where severe degradation (vegetation degradation) had already occurred. Therefore, more attention should be given to protect vegetation communities near the areas along roads, for example, establishing vegetation conservation zones, and to construct road networks in remote areas.

Human disturbance, mainly indicated by the variable DS, is an important factor that stimulated vegetation degradation in the period of 1985~1998. However, this phenomenon was not obvious in the later period, 1998~2004, even when more settlement centers were established. The difference of DS impact upon vegetation degradation is attributed to better vegetation protection measures near villages in the second period. Therefore, a policy of protecting vegetation patches that are under severe degradation around older settlements is effective. The implication of this policy is that a rigorous effort should be made to limit unplanned expansions of new settlements.

4.3. Selection of the Model Variables. The selection of the candidate variables for modeling the grassland vegetation degradation has been guided by our literature review, our field survey, and the characteristics of our study area. Many studies have proven that the changes in the global natural environment have made significant impact on vegetation dynamics [56–58]. The critical changes in natural environment are typically exemplified by global climate deteriorations, among which are the rising temperature and the shortage of precipitation in grassland areas [59]. The climate changes were regrettably not taken into consideration in our current study due to the fact that the study area covers a total area of only 10,000 km^2, which strides about 100 km in longitude and 100 km in latitude. The climate data collected at a few weather stations over the study area showed little variations.

5. Conclusions

Vegetation degradation is an important indicator of grassland ecosystem health. In this study, we examined the relationship between the occurrences of vegetation degradation and its driving factors in the Xilin River Basin, Inner Mongolia, China. As *Stipa grandis* and *Leymus chinensis* are the two most dominant grassland communities in the study area, their degradation patterns have important implications in terms of grassland ecology and management. Binary logistic regression (BLR) was used to fit the nonlinear correlations between the occurrences of vegetation degradation (dependent variable) and nine independent variables over two consecutive periods, 1985~1998 and 1998~2004. The independent variables include two indicating biotic conditions (the density of base vegetation and the normalized difference vegetation index), three reflecting human interferences (average grazing intensity, the distance to road network, and the distance to settlement center), and four denoting local habitat conditions elevation, slope, slope orientation, and the distance to water (river) body.

Several important findings were suggested by the BLR model. First, four variables, including the base vegetation density (for SG and LC), the average grazing intensity, the distance to road network, and the altitude, were important determinants of grassland degradation for both plant communities (SG and LC) and over both study periods. Secondly, some of our findings provide new insights into the causes of grassland degradation. For instance, severe degradation often happens in the most productive grassland; grassland is in general resilient to degradation when grazing intensity is kept under a certain threshold; the construction of road networks is the most destructive factor causing grassland degradation; the negative impact of human settlement on grassland degradation is much significant when the residents are primarily herdsmen. These findings should have clear policy implications in grassland management. Thirdly, the BLR models have moderate accuracy levels, over 75%, indicating that the BLR models are acceptable in studying grassland vegetation degradation, but some caution should be exercised by investigator. The models adopted in the current study should also work for other grassland regions when required data are available.

Finally, there are some limitations in current BLR models that should become the focus of future research topics or methods concerning grassland degradation. Due to the limitation of available data, current BLR models were implemented with periodic data (1985–1998 and 1998–2004). The results could be more convincing and accurate if the BLR models could be run with a yearly based dataset. The applicability of the BLR models could be extended to a regional scale if such variables as those which can tell spatial variations of climate changes could be added into the BLR models.

Acknowledgments

This research is partially supported by a grant from the Land-Cover/Land-Use Program at NASA (Grant no.

NNX09AK87G). The authors are also grateful to The Center for Ecological Research, Institute of Botany, Chinese Academy of Sciences, for the assistance of collecting the ground reference samples, and to Professor Tom Wagner at Eastern Michigan University for English proofreading.

References

[1] C. W. E. Moore, "Distribution of grasslands," in *Grasses and Grasslands*, C. Barnard, Ed., pp. 182–205, Macmillan, New York, NY, USA, 1966.

[2] F. S. Chapin, O. E. Sala, and E. Huber-Sannwald, *Global Biodiversity in a Changing Environment: Scenarios for the 21st Century*, Springer, New York, NY, USA, 2001.

[3] W. K. Lauenroth, I. C. Burke, and M. P. Gutmann, "The structure and function of ecosystem in the Central North American grassland region," *Great Plains Research*, vol. 9, no. 2, pp. 223–259, 1999.

[4] S. Kröpelin, D. Verschuren, A. M. Lézine et al., "Climate-driven ecosystem succession in the Sahara: the past 6000 years," *Science*, vol. 320, no. 5877, pp. 765–768, 2008.

[5] C. He, Q. Zhang, Y. Li, X. Li, and P. Shi, "Zoning grassland protection area using remote sensing and cellular automata modeling—a case study in Xilingol steppe grassland in northern China," *Journal of Arid Environments*, vol. 63, no. 4, pp. 814–826, 2005.

[6] N. A. Cutler, L. R. Belyea, and A. J. Dugmore, "The spatiotemporal dynamics of a primary degradation," *Journal of Ecology*, vol. 96, no. 2, pp. 231–246, 2007.

[7] M. D. Luis, J. Raventós, and J. C. González-Hidalgo, "Postfire vegetation succession in Mediterranean gorse shrublands," *Acta Oecologica*, vol. 30, no. 1, pp. 54–61, 2006.

[8] M. M. Heras, J. M. Nicolau, and T. Espigares, "Vegetation succession in reclaimed coal-mining slopes in a Mediterranean-dry environment," *Ecological Engineering*, vol. 34, no. 2, pp. 168–178, 2008.

[9] W. Thuiller, C. Albert, M. B. Araújo et al., "Predicting global change impacts on plant species' distributions: future challenges," *Perspectives in Plant Ecology, Evolution and Systematics*, vol. 9, no. 3-4, pp. 137–152, 2008.

[10] R. B. Harris, "Rangeland degradation on the Qinghai-Tibetan plateau: a review of the evidence of its magnitude and causes," *Journal of Arid Environments*, vol. 74, no. 1, pp. 1–12, 2010.

[11] V. A. Kovdaa, "Soil loss: an overview," *Agro-Ecosystems*, vol. 3, no. 1, pp. 205–224, 1976.

[12] X. R. Li, X. H. Jia, and G. R. Dong, "Influence of desertification on vegetation pattern variations in the cold semi-arid grasslands of Qinghai-Tibet Plateau, North-West China," *Journal of Arid Environments*, vol. 64, no. 3, pp. 505–522, 2006.

[13] M. Pellant, P. L. Shaver, D. A. Pyke, and J. E. Herrick, "Interpreting indicators of rangeland health," Technical Reference 1734–1736, U.S. Department of the Interior, Bureau of Land Management, National Science and Technology, 2005.

[14] W. G. Whitford, A. G. De Soyza, J. W. Van Zee, J. E. Herrick, and K. M. Havstad, "Vegetation, soil, and animal indicators of rangeland health," *Environmental Monitoring and Assessment*, vol. 51, no. 1-2, pp. 179–200, 1998.

[15] B. T. Bestelmeyer, J. E. Herrick, J. R. Brown, D. A. Trujillo, and K. M. Havstad, "Land management in the American southwest: a state-and-transition approach to ecosystem complexity," *Environmental Management*, vol. 34, no. 1, pp. 38–51, 2004.

[16] T. K. Stringham, W. C. Krueger, and P. L. Shaver, *States, Transitions and Thresholds: Further Refinement for Rangeland Applications*, Oregon State University Agricultural Experiment Station, Corvallis, Ore, USA, 2001.

[17] A. Hanafi and S. Jauffret, "Are long-term vegetation dynamics useful in monitoring and assessing desertification processes in the arid steppe, southern Tunisia," *Journal of Arid Environments*, vol. 72, no. 4, pp. 557–572, 2008.

[18] J. Cortina, F. T. Maestre, R. Vallejo, M. J. Baeza, A. Valdecantos, and M. Pérez-Devesa, "Ecosystem structure, function, and restoration success: are they related?" *Journal for Nature Conservation*, vol. 14, no. 3-4, pp. 152–160, 2006.

[19] F. X. Wang, Z. Y. Wang, and J. H. W. Lee, "Acceleration of vegetation succession on eroded land by reforestation in a subtropical zone," *Ecological Engineering*, vol. 31, no. 4, pp. 232–241, 2007.

[20] T. Nakayama, "Factors controlling vegetation succession in Kushiro Mire," *Ecological Modelling*, vol. 215, no. 1–3, pp. 225–236, 2008.

[21] D. F. Joubert, A. Rothauge, and G. N. Smit, "A conceptual model of vegetation dynamics in the semiarid Highland savanna of Namibia, with particular reference to bush thickening by Acacia mellifera," *Journal of Arid Environments*, vol. 72, no. 12, pp. 2201–2210, 2008.

[22] A. Kühner and M. Kleyer, "A parsimonious combination of functional traits predicting plant response to disturbance and soil fertility," *Journal of Vegetation Science*, vol. 19, no. 5, pp. 681–692, 2008.

[23] L. Kooistra, W. Wamelink, G. Schaepman-Strub et al., "Assessing and predicting biodiversity in a floodplain ecosystem: assimilation of net primary production derived from imaging spectrometer data into a dynamic vegetation model," *Remote Sensing of Environment*, vol. 112, no. 5, pp. 2118–2130, 2008.

[24] G. L. W. Perry and N. J. Enright, "Spatial modelling of vegetation change in dynamic landscapes: a review of methods and applications," *Progress in Physical Geography*, vol. 30, no. 1, pp. 47–72, 2006.

[25] B. Zhao, Y. Yan, H. Guo, M. He, Y. Gu, and B. Li, "Monitoring rapid vegetation succession in estuarine wetland using time series MODIS-based indicators: an application in the Yangtze River Delta area," *Ecological Indicators*, vol. 9, no. 2, pp. 346–356, 2009.

[26] D. Manuel-Navarrete, G. C. Gallopín, M. Blanco et al., "Multicausal and integrated assessment of sustainability: the case of agriculturization in the Argentine Pampas," *Environment, Development and Sustainability*, vol. 11, no. 3, pp. 621–638, 2009.

[27] J. G. Canadell, D. E. Pataki, and L. Pitelka, *Terrestrial Ecosystems in a Changing World*, Springer, Berlin, Germany, 2007.

[28] L. A. Garibaldi, M. Semmartin, and E. J. Chaneton, "Grazing-induced changes in plant composition affect litter quality and nutrient cycling in fooding Pampa grasslands," *Ecosystem Ecology*, vol. 151, pp. 650–662, 2007.

[29] H. Yang, X. Li, Y. Zhang, and A. J. B. Zehnder, "eEnvironmental—economic interaction and forces of migration: a case study of three counties in Northern China," in *Environmental Change and Its Implications for Population Migration*, J. D. Unruh, M. S. Krol, and N. Kliot, Eds., pp. 267–288, Kluwer Academic Publishers, Dodrecht, The Netherlands, 2004.

[30] U. Safriel and Z. Adeel, "Development paths of drylands: thresholds and sustainability," *Sustainability Science*, vol. 3, no. 1, pp. 117–123, 2008.

[31] R. B. Norgaard, "The case for methodological pluralism," *Ecological Economics*, vol. 1, no. 1, pp. 37–57, 1989.

[32] S. J. Rahlao, M. T. Hoffman, S. W. Todd, and K. McGrath, "Long-term vegetation change in the Succulent Karoo, South Africa following 67 years of rest from grazing," *Journal of Arid Environments*, vol. 72, no. 5, pp. 808–819, 2008.

[33] Y. Xie, Z. Sha, and M. Yu, "Remote sensing imagery in vegetation mapping: a review," *Journal of Plant Ecology*, vol. 1, pp. 9–23, 2008.

[34] J. Gao, "Quantification of grassland properties: how it can benefit from geoinformatic technologies?" *International Journal of Remote Sensing*, vol. 27, no. 7, pp. 1351–1365, 2006.

[35] S. Moreau, R. Bosseno, X. F. Gu, and F. Baret, "Assessing the biomass dynamics of Andean bofedal and totora high-protein wetland grasses from NOAA/AVHRR," *Remote Sensing of Environment*, vol. 85, no. 4, pp. 516–529, 2003.

[36] M. L. Nordberg and J. Evertson, "Monitoring change in mountainous dry-heath vegetation at a regional scale using multitemporal landsat TM data," *Ambio*, vol. 32, no. 8, pp. 502–509, 2003.

[37] R. Proulx and L. Parrott, "Measures of structural complexity in digital images for monitoring the ecological signature of an old-growth forest ecosystem," *Ecological Indicators*, vol. 8, no. 3, pp. 270–284, 2008.

[38] Y. Bai, X. Han, J. Wu, Z. Chen, and L. Li, "Ecosystem stability and compensatory effects in the Inner Mongolia grassland," *Nature*, vol. 431, no. 9, pp. 181–184, 2004.

[39] C. Tong, J. Wu, S. Yong, J. Yang, and W. Yong, "A landscape-scale assessment of steppe degradation in the Xilin River Basin, Inner Mongolia, China," *Journal of Arid Environments*, vol. 59, no. 1, pp. 133–149, 2004.

[40] B. Li, S. YON, and Z. Li, "The vegetation of the Xilin River Basin and its utilization," in *Research on Grassland Ecosystem*, Inner Mongolia Grassland Ecosystem Research Station, Ed., vol. 3, pp. 84–183, Science Press, Beijing, China, 1988.

[41] M.-Y. Xu, K. Wang, and F. Xie, "Effects of grassland management on soil organic carbon density in agro-pastoral zone of Northern China," *African Journal of Biotechnology*, vol. 10, pp. 4844–4850, 2011.

[42] A. Kang, "Reflections on the degradation, desertification and soil erosion in Xilingol grassland," *Water Resource Development Research*, vol. 2, pp. 36–38, 2002.

[43] K. Kawamura, T. Akiyama, H. O. Yokota et al., "Quantifying grazing intensities using geographic information systems and satellite remote sensing in the Xilingol steppe region, Inner Mongolia, China," *Agriculture, Ecosystems and Environment*, vol. 107, no. 1, pp. 83–93, 2005.

[44] Z. Sha, Y. Bai, Y. Xie, M. Yu, and L. Zhang, "Using a hybrid fuzzy classifier (HFC) to map typical grassland vegetation in Xilin River Basin, Inner Mongolia, China," *International Journal of Remote Sensing*, vol. 29, no. 8, pp. 2317–2337, 2008.

[45] M. M. Borman and D. A. Pyke, "Degradational theory and the desired plant community approach," *Rangelands*, vol. 16, pp. 82–85, 1994.

[46] Y. Xie, Z. Sha, and Y. Bai, "Classifying historical remotely sensed imagery using a tempo-spatial feature evolution (T-SFE) model," *ISPRS Journal of Photogrammetry and Remote Sensing*, vol. 65, no. 2, pp. 182–190, 2010.

[47] E. Meijering and M. Unser, "A note on cubic convolution interpolation," *IEEE Transactions on Image Processing*, vol. 12, no. 4, pp. 477–479, 2003.

[48] W. Tobler, "A computer movie simulating urban growth in the Detroit region," *Economic Geography*, vol. 46, pp. 234–240, 1970.

[49] G. S. Bilotta, R. E. Brazier, and P. M. Haygarth, "The impacts of grazing animals on the quality of soils, vegetation, and surface waters in intensively managed grasslands," *Advances in Agronomy*, vol. 94, pp. 237–280, 2007.

[50] A. Röder, T. Udelhoven, J. Hill, G. del Barrio, and G. Tsiourlis, "Trend analysis of Landsat-TM and -ETM+ imagery to monitor grazing impact in a rangeland ecosystem in Northern Greece," *Remote Sensing of Environment*, vol. 112, no. 6, pp. 2863–2875, 2008.

[51] L. J. Blanco, M. O. Aguilera, J. M. Paruelo, and F. N. Biurrun, "Grazing effect on NDVI across an aridity gradient in Argentina," *Journal of Arid Environments*, vol. 72, no. 5, pp. 764–776, 2008.

[52] Y. Bai, J. Wu, Q. Xing et al., "Primary production and rain use efficiency across a precipitation gradient on the Mongolia Plateau," *Ecology*, vol. 89, no. 8, pp. 2140–2153, 2008.

[53] D. Y. Xu, X. W. Kang, D. F. Zhuang, and J. J. Pan, "Multi-scale quantitative assessment of the relative roles of climate change and human activities in desertification—a case study of the Ordos Plateau, China," *Journal of Arid Environments*, vol. 74, no. 4, pp. 498–507, 2010.

[54] K. J. Wessels, S. D. Prince, P. E. Frost, and D. Van Zyl, "Assessing the effects of human-induced land degradation in the former homelands of northern South Africa with a 1 km AVHRR NDVI time-series," *Remote Sensing of Environment*, vol. 91, no. 1, pp. 47–67, 2004.

[55] X. Cui, Y. Wang, H. Niu et al., "Effect of long-term grazing on soil organic carbon content in semiarid steppes in Inner Mongolia," *Ecological Research*, vol. 20, no. 5, pp. 519–527, 2005.

[56] D. D. Breshears, N. S. Cobb, P. M. Rich et al., "Regional vegetation die-off in response to global-change-type drought," *Proceedings of the National Academy of Sciences of the United States of America*, vol. 102, no. 42, pp. 15144–15148, 2005.

[57] S. Sitch, C. Huntingford, N. Gedney et al., "Evaluation of the terrestrial carbon cycle, future plant geography and climate-carbon cycle feedbacks using five Dynamic Global Vegetation Models (DGVMs)," *Global Change Biology*, vol. 14, no. 9, pp. 2015–2039, 2008.

[58] D. J. Beerling, "Long-term responses of boreal vegetation to global change: an experimental and modelling investigation," *Global Change Biology*, vol. 5, no. 1, pp. 55–74, 1999.

[59] L. Yahdjian and O. E. Sala, "Climate change impacts on South American rangelands," *Rangelands*, vol. 30, no. 3, pp. 34–39, 2008.

6

Caatinga Revisited: Ecology and Conservation of an Important Seasonal Dry Forest

Ulysses Paulino de Albuquerque,[1] **Elcida de Lima Araújo,**[1] **Ana Carla Asfora El-Deir,**[1]
André Luiz Alves de Lima,[2] **Antonio Souto,**[3] **Bruna Martins Bezerra,**[3]
Elba Maria Nogueira Ferraz,[4] **Eliza Maria Xavier Freire,**[5]
Everardo Valadares de Sá Barreto Sampaio,[6] **Flor Maria Guedes Las-Casas,**[7]
Geraldo Jorge Barbosa de Moura,[1] **Glauco Alves Pereira,**[1] **Joabe Gomes de Melo,**[1]
Marcelo Alves Ramos,[1] **Maria Jesus Nogueira Rodal,**[1] **Nicola Schiel,**[1]
Rachel Maria de Lyra-Neves,[8] **Rômulo Romeu Nóbrega Alves,**[9]
Severino Mendes de Azevedo-Júnior,[1] **Wallace Rodrigues Telino Júnior,**[8]
and William Severi[10]

[1] *Departamento de Biologia, Universidade Federal Rural de Pernambuco, Rua Dom Manoel de Medeiros, s/n,*
Dois Irmãos, 52171-900 Recife, PE, Brazil
[2] *Unidade Acadêmica de Serra Talhada (UAST), Universidade Federal Rural de Pernambuco (UFRPE), Fazenda Saco s/n,*
56.900-000, Serra Talhada, PE, Brazil
[3] *Departamento de Zoologia, Centro de Ciências Biológicas Universidade Federal de Pernambuco (UFPE),*
Avenida Professor Moraes Rego, 1235-Cidade Universitária, 50670-901 Recife, PE, Brazil
[4] *Direção de Ensino/Gerência de Pesquisa e Pós-Graduação-Cidade Universitária, Instituto Federal de Pernambuco-Reitoria,*
Campus Recife, Avenida Professor Luis Freire 500, 50740-540 Recife, PE, Brazil
[5] *Laboratório de Herpetologia, Departamento de Botânica, Ecologia e Zoologia, Centro de Biociências,*
Universidade Federal do Rio Grande do Norte, Lagoa Nova, 59072-900 Natal, RN, Brazil
[6] *Departamento de Energia Nuclear, Centro de Tecnologia, Universidade Federal de Pernambuco (UFPE),*
Avenida Professor Luís Freire 1000, Cidade Universitária, 50740-540 Recife, PE, Brazil
[7] *Programa de Pós-Graduação em Ecologia e Recursos Naturais, Centro de Ciências Biológicas e da Saúde,*
Departamento de Ecologia e Biologia Evolutiva, Universidade Federal de São Carlos, Rodovia Washington Luiz Km 235,
13565-905 São Carlos, SP, Brazil
[8] *Unidade Acadêmica de Garanhuns, Universidade Federal Rural de Pernambuco, Avenida Bom Pastor, s/n, Boa Vista, Heliópolis,*
55296-901 Garanhuns, PE, Brazil
[9] *Departamento de Biologia, Universidade Estadual da Paraíba, Avenida das Baraúnas 351, Bodocongó,*
58109-753 Campina Grande, PB, Brazil
[10] *Departamento de Pesca e Aquicultura, Universidade Federal Rural de Pernambuco, Rua Dom Manoel de Medeiros, s/n,*
Dois Irmãos, 52171-900 Recife, PE, Brazil

Correspondence should be addressed to Ulysses Paulino de Albuquerque, upa677@hotmail.com

Academic Editors: B. B. Castro and H. Hasenauer

Besides its extreme climate conditions, the Caatinga (a type of tropical seasonal forest) hosts an impressive faunal and floristic biodiversity. In the last 50 years there has been a considerable increase in the number of studies in the area. Here we aimed to present a review of these studies, focusing on four main fields: vertebrate ecology, plant ecology, human ecology, and ethnobiology. Furthermore, we identify directions for future research. We hope that the present paper will help defining actions and strategies for the conservation of the biological diversity of the Caatinga.

1. Introduction

What is the current status of biodiversity research in the Brazilian Caatinga? In an attempt to resolve this issue, a team of researchers recently used the biological group "insects" as a proxy to estimate the status of biodiversity research in this environment [1]. Although this paper is very relevant and interesting, it attempts to extrapolate the condition of biodiversity research in a large region from the information available for insects alone.

In Brazil, the word Caatinga is used to designate a large geographic area comprising a variety of different types of vegetation. It is also used to name the semiarid region that occupies the largest portion of the northeast of Brazil. Its temperature and rainfall qualify the area as a tropical seasonal forest. Most of the information collected in studies of the Caatinga vegetation only applies to a small number of areas. It is difficult to formulate generalizations on the vegetation dynamics of this region because of the lack of replication, which is common for other vegetation types throughout the world. The study of the vertebrates inhabiting the Caatinga follows the same trend found in the study of the vegetation. Overall, sample size and sampling effort are often not satisfactory. For this reason, many researchers worldwide have adhered to the idea of maintaining an information network, such as TROPI-DRY, to aid in the preservation of Neotropical dry forests, such as the Caatinga. This particular region includes very large disturbed areas as well as areas undergoing desertification processes, both of which have been given high priority for preservation and/or restoration [2]. Information networks such as TROPI-DRY combine ecological research, remote sensing, and social sciences through the use of standardized protocols that allow for comparisons between different tropical dry forest areas.

Despite the recognized gaps in our knowledge of this region [3–5], the number of studies on Caatinga biodiversity has increased considerably over the past decades. In this paper we have made a comprehensive analyzes of a number of important studies carried out in the last 50 years about flora, vertebrate fauna, human ecology, and ethnobiology in the Caatinga. We aimed to summarize information on these subjects as well as identify key directions for future research. To compile data on the aforementioned topics, we used seven main databases (Biological Abstracts, Google Scholar, SciELO, Scirus, Scopus, JSTOR, and Web of Science) as well as several book chapters published on these topics. We hope that the present paper will help defining actions and strategies for the conservation of the biological diversity of the Caatinga.

2. Vertebrate Ecology

2.1. Ichthyofauna. Ichthyofauna are found in all aquatic environments across a wide range of habitats. It is the group that possesses the greatest richness among vertebrates. In this regard, there are 2,587 species of fish that are endemic to freshwater environments in Brazil [6].

Rosa et al. [7] have suggested that the fish fauna in the Caatinga have not yet been well documented; however, the authors highlight that various studies have been conducted in a few locations and this information is still restricted to monographs, dissertations, and environmental diagnostics for the construction of dams, which cannot be found online. In addition, several of these studies are rather specific, which does not allow for a general view of the diversity of fish fauna in this environment.

In recent years, several studies have increased our knowledge of the diversity of fish species in the Caatinga by providing descriptive studies, such as those concerning the annual species of the Rivulidae family, including Costa [8, 9], Costa and Brasil [10], and Costa et al. [11, 12]. Rosa et al. [7] regarded the Caatinga as the site with the greatest biodiversity of fish from the Rivulidae family in relation to other ecosystems. Rivulids, or annual fish, are known by their distinct life cycle; the adults develop rapidly and deposit their eggs during floods to avoid desiccation during the dry season [9].

Rosa et al. [7] and Rosa [13] reported the occurrence of 240 fish species in the Caatinga distributed across seven orders, with Siluriformes and Characiformes as the most diverse ones (101 and 89 species, resp.). Compiling the most recently published studies, this richness increases with the addition of four new species in the northeastern portion of the São Francisco River basin: *Aspidoras psammatides* [14], *Pimelodus pohli* [15], *Simpsonichthys harminicus* [16], and *Salminus franciscanus* [17]. Some additional species from the Siluriformes order have been recorded, such as *Platydoras brachylecis* in the Parnaíba River [18] and *Rhamdiopsis krugi* [19] in the caves of Chapada Diamantina National Park, demonstrating the preferences of some fish species for more sheltered habitats.

Medeiros and Maltchik [20] and Luz et al. [21, 22] noted that temporary environments such as lakes and tributaries are important marginal environments for shelter, feeding and rest for several fish species. There is an urgent need for preservation of the integrity of these habitats and their functionality, owing to their role in conservation of fish diversity in the Caatinga.

Based on this panorama of findings, the following challenges have been detected in relation to our knowledge of the ichthyofauna of the Caatinga: (i) long-term research to determine the actual diversity of fish species in this ecosystem; (ii) a taxonomic review of the systematically ambiguous groups; (iii) the characterization of the natural history of the fish species of the Caatinga, especially those that are involved in local economy; (iv) more precise monitoring of the consequences of introductions of exotic species, which requires the design of more efficient conservation projects to protect these species and the ecosystems within which they live.

2.2. Herpetofauna. The herpetofauna includes Anurans, Gymnophiona, Testudines, Squamata, and Alligators. These groups display a significant amount of variability in their morphotypes, that is, associated with specific ecological niches [23]. These organisms are also highly susceptible and sensitive to environmental changes [24, 25]. The first continuous efforts to monitor the herpetofauna of the

Caatinga were initiated in the 1970s and 1980s [26–31] and were especially concerned with the herpetofauna in the states of Pernambuco and Paraíba.

Since then, the number of studies on the herpetofauna of the Caatinga has increased. However, the group is still underrepresented because it encompasses a large diversity of unknown morphotypes [32–34], and those studies that have been performed have concentrated on only a few locations [32, 33, 35–38]. Research from the past three decades has revealed increases in the number of occurrences of new herpetofaunal species [36, 37, 39–44] as well as in the occurrences of described species [26, 45–62]. This work has helped to fill in geographical gaps but confirms the need for further studies to elucidate the full diversity of the group [32].

Despite its underrepresentation, the known herpetofauna of the Caatinga currently comprises 175 species (53 anurans, 3 Gymnophiona, 7 testudines, 47 lizards, 10 Amphisbaenia, 52 serpents, and 3 alligators). Approximately 12% of these species are endemic [33–38].

It is worth stressing that public policies providing financial support for multidisciplinary projects involving other taxa have also contributed to the increase in the number of publications concerning various aspects of the herpetofauna of the Caatinga, including studies providing specific species inventories ($n \approx 35$), studies on the ecology ($n \approx 22$) or conservation ($n \approx 22$) of herpetofauna, descriptions of new species ($n \approx 20$), reports of the geographic distribution of certain populations ($n \approx 12$), and, much less frequently, the behavioral biology and thermoregulation of herpetofaunal species ($n \approx 5$).

Lizards and serpents are the subjects of more than 50% of the scientific articles on the herpetofauna. Members of the Testudines order and crocodilians have been poorly studied, with occasional records of these species occurring in the main drainage basins of the Caatinga [36, 63–66]. Ecological research on the Gymnophiona order (amphibians) and the Amphisbaenia (Squamata), exclusively fossorial animals, is nonexistent, and there are few studies on frogs in the larval (tadpole) stage [35]. The large gap in knowledge concerning these animals is perhaps the result of the very low number of specialists in Brazil and/or the methodological difficulties involved in performing systematic research [36]. There is a clear need for financial incentives and training of herpetological specialists. The Caatinga herpetofauna is poorly understood compared to those found in other parts of Northeast Brazil, such as the Atlantic Forest [67, 68].

2.3. Avifauna. The Caatinga has been identified as an important center of endemism for South American birds [69–72]. Although the avifauna is the most well-known animal group with respect to its taxonomy, geographic distribution, and natural history, there are still many gaps in our understanding of the distribution, evolution, and ecology of the avifauna of the Caatinga as compared with other natural regions of Brazil, such as the Atlantic Forest and the Cerrado [73–76]. In fact, these latter regions have been disproportionately investigated because the majority of the universities in Brazil are concentrated in the southeast and

south regions of the country; a greater number of researchers are concentrated in the southeast region, and the majority of theses and dissertations are conducted in central south Brazil [77].

Most manuscripts that have been published on the avifauna of the Caatinga are inventories containing annotated lists of important species (endemic and threatened species); few studies have addressed issues at the community ecology level or analyzed the biology of specific bird species [74, 76, 78–83]. Published studies that investigated the evolutionary, behavioral, or reproductive biology of avifaunal species or examined their conservation potential are rare. In addition, few studies have analyzed the seasonal movements of species or the interactions among the avifauna of this region and the availability of resources, which encompass topics such as pollination, dispersion, and foraging ecology [84–89].

For decades, the avifauna of northeast Brazil was described only in the writings of foreign naturalists and preserved as material deposited in scientific collections. The majority of this material, however, was derived from the avifauna of the Atlantic Forest [90]. In the past 50 years, Brazilian ornithology has made great advances with respect to our knowledge of the birds of the Caatinga; however, the most significant advancements occurred in the past two decades. Altogether, 60 ornithological studies were conducted in the Caatinga between 1961 and 2011. The first study, published exactly 50 years ago, was the result of four expeditions performed in the states of Ceará, Alagoas, Paraíba, Bahia, and Piauí during 1957 and 1958 [91]. These expeditions surveyed portions of the Caatinga and the Atlantic Forest.

In the 1970s, three studies were performed in the Caatinga. The first was conducted by Coelho [92], who recorded 273 bird species in northeast Brazil, mainly in the state of Pernambuco. The next two studies were performed in the states of Paraíba [93] and Bahia [94]. In the 1980s and 1990s, 13 additional studies were performed, covering the states of Pernambuco, Ceará, Paraíba, and Bahia [95–100]. Among these studies, the study by Coelho [101], which was conducted in the Biological Reserve of the Serra Negra (in a semiarid region of Pernambuco), is notable because it examined bird species of both the Atlantic Forest (damp vegetation) and the Caatinga (dry vegetation). The noteworthy studies by Azevedo Júnior et al. [102] and Azevedo Júnior and Antas [103, 104] also contributed to our understanding of the biology of the game species *Zenaida auriculata*, which suffers a great deal of pressure from hunting in the semiarid areas of the northeast region.

Since the early 2000s, with the increases in ornithologists, research incentives, and requirements that environmental licensing organizations monitor avifauna, new areas of the Caatinga have been studied in the states of Pernambuco (Vale do Catimbau National Park and other parts of the Legal Reserve), Ceará (Chapada do Araripe and the Aiuaba Ecological Station), Bahia (Raso da Catarina), and Rio Grande do Norte (Seridó Ecological Station), resulting in more than 34 scientific studies [105–109]. These studies, above all, record new occurrences of bird species, expanding the known distributions of several species across the Caatinga [110–116].

The studies published on the avifauna of the Caatinga facilitated the publication of three lists of avifaunal species between 1995 and 2003. In the first two lists, 338 and 347 species were reported, respectively [117, 118]. However, the bird species recorded in other vegetation enclaves that occur in isolation in the Caatinga region, such as *brejos de altitude* (highland forests), alpine pastures, and areas of ecological tension, were not included in the first two lists. Therefore, Silva et al. [75] compiled a third, more exhaustive, list for the region; this list contains 510 bird species, distributed across 62 families, and includes the avian species of the aforementioned enclaves so as to improve estimates of regional species density and because of their influence on ecological processes such as dispersion, which helps to maintain the distribution of the avifauna across the Caatinga. Since the last of these three lists was published, only *Petrochelidon pyrrhonota* has been added to the list of birds of the Caatinga [114].

Although the diversity of the Caatinga avifauna may be relatively well established, there remains a need for further scientific research to determine the influence of seasonal variations in resource availability on the dynamics of bird populations and communities, both terrestrial and aquatic, particularly with regard to the mechanisms that contribute to the spatial-temporal heterogeneity that exists across the Caatinga. Filling in these knowledge gaps is extremely important for conservation initiatives and the management of Caatinga bird populations, which suffer from increasing anthropogenic pressures and are safeguarded in only a small number of protected areas [4, 120].

2.4. Mammals. Over the last four decades, notable progress has been made in expanding our knowledge of mammals of the Caatinga. The notable research efforts of the 1980s contributed information on the species richness, ecology, behavior, physiology, and distribution of mammalian fauna in the Caatinga [119, 121–124]. From this series of studies, the following five initial assumptions concerning the biology and ecology of the mammals of the Caatinga were made: (i) the number of species is relatively low, (ii) there is a low degree of endemism, (iii) the mammals of the Caatinga represent a subset of the mammals of the Cerrado, (iv) the mammals living in the Caatinga do not possess pronounced physiological adaptations to life in this semiarid region, and (v) these mammals exhibit behavioral adaptations that allow them to live in this environment.

Although extremely important, some of the results obtained by these authors have been amended by recent studies. For instance, the total number of species described by Willig and Mares in 1989 [119] increased from 80 to 101 in a study published by Fonseca and colleagues in 1996 [125]. More recently, Oliveira [126] added 47 new species to the list. This number increases even more if we consider the studies performed by Gregorin and Ditchfield [127], Gregorin et al. [128], Feijo et al. [129], and Moratelli et al. [130]. Together, these studies added six new species of bats to the Caatinga. Furthermore, Canale et al. [131] and Ferreira et al. [132]

recorded the presence of two new primates inhabiting the Caatinga. Collectively, these studies added nine new species that were not listed in the Oliveira [126] study, thus bringing the current total number of mammalian species living in the Caatinga to 156.

In 1989, the rodent *Kerodon rupestris* was the only mammalian species known to be endemic to the Caatinga [119]. However, in recent years, an additional 12 species have been defined as endemic, including ten rodents (*Wiedomys pyrrhorhinos, Trinomys yonenagae, Trinomys albispinus minor, Trinomys albispinus sertonius, Thylamys karimii, Dasyprocta* sp. n., *Oryzomys* sp. n., *Oxymycterus* sp. n., *Rhipidomys* sp. n. ssp. 1, and *Rhipidomys* sp. n. ssp. 2; Oliveira [126]); one primate (*Callicebus barbarabrownae*; Oliveira [126]); two bats (*Xeronycteris vieira* and *Chiroderma* sp. n; Gregorin and Ditchfield [127], and Gregorin et al. [128], resp.). As a result of these more recent findings, it is now accepted that there is a considerable richness of rodent species in the Caatinga [133]. As for bats, although studies remain scarce [134], it appears likely that the richness of bat species in this ecosystem will increase with further research.

The number of known mammalian species in the Caatinga (n = 156) is greater than that of the Pantanal (n = 113; Meserve [135]) or the Grande Chaco (n = 102; Meserve [135]), but it is less than that of the Amazon Forest (n = 350; Meserve [135]), the Atlantic Forest (n = 261; Ribeiro et al. [136]), and the Cerrado (n = 199; Klink and Machado [137]). There is a notably high level of endemism in the Amazon Forest (n = 205, 58.6%; Meserve [135]) relative to the Cerrado (n = 18, 9.3%; Marinho-Filho et al. [138]) and the Caatinga (n = 12, 7.7%). All of these recent results show that the Caatinga has a species richness that, though not comparable to the Amazon or Atlantic Forests, is higher than that found in other ecosystems and clearly higher than previously thought. The new data also highlight the inadequacy of the initial assumption regarding the mammals of the Caatinga as a subset of the mammalian species found in the Cerrado (e.g., Oliveira et al. [139]).

Life in semiarid conditions imposes constraints and limitations on many mammalian species. However, by means of water deprivation experiments, Streilein [123] discovered that there were no pronounced physiological adaptations in several small mammalian species living in the Caatinga. More recent studies support the earlier findings (e.g., Mendes et al. [140]; Ribeiro et al. [141]). According to Streilein [123], small mammals exhibit behavioral responses to compensate for limiting factors in the semiarid environment. As an example, Streilein [123] mentions the preference of some species for mesic and rocky areas (see, however, Freitas et al. [142]). Recent studies have confirmed the importance of behavior in the adaptation of a number of mammals to the Caatinga. For example, the endemic species *Trinomys yonenagae* digs holes in dunes to escape the hottest period of the day [143]. In addition to the thermoregulatory benefit of this behavior, Santos and Lacey [144] suggest that it also serves as protection against predators. Furthermore, Moura [145] proposed that *Sapajus libidinosus* (previously known as *Cebus libidinosus*; *C. apella* in Oliveira [139]) uses its high cognitive capacity to overcome the difficulties associated with obtaining food in the Caatinga.

Despite the advances in our knowledge of the mammals of the Caatinga since the first important assumptions were made decades ago, many geographical areas of the Caatinga have not yet been studied. This deficiency indicates a high potential for a further increases in knowledge concerning the richness and geographical distributions of mammalian species, not to mention their ecology and behavior. It has been recommended that researchers avoid examining locations close to cities and villages to prevent any methodological bias associated with these areas (Oliveira [139]). Moreover, Bernard et al. [134] made other recommendations that, although proposed for bat fauna studies, may be extrapolated to studies of other mammals; these suggestions include long-term research incentives involving previously studied areas; increasing the number and frequency of studies on the mammalian species present in collections and museums; investing in the training of taxonomists; and creating an online database to record and organize the occurrences and distributions of mammalian species in the Caatinga. Naturally, these efforts will only be viable if there is an attendant commitment to efficiently protect the regions established as priority areas for the conservation of the mammals of the Caatinga [126].

3. Plant Ecology

To date, at least 248 studies have examined the Caatinga vegetation, 33% of which attempted to answer questions related to the flora and the phytosociology of the region. This focus reflects the nation's policies that support increases in scientific and technological knowledge. Before the 2000s, these policies predominantly encouraged research aimed at identifying the composition of woody species and characterizing the structure of the communities in the areas considered to be protected or subject to a very low amount of anthropogenic intervention [146–152].

Despite the differences that exist among the sampling criteria and the sampling efforts of the various floristic and phytosociological studies of the Caatinga [146, 147], these studies confirm that the vegetation displays various types of physiognomies, ranging from predominantly herbaceous vegetation to arboreal vegetation, with differences in floristic composition among the physiognomic types, and includes a considerable number of endemic species [147, 149, 152]. However, there is still a large gap in our understanding of the richness of both the herbaceous component and the climbing and epiphytic plants [147, 148, 153], which are estimated to number approximately 1500 species, or threefold as many than the known richness of woody species [154].

Beginning in the 2000s, there was a large increase in the number of studies aimed at understanding the ecological and ecophysiological processes that influence the establishment and renewal of plant populations. Among these studies, those related to pollination, the reproduction system, and plant phenology make up the greatest percentage (17%) [155–161].

In general, the plant species in tropical dry forests exhibit diverse phenological patterns [162, 163] that reflect the heterogeneity in the environmental factors that create gradients in resource availability, such as precipitation, temperature, photoperiod, and soil type [162]. These forests generally encompass a range of deciduousness, from fully deciduous species to evergreen species [164]. However, despite the rich phenological diversity of tropical dry forests, little is actually known about the driving factors and mechanisms by which plants adapt their phenological patterns. This situation is even more complicated in the Caatinga, a dry tropical forest that occupies one of the largest land areas in the world and exists at the extreme of water resource availability for forests [165]. The necessity of collecting phenological information for these areas is reinforced by the high degree of anthropogenic disturbance to which these forests have been subjected [166] and the potential risks of probable climate changes and further reductions in water resource availability [165].

The timing of phenological events can be critical to the reproduction, establishment, and survival of plants [167]. Therefore, an understanding of the mechanisms and factors that influence these events and their relationships to pollinating and dispersing agents may help to inform better biodiversity management and conservation planning efforts. According to Quesada et al. [168], tropical dry forests are highly dependent on pollinators (54–80% of the species are self-incompatible), and in the case of the Caatinga vegetation, Machado and Lopes [169] found that 98% ($n = 147$) of the plant species they examined were pollinated by animals, with entomophily as the predominant means of pollination.

Of the phenological studies published about the Caatinga, five examined groups of 10 to 20 woody species [156, 158, 170–172] and two other studies investigated one and a few woody species, respectively [173, 174]. Only one study examined two herbaceous species [175]. Several other studies dealt exclusively with aspects of reproduction [155, 157, 169, 176–180]. In total, these studies considered only a small proportion (<10%) of the Caatinga flora, which contains more than 1500 species, even when only considering the most typical vegetation types [154].

The few phenological studies undertaken in the Caatinga region have indicated that precipitation is the major driving factor of phenophases [181]. Precipitation controls the phenology of numerous species, although other species initiate their phenophases independent of the occurrence of rainfall [157]. Amorim et al. [156] demonstrated that different species show different responses to soil water availability, with some budding in response to sporadic rainfall during the dry season, while others remain dormant. Recently, Lima and Rodal [158] reported a close relationship between plant phenology and wood density (the quantity of water stored in tree trunks). These authors noted that deciduous plants with low wood densities are capable of storing more water in their trunks and that budding and reproduction often occur during the dry season in species with low wood densities. In contrast, the deciduous trees with high-density wood tend to initiate their phenophases more in accordance with soil water availability. Lima [182] likewise determined that species with low-density wood maintained high water potentials throughout the year, although they only budded or flowered when the photoperiod increased (not necessarily when soil water was available). There have been few studies published in this area, and the effects of the photoperiod on Caatinga plants still need to be examined along latitudinal gradients and in terms of the synchrony of phenological events, which has been investigated in other dry tropical forests [183–187].

Future phenological studies in the Caatinga should not only focus on the relationships between the timing of phenological occurrences and rainfall, which have been examined numerous times [171, 181], but also investigate the biotic mechanisms responsible for the occurrence of various phenophases and attempt to integrate the diverse characteristics of a given plant species into a whole entity [188], with the aim of predicting possible phenological patterns from assemblages of such characteristics [168, 189, 190]. These types of studies would allow Caatinga species to be classified in terms of their functional phenological types, that is, as groups of species that demonstrate certain sets of characteristics and functions and respond in similar manners to multiple environmental factors [191], independent of their phylogenetic or taxonomic relationships [191, 192]. One paper [193] analyzed a large number of leaf, stem, and life-form characteristics to distinguish functional groups of species in the Caatinga, but phenological traits were not incorporated into the model.

Studies of phenological types can greatly contribute to our understanding of ecosystem functioning and biodiversity maintenance and should include (i) long-term studies [194]; (ii) environmental and latitudinal gradients and multiple successional stages [168]; (iii) experimental manipulations that simulate climatic variations, especially those related to soil water availability and longer dry periods; (iv) relationships between the physiology and phenology of plants [195]; (v) phenology and storage mechanisms; (vi) fruiting season and dispersion syndromes. Finally, (vii) the different Caatinga physiognomies should be characterized in terms of functional groups based on phenological and other plant traits.

In addition to the studies on the phenology and reproduction of plants in the Caatinga, other significant processes that influence the establishment of plants and the renewal of plant populations in the Caatinga have been investigated. Studies on seed germination constitute the greatest percentage (10%) of such studies [196–199], followed by studies of plant physiology and ecophysiology (8%) [200–203] and studies examining the dispersion and dynamics of the seed bank (6%) [204–207]. These studies demonstrate the following: (i) many species have seeds with a rigid tegument and seed coat dormancy, which help the plant embryo to avoid germination in the dry season and protect the seed from predation by small animals; (ii) leaf fall is the main survival strategy during the dry period, but leaf replacement can take place in the dry season given the occurrence of thunderstorms with intense rains; (iii) many adult plants of the Caatinga contain chlorophyll covered by a

fine layer of rhytidome in their stems, leading to the hypothesis that in some woody (noncactus) species, photosynthesis is performed in the stem organ during the dry season, explaining the positive growth rate of the plants during the dry season; (iv) the dispersion of seeds is influenced by the seasonality of the region, with a predominance of zoochory recorded in the moister Caatinga areas; and (v) the seed bank dynamics suggest the existence of opposite tendencies among Caatinga areas, with greater seedling emergence occurring in either the dry season or the rainy season, depending on the area.

All of the aforementioned processes influence the dynamics of plant populations. However, few studies have been designed to model the dynamics that exist in the Caatinga. To date, only 12% of studies have provided information on the birth and mortality rates of the plant populations of the Caatinga; these studies mainly focused on the influence of regional climatic seasonality on herbaceous species [175, 196, 208–210]. No studies have been performed on the population dynamics of epiphytic or climbing plants.

It is important to stress that approximately 80% of the annual precipitation in the Caatinga is distributed irregularly and unpredictably within the rainy season [147, 152, 211], and any of the following scenarios may occur: (i) a prolongation of the duration of the rainy season or the dry season; (ii) erratic rainfall events during the dry season; (iii) droughts (dry spells) during the rainy season; though very rarely, (iv) years that lack a dry season entirely, particularly in the hypoxerophilous areas of the Caatinga nearest to the coast [200]. These rain distribution models act as a selective force because they influence the reproductive behavior of the plants and cause differential mortality in the plant populations, thereby constituting key factors that influence the dynamics of the ecosystem.

Plant birth is predictable over time both for herbaceous and woody species, with the predominant period of birth occurring during the rainy season or after the erratic rains occurring in the dry season [196, 212]. The latter incidents have a negative effect on the population because the seedlings die when the dry season resumes [175, 210]. Some populations display synchronous and concentrated births at the beginning of the rainy season, which may be disadvantageous if dry spells occur right after the beginning of the rainy period. Other populations distribute births among the months of the rainy season, a model that minimizes the mortality caused by the occurrence of water deficits [148, 196]. Thus, various birth distribution models may have evolved as strategies to escape the unpredictability of the rain distribution from year to year, indicating that any forecasts of the size of a given population, especially those species with a long-life cycle, must be based on data from a long-time series that incorporates the effect of this unpredictability.

Plant mortality in the Caatinga is unpredictable over time. Mortality may occur in any of the climatic seasons, and mortality rates vary among years [208, 210]. Nevertheless, marked differences in the timing of mortality have been observed between woody and herbaceous populations in the

Caatinga. Herbaceous populations display heightened mortality in the dry season because of the predominance of the therophyte life-form [212]. The opposite has been recorded for woody plants because of the heightened recruitment of seedlings, a seasonal state that is delimited by the duration of the rainy period [148, 213].

The direct and indirect impacts of the rainfall distribution appear to function as the main causes of population mortality. At the transition between climatic seasons, when individual adults still display deciduous characteristics or have only undergone incomplete leaf replacement [156, 171], the force of the water can directly uproot young individuals in the soil [148]. Rainfall events can also indirectly cause mortality because, due to the weight of the rain, dry branches of woody plants may fall onto fragile seedlings, uprooting them from the soil. Vines, which grow rapidly after the arrival of the rainy season, may also uproot fragile plants by using the seedlings as support [211]. Another cause of mortality recorded during the rainy season is herbivory of newly recruited seedlings by wild fauna (populations of beetles, ants, and other organisms that grow in size during the rainy season).

Little importance has been placed on the influence of microhabitats on the vegetation of the Caatinga, but recent studies indicate that the microhabitats formed by cracks in rocks, at the edges of streams, or along stretches shaded by nondeciduous woody tree canopies can influence the dynamics of the vegetation of semiarid environments. These microhabitats serve as refuges, minimizing the severity of the dry season and facilitating the survival of individuals from various plant populations [175, 211]. These microhabitats may also influence plant growth rates, but this potential impact needs to be directly assessed in the Caatinga.

Few studies have characterized the processes of litterfall, nutrient cycling, biological fixation, and the environmental services provided by the vegetation of the Caatinga. Those studies that have been performed indicate that climatic seasonality is a highly significant factor that influences these processes and that not all legume species fix nitrogen [214–217].

Similarly, few studies have analyzed the impacts of different forms of management on the dynamics of the Caatinga plant populations or on the resiliency of disturbed areas. According to Castelleti et al. [218], management has increased in the Caatinga, despite the role of the plant populations in the local and regional conservation of the climate [219]. These studies have shown that plant management is mainly aimed at agropastoral activities and that abandoned areas exhibit a capacity for natural recovery, although the composition of the native species may be altered as a result of biological invasion, which can alter the abundances of various plant populations [220–224]. The land use history and the duration of use influence the recovery speed of habitats modified by human activity [225], but the parameters that can be used to indicate that a disturbed area has completed its recovery through natural regeneration processes have not yet been defined.

4. Human Ecology and Ethnobiology

By virtue of the adverse environmental conditions, a large portion of the human population living in semiarid regions develops strong relationships with the local floristic and faunistic resources [226]. Animals and plants of the Caatinga provide sources of food but also serve many other needs, such as medicinal remedies (medicinal plants and animals), leather, hide, and ornamental pieces (horns, hooves, eggs, and furs) as well as providing pleasure and decoration (e.g., canaries and other pets) [226–232]. Additionally, some animal species are persecuted and killed because of their conflicting relationships with the human population [226, 230, 233]. In this context, hunting in the Caatinga region has long been practiced and represents a traditional form of wildlife management.

Because of the cultural richness of the local population and their diverse interactions with the local fauna, the Caatinga is an extremely suitable area to conduct ethnozoological studies. These studies are of fundamental importance from a socioenvironmental perspective because factors such as excessive exploitation, hunting, and illegal trading of wild animals have been designated as threats to some vertebrate species. Nonetheless, in the past few decades, researchers have begun to systematically investigate the relationships between the local inhabitants and the wild fauna of the Caatinga region. Our review indicates that 92 studies on the ethnozoology of the area have been published to date (89% of these in the last ten years). The majority of these studies have examined the popular uses of medicinal animals ($n = 25$) and ethnoentomology ($n = 22$). The other topics examined include hunting ($n = 9$), ethnoornithology ($n = 6$), ethnoherpetology ($n = 5$), ethnoichthyology ($n = 4$), the ethnozoology of noninsect arthropods ($n = 4$), religious uses of fauna ($n = 4$), ethnocarcinology ($n = 3$), and ethnomastozoology ($n = 3$). Eight other studies covered topics that are not restricted to one zoological group. The majority of the investigations were performed in two states: Bahia ($n = 52$) and Paraiba ($n = 21$). It should be noted that, even within these two states, this area of research has been restricted to only a few areas. For example, in Bahia, more than 90% of the studies were conducted in two municipalities, Feira de Santana ($n = 12$) and Santa Terezinha ($n = 20$). These patterns provide evidence that the scarce ethnozoological research on the Caatinga has been restricted to a small number of themes and geographic areas.

Despite the scarcity of information, some patterns can be identified in the forms of the interactions between the local populations and various animal taxa. Among the invertebrates, bees stand out as an important group because of their honey production and medicinal uses. Vertebrates are the main hunting targets in the region. Mammals comprise the preferred sources of food because of their size and the possibility of a greater yield for the energy invested in hunting. The species hunted most commonly are *Dasypus novemcinctus* (nine-banded armadillo), *Euphractus sexcinctus* (six-banded armadillo), *Tamandua tetradactyla* (southern tamandua), and *Conepatus semistriatus* (striped hog-nosed skunk) [226, 231, 234, 235]. The populations of some animals that were previously common in the Caatinga, such as the deer *Mazama guazoubira* and *M. americana* (important game species in the region), appear to have declined in many areas [226, 228]. Despite the general preference for mammals as game, when considering the diversity of species used for food, the avifauna is actually the most notable group, particularly species of the families Columbidae and Tinamidae. These results appear to reflect the richness of these groups in the Caatinga, where 511 species of birds and 156 species of mammals have been recorded (see comments above).

Despite their value as a source of protein, the high frequency of game birds targeted is primarily related to their use as pets [229, 232, 236, 237]. This value represents a strong stimulatory factor for the illegal trade of birds in the Caatinga. Various cities in the interior of northeast Brazil have public markets and open fairs where birds and other wild animals are sold [237].

With regard to reptiles, few species are used as food. The lizard *Tupinambis merianae* has been identified as the species most frequently consumed. Although the practice is not common, other reptiles, such as *B. constrictor, Iguana iguana*, and *Caudisona durissa*, can also be eaten [230, 238]. Nonetheless, the principal practical value of reptiles appears to be related to the medicinal value of the products derived from these animals, and various species of chelonians, snakes, and lizards have been used as remedies by local populations [77, 227, 239–243].

Conflicts represent another important type of interaction between humans and animals in semiarid northeast Brazil [230]. The reasons for the conflicts, which lead to the killing of wild animals, include attacks on livestock, risk to human lives, destruction of crops, and risk of disease transmission. The principal taxa involved in conflicts with local inhabitants are reptiles (particularly snakes), mammals (particularly carnivores), and, to a lesser extent, birds (granivorous or falconiformes). Of these animals, reptiles tend to be the group most frequently considered dangerous and persecuted. Snakes, even the nonpoisonous species, are often beaten or killed when encountered. Medium- and large-sized carnivores, such as *Leopardus tigrinus, Puma yagouaroundi, Puma concolor*, and *Cerdocyon thous*, are also killed because they prey on domestic animals [226].

All forms of interaction between the fauna and inhabitants of the Caatinga require further investigation, particularly considering their importance for the development of management and conservation plans. Limited ethnozoological information is available for important groups of vertebrates, likely because of the legal implications surrounding the principal groups of game animals, which are protected by law. This situation influences the choice of topics for ethnozoological studies, most of which involve groups such as insects and fishes. In addition to the scarcity of ethnozoological studies concerned with important animal groups, there are gaps in knowledge from the geographical perspective because, even in the states for which a large number of studies have been performed, the studies have been restricted to two or three cities and then to the same community within these cities.

Most of the studies on plants in Brazil have focused on ethnobiology. For example, we found 156 articles on ethnobotany in Brazil published between 1992 and 2011. Of these, 75% were published between the years 2006 and 2011. When classified according to the ecosystem, a very large proportion of these studies focused on the Atlantic Forest (42.95%), followed by the Caatinga (30.13%), the Cerrado (16.67%), the Amazon (12.18%), the Pampas (4.49%), and the Pantanal (1.28%). This scenario, in which more studies have been performed in the Caatinga than the Cerrado or Amazon, is surprising in a way. Along with the Atlantic Forest, we expected these ecosystems to have hosted the greatest number of studies. In the case of the Cerrado, it is a biodiversity "hotspot" (as is the Atlantic Forest) and an environmental priority for conservation, given that it is the largest tropical forest on the planet, covering nearly half of Brazil.

Considering only the studies focused on the Caatinga, the scenario is similar to the national trend; that is, the majority of the studies (89.36%) were published during the same time interval mentioned above. This fact was also recorded by Oliveira et al. [244], who found a greater volume of ethnobotanical studies published in recent years, possibly due to the growing number of researchers (senior and junior) who have started working on these topics over the years. Furthermore, the Caatinga covers nine Brazilian states (eight in the northeast region and one in the southeast region), but the ethnobotanical studies have mainly concentrated on the state of Pernambuco, in which more than 65% of the studies on this region were conducted.

The majority of the ethnobotanical studies performed in the Caatinga are either directly or indirectly concerned with medicinal plants. Studies that express the intention of contributing to a foundation by searching for new biologically active molecules particularly stand out [245–248], as do studies focused on conservation and sustainable management [245–248] and the cognitive effects of medicinal plants [249, 250]. An example of a study that aimed to contribute to the search for new biomolecules is that of Araújo et al. [245], who applied syndromic importance as a tool in the search for plants in the Caatinga with high levels of tannins, optimizing the search for plant resources with high yields of this molecule. This method was made possible by the use of tools and/or concepts derived from other branches of knowledge, such as phytochemistry, chemical ecology, and pharmacology. Although medicinal plants have been considered with the greatest frequency, other ethnobotanical topics have been gaining ground, such as the research on wood forest products that emphasizes their uses and the impacts of harvesting on the native vegetation [251, 252].

In general, the ethnobotanical research conducted in the region has helped not only to clarify the biodiversity of useful plants of the Caatinga [246, 253] but also to determine the possible impacts that the collection practices may exert on the environment [251, 254]. One of the studies that compiled the medicinal plants used for traditional and nontraditional purposes in the Caatinga revealed that more than 350 taxa (including native and exotic plants) compose the popular pharmacopoeia of the local human populations

[253]. With regard to the conservation of plant resources, ethnobotany has contributed by measuring the usage and availability of the plant species in the environment or rather by seeking to improve our understanding of how much the native local vegetation may provide without harming the wild populations [254–258]. In this regard, ethnobotanical research has made many contributions to this field. An example is the study of Oliveira et al. [254], who suggested priority species for management and conservation programs based on ecological data along with information on plant usage.

Another point worth stressing is that the majority of the ethnobotanical studies performed in the Caatinga have been performed in communities that are located close to the native vegetation areas. These populations clearly collect and use species from the neighboring vegetated areas for various purposes, such as medicine, wood (e.g., housing construction or to delineate territories), food, and fuel (firewood and charcoal). These studies have contributed to both an understanding of the local demands for the resources necessary for survival and the identification of the species/populations that suffer from the greatest anthropogenic pressure. These studies can further contribute to the formulation of proposals for the rational use and sustainable management of these species, thus contributing to the conservation of the local biodiversity.

Another notable research topic, though not relevant to the majority of the existing studies, is the testing of hypotheses and predictive models. In this regard, researchers from the region have tested various hypotheses/models, such as those associated with the appearance, diversification, and functional redundancy of species, in an attempt to explain patterns in the selection and usage of plants by humans. An example of a hypothesis that was conceived of and tested in an area of the Caatinga is the hypothesis of diversification, which asserts that exotic plants are inserted into a given community to supply therapeutic treatments for diseases for which the native plants are not effective. This hypothesis was corroborated in a study conducted in the Caatinga by the observation of significant differences in the presence of particular classes of secondary metabolites between native and exotic plants [259].

5. Perspectives

The heterogeneity in our understanding of the Caatinga is possibly what led Santos et al. [1] to assert that "objectively, society must rapidly reduce the institutional anemia experienced by some SDTFs [seasonally dry tropical forests] and other seasonal ecosystems by expanding local institutional capacity and research networks (i.e., aggressive capacity-building) with the task of (1) informing stakeholders [of] the costs and benefits from general land use patterns and those imposed by public policies, and (2) developing and transferring the better-practices required for using natural resources sustainably."

Clearly, more studies need to examine the communities and populations of plants and animals of the Caatinga because generalizations on the dynamics of this type of

vegetation must consider the differentiation of existing habitats, which in turn influences the ecophysiological behavior, reproductive dynamics, and survival capacity of the individuals of different populations. Much of this research has been financed by public organizations, but no policy exists to establish a central storage bank of field data. Such a database would enable subsequent analyses considering longer time series.

With regard to the flora of the Caatinga, many questions and gaps have been acknowledged in previous studies [4, 5, 213] but remain unanswered. For example, what is the annual contribution of seeds to the renewal of the seed bank in the Caatinga? Within a single population of Caatinga plants, is there temporal variation in sexual characteristics? What is the biological significance of the variation in the phenological and reproductive behaviors of plants for the population dynamics and evolution of the plants in this environment? What Caatinga species display interactive dynamic models, and how might these interactions facilitate our understanding of the ecosystem functions of the Caatinga? What survival mechanisms are employed by plants in the Caatinga? What species can be used to assist with the recovery of anthropogenically disturbed areas?

With regard to the fauna, the main challenges to be overcome in the next decades are as follows: (i) directing efforts for multidisciplinary studies with long-term research goals; (ii) promoting taxonomic reviews of species with many morphotypes; (iii) recording the real geographic distribution, diet, spatial-temporal specifics, and reproductive aspects of the species that occur in the Caatinga.

With regard to the conservation status of Caatinga species, there are a number of relevant issues, such as the consequences of climatic modifications and desertification processes for population structures, particularly given that while Caatinga species may be adapted to long periods of drought, certain species may be extremely vulnerable to rapid climate changes.

The gap in our knowledge concerning the resiliency of this ecosystem is substantial. It will be necessary to invest in studies examining the dynamics of the recovery of anthropogenically disturbed areas with regard to both the land use history and duration of use. Currently, the similarities and differences in the natural regeneration process between preserved and disturbed areas are poorly understood. Thus, it is difficult to contemplate and discuss strategies to recover disturbed areas, which are continuously increasing in area in this type of ecosystem because of the social demands and the technological development of the country.

In view of the current panorama of research findings, it is evident that future ethnobiological studies should be extended to a greater diversity of taxa. Such data would make it possible to confirm the ethnozoological and ethnobotanical patterns recorded to date and to search for answers to questions not yet resolved, including the following: are socioeconomic parameters (e.g., schooling, income, and age) the main factors influencing the interactions between animals and the local population? Or is hunting motivated by leisure and culture? Are cultural aspects more important than either of these factors? Does subsistence hunting persist only in more isolated sites? Are hunting activities more intense in well-preserved areas of the Caatinga? Do the types of uses of the fauna vary according to taxonomic group? Are hunting and the use of wild animals similar in urban and rural areas? Which forms of interactions/uses have the most negative effects on the species exploited? Is there a relationship between the uses of particular species and the availability of the local fauna? Does the illegal trade of wild animals persist because of cultural or socioeconomic aspects? We believe that these are some of the questions that should guide new ethnobiological studies. These studies should be performed with a greater taxonomic rigor, an aspect that is lacking in a large number of ethnozoological studies carried out to date.

Despite all of these knowledge gaps, the Ministry of the Environment has gathered together researchers to consider and propose priorities for the conservation of the biological diversity of the Caatinga [3, 219]. If, however, there remains too great of a gap in the scientific knowledge of the Caatinga [1], the proposed public policies may not achieve the stated goals and may need to be rethought.

Acknowledgments

This paper is the P010 contribution of the Rede de Investigação em Biodiversidade e Saberes Locais (REBISA-Network of Research in Biodiversity and Local Knowledge), with financial support from FACEPE (Foundation for Support of Science and Technology) to the Project Nùcleo de Pesquisa em Ecologia, conservação e Potencial de Uso de Recursos Biológicos no Semiárido do Nordeste do Brasil (Center for Research in Ecology, Conservation and Potential Use of Biological Resources in the semiarid region of northeastern Brazil-APQ-1264-2.05/10).

References

[1] J. C. Santos, I. R. Leal, J. S. Almeida-Cortez, G. W. Fernandes, and M. Tabarelli, "Caatinga: the scientific negligence experienced by a dry tropical forest," *Tropical Conservation Science*, vol. 4, no. 3, pp. 276–286, 2011.

[2] G. A. Sánchez-Azofeifa, M. Quesada, J. P. Rodríguez et al., "Research priorities for neotropical dry forests," *Biotropica*, vol. 37, no. 4, pp. 477–485, 2005.

[3] M. Barbosa, R. Castro, F. Araújo, and M. Rodal, "Estratégias para conservação da biodiversidade e pesquisas futuras no bioma Caatinga," in *Análise das Variações da Biodiversidade do Bioma com Apoio de Sensoriamento Remoto e Sistema de Informações Geográficas para Suporte de Estratégias Regionais de Conservação*, F. S. Araújo, M. J. N. Rodal, and M. R. V. Barbosa, Eds., pp. 417–434, Ministério do Meio Ambiente, Fortaleza, Brazil, 2005.

[4] M. Tabarelli and A. Vicente, "Lacunas de conhecimento sobre as plantas lenhosas da caatinga," in *Caatinga: Vegetação e Flora*, E. V. S. B. Sampaio, A. M. Giulliette, J. Virgírio, and C. F. L. Gamarra-Rojas, Eds., pp. 25–40, Associação Plantas do Nordeste e Centro Nordestino de Informações sobre Plantas, Recife, Brazil, 2002.

[5] M. Tabarelli and A. Vicente, "Conhecimento sobre plantas lenhosas da Caatinga: lacunas geográficas e ecológicas," in *Biodiversidade da Caatinga: Áreas e Ações Prioritárias para a Conservação*, J. M. C. Silva, M. Tabarelli, M. T. Fonseca, and L. V. Lins, Eds., pp. 101–111, Ministério do Meio Ambiente e Universidade Federal de Pernambuco, Recife, Brazil, 2004.

[6] P. Buckup, N. Menezes, and M. Ghazzi, *Catálogo das Espécies de Peixes de Água Doce do Brasil*, Museu Nacional, Rio de Janeiro, Brazil, 2007.

[7] R. Rosa, N. Menezes, H. Britski, W. Costa, and F. Groth, "Diversidade, padrões de distribuição e conservação dos peixes da Caatinga," in *Ecologia e Conservação da Caatinga*, I. R. Leal, M. Tabarelli, and J. M. C. Silva, Eds., pp. 135–162, Editora Universitária da UFPE, Recife, Brazil, 2003.

[8] W. Costa, "The neotropical annual fish genus Cynolebias: philogenetic relationships, taxonomic revision and biogeography," *Icththyological Exploration of Freshwaters*, vol. 12, no. 1, pp. 333–383, 2001.

[9] W. Costa, *Peixes Anuais Brasileiros: Diversidade e Conservação*, Editora da UFPR, Curitiba, Brazil, 2002.

[10] W. Costa and G. Brasil, "Two new species of Cynolebias (Cyprinodontiformes: Rivulidae) from the São Francisco basin, Brazil with notes on phylogeny and biogeography of annual fishes," *Icththyological Exploration of Freshwaters*, vol. 4, pp. 193–200, 1993.

[11] W. Costa, A. Cyprino, and A. Nielsen, "Description d'une nouvelle espece de poisson du gene Simpsonichthys (Cyprinodontiforrmes: Rivulidae) du Dassin du rio São Francisco, Brazil," *Revue Française d'aquariologie*, vol. 23, pp. 17–20, 1996.

[12] W. J. E. M. Costa, T. P. A. Ramos, L. C. Alexandre, and R. T. C. Ramos, "Cynolebias parnaibensis, a new seasonal killifish from the Caatinga, Parnaíba River basin, northeastern Brazil, with notes on sound producing courtship behavior (Cyprinodontiformes: Rivulidae)," *Neotropical Ichthyology*, vol. 8, no. 2, pp. 283–288, 2010.

[13] R. Rosa, "Diversidade e conservação dos peixes da caatinga," in *Biodiversidade da Caatinga: Áreas e Ações Prioritárias para a Conservação*, J. M. C. Silva, M. Tabarelli, M. T. Fonseca, and M. T. Lins, Eds., Ministério do Meio Ambiente, Brasília, Brazil, 2004.

[14] M. Britto, F. Lima, and A. Santos, "A new Aspidoras (Siluriformes: Callichthyidae) from rio paraguaçu basin, chapada diamantina, Bahia, Brazil," *Neotropical Ichthyology*, vol. 3, no. 4, pp. 473–479, 2005.

[15] F. R. V. Ribeiro and C. A. S. De Lucena, "A new species of Pimelodus LaCépède, 1803 (Siluriformes: Pimelodidae) from the rio São Francisco drainage, Brazil," *Neotropical Ichthyology*, vol. 4, no. 4, pp. 411–418, 2006.

[16] W. J. E. M. Costa, "Simpsonichthys harmonious, a new seasonal killifish from the são francisco river basin, northeastern Brazil (Cyprinodontiformes: Rivulidae)," *Ichthyological Exploration of Freshwaters*, vol. 21, no. 1, pp. 73–78, 2010.

[17] F. C. T. Lima and H. A. Britski, "Salminus franciscanus, a new species from the rio São Francisco basin, Brazil (Ostariophysi: Characiformes: Characidae)," *Neotropical Ichthyology*, vol. 5, no. 3, pp. 237–244, 2007.

[18] N. M. Piorski, J. C. Garavello, M. Arce, and M. H. Sabaj Pérez, "Platydoras brachylecis, a new species of thorny catfish (Siluriformes: Doradidae) from northeastern Brazil," *Neotropical Ichthyology*, vol. 6, no. 3, pp. 481–494, 2008.

[19] F. A. Bockmann and R. M. C. Castro, "The blind catfish from the caves of Chapada Diamantina, Bahia, Brazil (Siluriformes: Heptapteridae): description, anatomy, phylogenetic relationships, natural history, and biogeography," *Neotropical Ichthyology*, vol. 8, no. 4, pp. 673–706, 2010.

[20] E. S. F. Medeiros and L. Maltchik, "Diversity and stability of fishes (Teleostei) in a temporary river of the Brazilian semiarid region," *Iheringia*, vol. s, no. 90, pp. 157–166, 2001.

[21] S. Luz, A. El-Deir, E. França, and W. Severi, "Estrutura da assembléia de peixes de uma lagoa marginal desconectada do rio, no submédio Rio São Francisco, Pernambuco," *Biota Neotropica*, vol. 9, no. 3, pp. 117–129, 2009.

[22] S. Luz, H. Lima, and W. Severi, "Composição da ictiofaunaem ambientes marginais e tributários do médio-submédio rio São Francisco," *Revista Brasileira e Ciências Agrárias*. In press.

[23] M. Rodrigues, "Fauna de anfíbios e répteis das caatingas," in *Biodiversidade da Caatinga: Áreas de Ações Prioritárias para a Conservação*, J. M. C. Silva, M. Tabarelli, M. T. Fonseca, and L. V. Lins, Eds., pp. 173–179, Ministério do Meio Ambiente, Brasília, Brazil, 2004.

[24] W. Duellman and L. Trueb, *Biology of Amphibians*, Johns Hopkins University Press, Baltimore, Md, USA, 1994.

[25] L. Vitt and J. P. Caldwell, "The effects of logging on reptiles and amphibians of tropical forests," in *The Cutting Edge: Conservation Wildlife in Logged Tropical Forest*, R. A. Fimbel, A. Grajal, and J. G. Robinson, Eds., pp. 239–260, Columbia University Press, New York, NY, USA, 2001.

[26] M. Rodrigues, "Uma nova espécie de Phyllopezus de Cabaceiras: paraiba: Brasil, com comentarios sobre a fauna de lagartos da área (Sauria, Gekkonidae)," *Papéis Avulsos de Zoologia*, vol. 36, no. 20, pp. 237–250, 1986.

[27] P. Vanzolini, "Miscellaneous notes on the ecology of some Brasílian lizards (Sauria)," *Papéis Avulsos de Zoologia, São Paulo*, vol. 26, no. 8, pp. 83–115, 1972.

[28] P. Vanzolini, "Ecological and geographical distribution of lizards in Pernambuco, northeastern Brasil (Sauria)," *Papéis Avulsos de Zoologia, São Paulo*, vol. 28, no. 4, pp. 61–90, 1974.

[29] P. Vanzolini, "On the lizards of a cerrado-Caatinga contact: evolutionary and zoogeographical implications (Sauria)," *Papéis Avulsos de Zoologia, São Paulo*, vol. 29, no. 16, pp. 111–119, 1976.

[30] L. Vitt, "Ecological observations on sympatric Philodryas (Colubridae) in northeastern Brasil," *Papéis Avulsos de Zoologia, São Paulo*, vol. 34, no. 5, pp. 87–98, 1980.

[31] L. Vitt, "Reproduction and sexual dimorphism in tropical Teiid Lizard Cnemidophorus ocelifer," *Copeia*, vol. 2, pp. 359–366, 1983.

[32] M. Camardelli and M. F. Napoli, "Amphibian conservation in the caatinga biome and semiarid region of Brazil," *Herpetologica*, vol. 68, no. 1, pp. 31–47, 2012.

[33] E. Freire, G. Sugliano, M. Kolodiuk et al., "Répteis squamata das caatingas do seridó do Rio Grande do Norte e do Cariri da Paraíba: síntese do conhecimento atual e perspectivas," in *Recursos Naturais das Caatingas: Uma Visão Multidisciplinar*, E. M. X. Freire, Ed., pp. 51–84, Editora da UFRN-EDUFRN, Natal, Brazil, 1st edition, 2009.

[34] M. Rodrigues, "Herpetofauna da caatinga," in *Ecologia e Conservação da Caatinga*, I. R. Leal, M. Tabarelli, and J. M. C. Silva, Eds., vol. 4, pp. 181–236, Universidade Federal de Pernambuco, Recife, Brazil, 2003.

[35] D. Borges-Nojosa and C. Arzabe, "Diversidade de anfíbios e répteis em áreas prioritárias para a conservação da Caatinga," in *Análise das Variações da Biodiversidade do Bioma Caatinga*, F. S. Araújo, M. J. N. Rodal, and M. R. V. Barbosa, Eds., pp. 227–241, Ministério do Meio Ambiente, Brasília, Brazil, 2005.

[36] G. Moura, E. Freire, E. Santos et al., "Distribuição geográfica e caracterização ecológica dos répteis do Estado de Pernambuco," in *Herpetologia do Estado de Pernambuco*, G. J. B. de Moura, E. M. dos Santos, M. A. B. Oliveira, and M. C. C. Cabral, Eds., vol. 1, pp. 229–290, Ministério do Meio Ambiente, Brasília, Brazil, 1st edition, 2011.

[37] G. Moura, E. Santos, E. V. E. Andrade, and E. Freire, "Distribuição geográfica e caracterização ecológica dos anfíbios do Estado de Pernambuco," in *Herpetologia do Estado de Pernambuco*, G. J. B. de Moura, E. M. dos Santos, M. A. B. Oliveira, and M. C. C. Cabral, Eds., vol. 1, pp. 51–84, Ministério do Meio Ambiente, Brasília, Brazil, 1st edition, 2011.

[38] M. Rodrigues, C. Carvalho, D. Borges et al., "Anfíbios e Répteis: áreas e ações prioritárias para a conservação da Caatinga," in *Biodiversidade da Caatinga: Áreas e Ações Prioritárias para a Conservação*, J. M. C. Silva, M. Tabarelli, M. T. Fonseca, and L. V. Lins, Eds., pp. 181–188, Ministerio do Meio Ambiente, Universidade Federal de Pernambuco, Brasília, Brazil, 2004.

[39] F. Delfim, E. Gonçalves, and S. Silva, "Squamata, gymnophthalmidae, psilophthalmus paeminosus: distribution extension, new state record," *Check List*, vol. 2, no. 3, pp. 89–92, 2006.

[40] D. Loebmann, *First State Record Extends the Species Distribution*, vol. 39, Herpetological Review, Berkeley, Calif, USA, 2008.

[41] D. Loebmann, "Notes on geographic distribution: reptilia, squamata, serpentes, viperidae, bothrops lutzi: distribution extension, geographic distribution map," *Check List*, vol. 5, no. 3, pp. 373–375, 2009.

[42] D. Loebmann, "Notes on geographic distribution: reptilia, squamata, serpentes, scolecophidia, anomalepididae, liotyphlops cf. ternetzii (Boulenger, 1896): first family record for the state of Ceará, Brazil," *Check List*, vol. 5, no. 2, pp. 249–250, 2009.

[43] I. Roberto, S. Ribeiro, M. Delfino, and W. Almeida, "Notes on geographic distribution: reptilia, colubridae, helicops angulatus: distribution extension and rediscovery in the state of Ceará," *Check List*, vol. 5, no. 1, pp. 118–121, 2009.

[44] G. J. B. Moura, E. V. E. Andrade, and E. M. X. Freire, "Amphibia, anura, microhylidae, stereocyclops incrassatus Cope, 1870: Distribution extension," *Check List*, vol. 6, no. 1, pp. 71–72, 2010.

[45] F. Arias, C. Carvalho, M. Rodrigues, and H. Zaher, "Two new species of cnemidophorus of the ocellifer group, from Bahia, Brazil," *Zootaxa*, vol. 3022, pp. 1–21, 2011.

[46] F. Arias, C. M. D. Carvalho, M. T. Rodrigues, and H. Zaher, "Two new species of Cnemidophorus (Squamata: Teiidae) from the Caatinga, Northwest Brazil," *Zootaxa*, no. 2787, pp. 37–54, 2011.

[47] R. Bour and H. Zaher, "A new species of mesoclemmys, from the open formations of Northeastern Brazil (Chelonii, Chelidae)," *Papéis Avulsos de Zoologia*, vol. 45, no. 24, pp. 295–311, 2005.

[48] J. Cassimiro and M. T. Rodrigues, "A new species of lizard genus Gymnodactylus Spix, 1825 (squamata: gekkota: phyllodactylidae) from Serra do Sincorá, Northeastern Brazil, and the status of G. carvalhoi Vanzolini, 2005," *Zootaxa*, no. 2008, pp. 38–52, 2009.

[49] P. Manzani and A. Abe, "A new species of tapinurus from the caatinga of Piauí, Northeastern Brazil (Squamata: Tropiduridae)," *Herpetologica*, vol. 46, no. 4, pp. 462–467, 1990.

[50] T. Mott, M. T. Rodrigues, M. A. De Freitas, and T. F. S. Silva, "New species of Amphisbaena with a nonautotomic and dorsally tuberculate blunt tail from state of Bahia, Brazil (Squamata, Amphisbaenidae)," *Journal of Herpetology*, vol. 42, no. 1, pp. 172–175, 2008.

[51] T. Mott, M. T. Rodrigues, and E. M. D. Santos, "A new Amphisbaena with chevron-shaped anterior body annuli from state of Pernambuco: Brazil (Squamata: Amphisbaenidae)," *Zootaxa*, no. 2165, pp. 52–58, 2009.

[52] M. Rodrigues, "Herpetofauna das dunas interiores do rio Sao Francisco: Bahia: Brasil II Psilophthalmus: um novo genero de microteiideos sem palpebra (Sauria: Teiidae)," *Papéis Avulsos de Zoologia (São Paulo)*, vol. 37, no. 20, pp. 321–327, 1991.

[53] M. Rodrigues, "Herpetofauna das dunas interiores do rio Sao Francisco: Bahia: Brasil IV Psilophthalmus: um novo genero de microteiideos sem palpebra (Sauria: Teiidae)," *Papéis Avulsos de Zoologia (São Paulo)*, vol. 37, no. 22, pp. 343–346, 1991.

[54] M. Rodrigues, "Herpetofauna das dunas interiores do rio Sao Francisco: Bahia: Brasil: I Introducao a área e descricao de um novo genero de microteiideos (Calyptommatus) com notas sobre sua ecologia, distribuição e especiacao (Sauria, Teiidae)," *Papéis Avulsos de Zoologia (São Paulo)*, vol. 37, no. 19, pp. 285–320, 1991.

[55] M. Rodrigues, "Herpetofauna das dunas interiores do rio Sao Francisco: Bahia: Brasil III Procellosaurinus: um novo genero de microteiideo sem palpebra, com a redefinicao do genero Gymnophthalmus (Sauria: Teiidae)," *Papéis Avulsos de Zoologia (São Paulo)*, vol. 37, no. 21, pp. 329–342, 1991.

[56] M. Rodrigues, G. V. Andrade, and J. Lima, "A new species of Amphisbaena (Squamata, Amphisbaenidae) from state of Maranhão, Brazil," *Phyllomedusa*, vol. 2, no. 1, pp. 21–26, 2003.

[57] M. T. Rodrigues, J. Cassimiro, M. A. De Freitas, and T. F. S. Silva, "A new microteiid lizard of the genus Acratosaura (Squamata: Gymnophthalmidae) from Serra do Sincorá, State of Bahia, Brazil," *Zootaxa*, no. 2013, pp. 17–19, 2009.

[58] M. T. Rodrigues, M. A. De Freitas, T. F. S. Silva, and C. E. V. Bertolotto, "A new species of lizard genus Enyalius (Squamata, Leiosauridae) from the highlands of Chapada Diamantina, state of Bahia, Brazil, with a key to species," *Phyllomedusa*, vol. 5, no. 1, pp. 11–24, 2006.

[59] M. T. Rodrigues, M. A. De Freitas, and T. F. S. Silva, "New species of earless lizard genus heterodactylus (Squamata: Gymnophthalmidae) from the highlands of chapada diamantina, state of Bahia, Brazil," *Journal of Herpetology*, vol. 43, no. 4, pp. 605–611, 2009.

[60] M. Rodrigues, H. Zaher, and F. Curcio, "A new species of lizard, genus Calyptommatus, from the caatingas of the state of Piauí, northeastern Brazil (Squamata, Gymnophthalmidae)," *Papéis Avulsos de Zoologia (São Paulo)*, vol. 41, no. 28, pp. 529–546, 2001.

[61] M. Rodrigues, "A new species of Mabuya (Squamata: Scincidae) from the semiarid caatingas of northeastern Brazil," *Papéis Avulsos de Zoologia (São Paulo)*, vol. 41, no. 21, pp. 313–328, 2000.

[62] M. F. Napoli and F. A. Juncá, "A new species of the Bokermannohyla circumdata group (Amphibia: Anura: Hylidae) from Chapada Diamantina, State of Bahia, Brazil," *Zootaxa*, no. 1244, pp. 57–68, 2006.

[63] D. Borges-Nojosa and P. Cascon, "Herpetofauna da área reserva da serra das Almas, Ceará," in *Análise das Variações da Biodiversidade do Bioma Caatinga, Brasília*, F. S. Araújo, M. J. N. de Rodal, M. R. V. Barbosa, and M. R. de Vasconcellos, Eds., pp. 245–260, Ministério do Meio Ambiente, Brasília, Brazil, 2005.

[64] D. Borges-Nojosa and D. Lima, "Caiman crocodilus (Common caiman) geographic distribution," *Herpetological Review*, vol. 39, pp. 480–481, 2008.

[65] D. Lima, F. Lima, and D. Borges-Nojosa, "Paleosuchus palpebrosus (Cuvier's Dawrf Caiman, Jacaré-paguá), geographic distribution," *Herpetological Review*, vol. 42, pp. 109–109, 2011.

[66] D. Loebmann and C. F. B. Haddad, "Amphibians and reptiles from a highly diverse área of the Caatinga domain: Composition and conservation implications," *Biota Neotropica*, vol. 10, no. 3, pp. 227–256, 2010.

[67] SBH, "Brazilian amphibians—list of species," http://www.sbherpetologia.org.br/, Sociedade Brasileira de Herpetologia, 2010.

[68] R. S. Bérnils, "Brazilian reptiles—list of species," http://www.sbherpetologia.org.br/, Sociedade Brasileira de Herpetologia, 2010.

[69] J. Cracraft, "Historical biogeography and patterns of differentiation within the South American avifauna: áreas of endemism," *Ornithological Monographs*, vol. 36, pp. 49–84, 1985.

[70] M. Ferri, *A Vegetação Brasileira*, Editora Itatiaia/EDUSP, São Paulo, Brazil, 1980.

[71] P. Gil, *Wilderness—earth's last wild places*, CEMEX, Distrito Federal, Mexico, 2002.

[72] D. Prado and P. Gibbs, "Patterns of species distirbutions in the dry seasonal forests of South America," *Annals of the Missouri Botanical Garden*, vol. 80, pp. 902–927, 1993.

[73] L. B. Mendonça and L. Dos Anjos, "Bird-flower interactions in Brazil: A review," *Ararajuba*, vol. 11, no. 2, pp. 195–205, 2003.

[74] M. Santos, "As comunidades de aves em duas fisionomias da vegetação de Caatinga no estado do Piauí, Brasil," *Ararajuba*, vol. 12, no. 2, pp. 113–123, 2004.

[75] J. Silva, M. Souza, A. Bieber, and C. Carlos, "Aves da Caatinga: status, uso do habitat e sensitividade," in *Ecologia e Conservação da Caatinga Editora Universitária*, I. R. Leal, M. Tabarelli, and J. M. C. Silva, Eds., pp. 237–274, Universidade Federal de Pernambuco, Recife, Brazil, 2003.

[76] W. Telino-Júnior, R. Lyra-Neves, and J. Nascimento, "Biologia e composição da avifauna em uma Reserva Particular de Patrimônio Natural da caatinga paraibana," *Ornithologia*, vol. 1, no. 1, pp. 49–57, 2005.

[77] R. R. N. Alves, H. N. Lima, M. C. Tavares, W. M. S. Souto, R. R. D. Barboza, and A. Vasconcellos, "Animal-based remedies as complementary medicines in Santa Cruz do Capibaribe, Brazil," *BMC Complementary and Alternative Medicine*, vol. 8, p. 44, 2008.

[78] R. Lyra-Neves and W. Telino-Júnior, *Aves da Fazenda Tamanduá*, Avis Brasilis, Brasília, Brazil, 2010.

[79] E. D. Neves and B. Viana, "Dispersão e predação de sementes de três espécies de Jatropha l, (Euphorbiaceae) da caatinga, semi-árido do Brasil," *Candombá—Revista Virtual*, vol. 4, no. 2, pp. 146–157, 2008.

[80] G. A. Pereira and S. Azevedo-Júnior, "Estudo comparativo entre as comunidades de aves de dois fragmentos florestais de caatinga em Pernambuco, Brasil," *Revista Brasileira de Ornitologia*, vol. 19, pp. 22–31, 2011.

[81] G. Pereira, "Avifauna associada a três lagoas temporárias no estado do Rio Grande do Norte, Brasil," *Atualidades Ornitológicas*, vol. 156, pp. 53–60, 2010.

[82] A. Roos, M. Nunes, E. Souza et al., "Avifauna da região do Lago de Sobradinho: composição, riqueza e biologia," *Ornithologia*, vol. 1, no. 2, pp. 135–160, 2006.

[83] J. Santos-Neto and M. Camandaroba, "Mapeamento dos sítios de alimentação da arara-azul-de-Lear Anodorhynchus leari (Bonaparte, 1856)," *Ornithologia*, vol. 3, no. 1, pp. 1–20, 2008.

[84] J. A. A. Barbosa, V. A. Nóbrega, and R. N. N. Alves, "Aspectos da caça e comércio ilegal da avifauna silvestre por populações tradicionais do semi-árido paraibano," *Revista de Biologia e Ciências da Terra*, vol. 10, no. 2, pp. 39–49, 2010.

[85] F. Las-Casas, S. Azevedo-Júnior, and M. Dias-Filho, "The community of hummingbirds (Aves: Trochilidae) and the assemblage of flowers in a Caatinga vegetation," *Brazilian Journal of Biology*, vol. 72, no. 1, pp. 51–58, 2012.

[86] F. Leal, A. Lopes, and I. Machado, "Polinização por beija-flores em uma área de Caatinga no Município de Floresta, Pernambuco, Nordeste do Brasil," *Revista Brasileira de Botânica*, vol. 29, no. 3, pp. 379–389, 2006.

[87] C. Machado, "Beija-flores (Aves: Trochilidae) e seus recursos florais em uma área de caatinga da Chapada Diamantina, Bahia, Brasil," *Zoologia*, vol. 26, no. 2, pp. 255–265, 2009.

[88] F. Olmos, "Aves ameaçadas, prioridades e políticas de conservação no Brasil," *Natureza & Conservação*, vol. 3, no. 1, pp. 21–42, 2005.

[89] E. Souza, W. Telino-Júnior, J. Nascimento et al., "Estima-tivas populacionais de avoantes Zenaida auriculata (Aves Columbidae, DesMurs, 1847) em colônias reprodutivas no Nordeste do Brasil," *Ornithologia*, vol. 2, no. 1, pp. 28–33, 2007.

[90] J. Pacheco, "As aves da Caatinga: uma análise histórica do conhecimento," in *Biodiversidade da Caatinga: Áreas e Ações Prioritárias para Conservação*, J. M. C. Silva, M. Tabarelli, M. T. Fonseca, and L. V. Lins, Eds., pp. 189–250, MMA, Brasília, Brazil, 2004.

[91] O. Pinto and E. Camargo, "Resultados ornitológicos de quatro recentes expedições do Departamento de Zoologia ao Nordeste do Brasil, com a descrição de seis novas subespécies," *Arquivos de Zoologia*, vol. 11, pp. 193–284, 1961.

[92] A. Coelho, "Lista de algumas espécies de aves do Nordeste do Brasil," *Notulae Scientia Biologicae*, vol. 1, pp. 1–7, 1978.

[93] P. Dekeyser, "Une contribution méconnue à l'ornithologie de l'état de la Paraíba," *Revista Nordestina de Biologia*, vol. 2, no. 1-2, pp. 127–145, 1979.

[94] H. Sick, "Notes on some brazilian birds," *Bulletin of the British Ornithologists' Club*, vol. 99, pp. 115–120, 1979.

[95] A. Fiuza, *A Avifauna da Caatinga do Estado da Bahia: Composição e Distribuição*, Texto e notas Adicionais de Deodato Souza Anor, Articulação Nordestina de Ornitologia, Bahia, Brazil, 1999.

[96] R. Lyra-Neves, W. Telino-Júnior, and J. Nascimento, *Aves da Fazenda Tamanduá, Santa Terezinha, Paraíba*, Dos Autores, Santa Terezinha, Brazil, 1999.

[97] J. Nascimento, *Aves da Floresta Nacional do Araripe, Ceará*, Instituto Brasileiro de Meio Ambiente e dos Recursos Naturais Renováveis, Brasília, Brazil, 1996.

[98] J. Nascimento and A. Schulz-Neto, *Aves da Estação Ecológica de Aiuaba, Ceará*, Instituto Brasileiro de Meio Ambiente e dos Recursos Naturais Renováveis, Brasília, Brazil, 1996.

[99] F. Olmos, "Birds of Serra da Capivara National Park, in the "caatinga" of northeastern Brazil," *Bird Conservation International*, vol. 3, no. 1, pp. 21–36, 1993.

[100] E. Willis and Y. Oniki, "Avifaunal transects across the open zones of northern Minas Gerais, Brazil," *Ararajuba*, vol. 2, pp. 41–58, 1991.

[101] A. Coelho, *Aves da Reserva Biológica de Serra Negra (Floresta, PE), Lista Preliminar*, Publicação avulsa, Recife, Brazil, 1987.

[102] S. Azevedo-Júnior, P. Antas, and J. Nascimento, "Censo da Zenaida auriculata Noronha fora da época de reprodução no nordeste," *Caderno Omega Universidade Federal de Pernam-buco*, vol. 2, pp. 157–168, 1989.

[103] S. Azevedo-Júnior and P. Antas, "Novas informações sobre a alimentação da Zenaida auriculata no Nordeste do Brasil," *Anais do IV Encontro Nacional de Anilhadores de Aves*, pp. 59–64, 1990.

[104] S. Azevedo-Júnior and P. Antas, "Observações sobre a reprodução da Zenaida auriculata no Nordeste do Brasil," *Anais do IV Encontro Nacional de Anilhadores de Aves*, pp. 65–72, 1990.

[105] G. Farias, "Aves do parque Nacional do Vale do Catimbau, Buique, Pernambuco," *Atualidades Ornitologicas*, vol. 147, pp. 36–39, 2009.

[106] G. D. Farias, "Avifauna em quatro áreas de caatinga strictu senso no centro-oeste de Pernambuco, Brasil," *Revista Brasileira de Ornitologia*, vol. 15, pp. 53–60, 2007.

[107] P. Lima, S. Santos, and R. Lima, "2003Levantamento e Anil-hamento da Ornitofauna na Patria da Arara-Azul-de-Lear (Anodorhynchus leari, Bonaparte, 1856): um complemento ao Levantamento realizado por H Sick LP Gonzaga e DM Teixeira," *Atualidades Ornitologicas*, vol. 112, p. 11, 1987.

[108] J. Nascimento, "Estudo comparativo da avifauna em duas Estações Ecologicas da Caatinga: aiuaba e Serido," *Mellopsit-tacus*, vol. 3, pp. 12–35, 2000.

[109] J. Nascimento, I. Nascimento, and S. Azevedo-Junior, "Aves da Chapada do Araripe (Brasil): biologia e Conservação," *Ararajuba*, vol. 8, pp. 115–125, 2000.

[110] S. Dantas, G. Pereira, G. Farias et al., "Registros relevantes de aves para o Estado de Pernambuco, Brasil," *Revista Brasileira de Ornitologia*, vol. 15, pp. 113–115, 2007.

[111] G. Farias and G. Pereira, "Aves de Pernambuco: o estado atual do conhecimento ornitologico," *Biotemas*, vol. 22, pp. 1–10, 2009.

[112] G. B. Farias, W. A. G. Silva, and C. G. Albano, "Diversidade de aves em áreas prioritárias para conservação da Caatinga," in *Análise das Variações da Biodiversidade do Bioma Caatinga: Suporte e Estratégias Regionais de Conservação*, F. S. Araújo, M. J. Rodal, and M. R. V. Barbosa, Eds., pp. 203–226, MMA, Brasília, BRAZIL, 2005.

[113] F. Las-Casas and S. Azevedo-Junior, "Ocorrencia de Knipole-gus nigerrimus (Vieillot, 1818) (Aves, Tyrannidae) no Dis-trito do Para, Santa Cruz do Capibaribe, Pernambuco, Brasil," *Ornithologia*, vol. 3, pp. 18–20, 2008.

[114] G. Pereira, A. Whittaker, B. Whitney et al., "Novos registros relevantes de aves para o estado de Pernambuco, Brasil, incluindo novos registros para o Estado," *Revista Brasileira de Ornitologia*, vol. 16, pp. 47–53, 2008.

[115] E. Souza, M. Nunes, I. Simao et al., "Ampliacao de área de ocorrencia do Beija-flor-de-gravatinha-vermelha Augastes lumachella (Lesson, 1838) (Trochilidae)," *Ornithologia*, vol. 3, pp. 145–148, 2009.

[116] W. Telino-Junior, R. Lyra-Neves, S. Azevedo-Junior, and M. Larrazabal, "First occurrence of the Saltator atricollis Vieillot 1817 (Aves, Cardinalidae) in the state of Pernambuco, Brazil," *Ornithologia*, vol. 3, pp. 34–37, 2008.

[117] J. Pacheco and C. Bauer, "As aves da caatinga—apreciacao historica do processo de conhecimento, In:," in *Avaliacao*

e identificacao de ações Prioritárias para a Conservação, utilizacao sustentavel e reparticao de beneficios da biodiversidade do bioma Caatinga, Documento tematico, Seminario Biodiversidade da Caatinga, Petrolina, Brazil, 2000.

[118] A. Souto and C. Hazin, "Diversidade animal e desertificacao no semi-árido nordestino," *Biologica Brasilica*, vol. 6, pp. 39–50, 1995.

[119] M. R. Willig and M. A. Mares, "Mammals from the Caatinga: an updated list and summary of recent research," *Revista Brasileira de Biologia*, vol. 49, no. 2, pp. 361–367, 1989.

[120] C. Castelletti, A. Santos, M. Tabarelli, and J. Silva, "Quanto ainda resta da Caatinga? Uma estimativa preliminar," in *Ecologia e Conservação da Caatinga Editora Universitaria*, I. R. Leal, M. Tabarelli, and J. M. C. Silva, Eds., pp. 719–734, Universidade Federal de Pernambuco, Recife, Brazil, 2003.

[121] M. A. Mares, M. R. Willig, and T. E. Lacher, "The Brazilian caatinga in South American zoogeography: tropical mammals in a dry region," *Journal of Biogeography*, vol. 12, no. 1, pp. 57–69, 1985.

[122] M. A. Mares, M. Willig, T. Streilen, and J. T. Lacher, "The mammals of northeastern Brazil: a preliminary assessment," *Annals of Carnegie Museum*, vol. 50, pp. 81–137, 1981.

[123] K. Streilein, "the ecology of small mammals in the semiarid Brazilian Caatinga. IV. Habitat selection," *Annals of Carnegie Museum*, vol. 51, pp. 331–343, 1982.

[124] M. Willig, "Bat community structure in South America: a tenacious chimera," *Revista Chilena de Historia Natural*, vol. 59, pp. 151–168, 1986.

[125] G. A. B. Fonseca, G. Herrman, Y. L. R. Leite, R. A. M. Mittermeier, A. B. Rylands, and J. L. Patton, *Lista Anotada dos Mamíferos do Brasil. Occasional Papers in Conservation Biology*, vol. 4, Conservation International, Washington, DC, USA, 1996.

[126] J. Oliveira, "Diversidade de mamíferos e o estabelecimento de áreas prioritárias para conservação da Caatinga," in *Biodiversidade da Caatinga: áreas e ações prioritárias para a conservação*, J. M. C. da Silva, M. Tabarelli, M. T. da Fonseca, and L. V. Lins, Eds., pp. 263–282, Ministerio do Meio Ambiente, Brasília, Brazil, 2004.

[127] R. Gregorin and A. D. Ditchfield, "New genus and species of nectar-feeding bat in the tribe lonchophyllini (Phyllostomidae: Glossophaginae) from northeastern Brazil," *Journal of Mammalogy*, vol. 86, no. 2, pp. 403–414, 2005.

[128] R. Gregorin, A. Carmignotto, and A. Percequillo, "Quiropteros do Parque Nacional da Serra das Confusoes, Piaui," *Chiroptera Neotropical*, vol. 14, pp. 366–383, 2008.

[129] J. Feijo, P. Araújo, M. Fracasso, and K. Santos, "New records of three bat species for the Caatinga of the state of Paraiba, northeastern Brazil," *Chiroptera Neotropical*, vol. 16, pp. 723–727, 2010.

[130] R. Moratelli, A. L. Peracchi, D. Dias, and J. A. de Oliveira, "Geographic variation in South American populations of Myotis nigricans (Schinz, 1821) (Chiroptera, Vespertilionidae), with the description of two new species," *Mammalian Biology*, vol. 76, no. 5, pp. 592–607, 2011.

[131] G. R. Canale, C. E. Guidorizzi, M. C. M. Kierulff, and C. A. F. R. Gatto, "First record of tool use by wild populations of the yellow-breasted capuchin monkey (Cebus xanthosternos) and new records for the bearded capuchin (Cebus libidinosus)," *American Journal of Primatology*, vol. 71, no. 5, pp. 366–372, 2009.

[132] R. G. Ferreira, L. Jerusalinsky, T. C. F. Silva et al., "On the occurrence of Cebus flavius (Schreber 1774) in the Caatinga, and the use of semi-arid environments by Cebus species in the Brazilian state of Rio Grande do Norte," *Primates*, vol. 50, no. 4, pp. 357–362, 2009.

[133] C. Tribe, "A new species of Rhipidomys (Rodentia, Muroidea) from North-Eastern Brazil," *Archivos do Museu Nacional do Rio de Janeiro*, vol. 63, no. 1, pp. 131–146, 2005.

[134] E. Bernard, L. M. S. Aguiar, and R. B. Machado, "Discovering the Brazilian bat fauna: A task for two centuries?" *Mammal Review*, vol. 41, no. 1, pp. 23–39, 2011.

[135] P. Meserve, "Zoogeography of South America," in *Physical geography of South America*, A. Orme, Ed., pp. 112–132, Oxford University Press, Oxford, UK, 2007.

[136] S. Ribeiro, F. Ferreira, S. Brito et al., "The squamata fauna of the Chapada do Araripe, Northeastern," *Cadernos de Cultura e Ciência*, vol. 1, no. 1, pp. 67–76, 2008.

[137] C. A. Klink and R. B. Machado, "Conservation of the Brazilian Cerrado," *Conservation Biology*, vol. 19, no. 3, pp. 707–713, 2005.

[138] J. Marinho-Filho, F. Rodrigues, and K. Juarez, "The Cerrado mammals: diversity, ecology, and natural history," in *The Cerrados of Brazil: Ecology and Natural History of a Neotropical Savanna*, P. S. Oliveira and R. J. Marquis, Eds., pp. 266–284, Columbia University Press, New York, NY, USA, 2002.

[139] J. Oliveira, P. Goncalves, and C. Bonvicino, "Mamiferos da caatinga," in *Ecologia e Conservação da Caatinga*, I. R. Leal, M. Tabarelli, and J. M. C. Silva, Eds., pp. 275–336, Universitaria da UFPE, Recife, Brazil, 1st edition, 2003.

[140] L. Mendes, P. Rocha, M. Ribeiro, S. Perry, and E. Oliveira, "Differences in ingestive balance of two populations of Neotropical Thrichomys apereoides (Rodentia, Echimyidae)," *Comparative Biochemistry and Physiology A*, vol. 138, pp. 327–332, 2004.

[141] M. Ribeiro, P. Rocha, L. Mendes, S. Perry, and S. Oliveira, "Physiological effects of the short-term water deprivation in the black-footed pygmy rice rat (Oligoryzomys nigripes) and the South American water rat (Nectomys quamipes) within a phylogenetic context," *Canadian Journal of Zoology*, vol. 82, no. 1, pp. 1–10, 2004.

[142] R. Freitas, R. P. da, and P. Simoes-Lopes, "Habitat structure and small mammals abundances in one semiarid landscape in the Brazilian Caatinga," *Revista Brasileira de Zoologia*, vol. 22, no. 1, pp. 119–129, 2005.

[143] P. L. B. Da Rocha, "Proechimys yonenagae, a new species of spiny rat (Rodentia: Echimyidae) from fossil sand dunes in the Brazilian Caatinga," *Mammalia*, vol. 59, no. 4, pp. 537–549, 1995.

[144] J. W. A. Santos and E. A. Lacey, "Burrow sharing in the desert-adapted torch-tail spiny rat, Trinomys yonenagae," *Journal of Mammalogy*, vol. 92, no. 1, pp. 3–11, 2011.

[145] A. Moura, "Primate group size and abundance in the Caatinga dry forest, Northeastern Brazil," *International Journal of Primatology*, vol. 28, no. 6, pp. 1279–1297, 2007.

[146] L. Andrade, J. Fabricante, and E. Araújo, "Estudo de Fitossociologia em Vegetação de Caatinga," *Fitossociologia no Brasil—Métodos e estudo de casos*, Editora UFV, Viçosa, Brazil, vol. 1, pp. 339–371, 1st edition, 2011.

[147] E. Araújo, C. Castro, and U. Albuquerque, "Dynamics of Brazilian caatinga a review concerning the plants, environment and people," *Functional Ecosystems and Communities*, vol. 1, no. 1, pp. 15–29, 2007.

[148] E. Araújo, F. Martins, and F. Santos, "Establishment and death of two dry tropical forest woody species in dry and rainy seasons in northeastern Brazil," in *Estresses Ambientais—Danos e Beneficios em Plantas*, R. J. M. Nogueira, E. L. Araújo, L. G. Willadino Uidede, and M. T.

Cavalcante, Eds., vol. 1, pp. 76–91, MXM Grafica e Editora, Recife, Brazil, 2005.

[149] A. Giulietti, A. Paula, D. Barbosa et al., "vegetação: áreas e ações Prioritárias para a Conservação da Caatinga," in *Biodiversidade da Caatinga: Áreas e ações Prioritárias para a Conservação*, M. Silva, M. T. Tabarelli, M. Fonseca, and J. M. C. Lins, Eds., vol. 1, pp. 113–131, MMA-UFPE-Conservation International-Biodiversitas-Embrapa semi-árido, Brasília, Brazil, 2004.

[150] M. Rodal, "Vegetação do semi-árido do nordeste do Brasil," in *Biodiversidade, Conservação e Uso Sustentado da Flora do Brasi*, E. D. L. Araújo, A. D. N. Moura, Barreto EVDSS, Gestinari LMDS, and Carneiro JDMT, Eds., vol. 1, Editora da Universidade Federal Rural de Pernambuco, Recife, Brazil, 1st edition, 2002.

[151] M. Rodal and E. Sampaio, "A vegetação do bioma Caatinga," in *vegetação e Flora da Caatinga*, E. V. S. B. Sampaio, A. M. Giulietti, J. Virginio, and C. F. L. Gamarra-Rojas, Eds., pp. 11–24, Associação Plantas do Nordeste, Recife, Brazil, 2002.

[152] E. Sampaio, "Overview of the Brasílian caatinga," in *Seasonally Dry Tropical Forest*, S. H. Bullock, H. A. Mooney, and E. Medina, Eds., pp. 35–63, Cambridge University Press, Cambridge, UK, 1995.

[153] J. Cruz, E. Araújo, E. Ferraz, and S. Silva, "Plantas trepadeiras da caatinga: aspectos da distribuição e usos baseados no checklist dos herbarios de Pernambuco-Brasil," in *Biodiversidade, Potencial econômico e Processos Eco-Fisiologicos em Ecossistemas Nordestinos*, U. P. Albuquerque, A. N. Moura, and E. L. Araújo, Eds., vol. 2, pp. 217–251, NUPEEA, Brasília, Brazil, 2010.

[154] A. Giullieti, A. Conceição, and L. Queiroz, *Diversidade e caracterização das fanerogamas do semi-árido brasileiro*, Associação Plantas do Nordeste, Recife, Brazil, 2006.

[155] A. L. S. Almeida, U. P. Albuquerque, and C. C. Castro, "Reproductive biology of Spondias tuberosa Arruda (Anacardiaceae), an endemic fructiferous species of the caatinga (dry forest), under different management conditions in northeastern Brazil," *Journal of Arid Environments*, vol. 75, no. 4, pp. 330–337, 2011.

[156] I. Amorim, E. Sampaio, and E. Araújo, "Fenologia de espécies lenhosas da caatinga do Seridó," *Revista Árvore*, vol. 33, no. 3, pp. 491–499, 2009.

[157] A. Virgínia de Lima Leite and I. C. Machado, "Reproductive biology of woody species in Caatinga, a dry forest of northeastern Brazil," *Journal of Arid Environments*, vol. 74, no. 11, pp. 1374–1380, 2010.

[158] A. L. A. Lima and M. J. N. Rodal, "Phenology and wood density of plants growing in the semi-arid region of northeastern Brazil," *Journal of Arid Environments*, vol. 74, no. 11, pp. 1363–1373, 2010.

[159] I. Machado, "Biologia floral e fenologia," in *Pesquisa Botânica Nordestina: Progresso e Perspectivas*, E. V. S. B. Sampaio, S. J. Mayo, and M. R. Barbosa, Eds., pp. 161–171, Sociedade Botânica do Brasil, Recife, Brazil, 1996.

[160] I. Machado and A. Lopes, "A polinização em ecossistemas de Pernambuco. uma revisão do estado atual do conhecimento," in *Diagnóstico da Biodiversidade do estado de Pernambuco*, J. M. Silva and M. Tabarelli, Eds., pp. 583–595, SECTMA, Recife, Brazil, 2002.

[161] I. Machado and A. Lopes, "polinização em espécies da caatinga," in *Desafios da Botânica Brasileira no Novo Milenio. Inventario, Sistematização e Conservação da Diversidade Vegetal*, E. A. G. Jardin, M. N. C. Bastos, and J. U. M. Santos, Eds.,

pp. 105–107, Sociedade Brasileira de Botânica, Belém, Brazil, 2003.

[162] R. Borchert, "Soil and stem water storage determine phenology and distribution of tropical dry forest trees," *Ecology*, vol. 75, no. 5, pp. 1437–1449, 1994.

[163] C. Kushwaha, S. Tripathi, B. Tripathi, and K. Singh, "Patterns of tree phenological diversity in dry tropics," *Acta Ecologica Sinica*, vol. 31, no. 4, pp. 179–185, 2011.

[164] C. P. Kushwaha and K. P. Singh, "Diversity of leaf phenology in a tropical deciduous forest in India," *Journal of Tropical Ecology*, vol. 21, no. 1, pp. 47–56, 2005.

[165] L. Miles, A. C. Newton, R. S. DeFries et al., "A global overview of the conservation status of tropical dry forests," *Journal of Biogeography*, vol. 33, no. 3, pp. 491–505, 2006.

[166] I. Leal, J. Silva, M. Tabarelli, and J. Lacher, "Mudando o curso da Conservação da biodiversidade na Caatinga do Nordeste do Brasil," *Megadiversidade*, vol. 1, pp. 140–146, 2005.

[167] B. Rathcke and E. P. Lacey, "Phenological patterns of terrestrial plants," *Annual review of ecology and systematics. Vol. 16*, pp. 179–214, 1985.

[168] M. Quesada, G. A. Sanchez-Azofeifa, M. Alvarez-Añorve et al., "Succession and management of tropical dry forests in the Americas: Review and new perspectives," *Forest Ecology and Management*, vol. 258, no. 6, pp. 1014–1024, 2009.

[169] I. C. Machado and A. V. Lopes, "Floral traits and pollination systems in the Caatinga, a Brazilian tropical dry forest," *Annals of Botany*, vol. 94, no. 3, pp. 365–376, 2004.

[170] D. Barbosa, J. Alves, S. Prazeres, and A. Paiva, "Dados fenológicos de 10 espécies arbáreas de uma área de Caatinga (Alagoinha-PE)," *Acta Botanica Brasilica*, vol. 3, no. 2, pp. 109–117, 1989.

[171] I. C. S. Machado, L. M. Barros, and E. V. S. B. Sampaio, "Phenology of caatinga species at Serra Talhada, PE, Northeastern Brazil," *Biotropica*, vol. 29, no. 1, pp. 57–68, 1997.

[172] R. Pereira, F. J. Araújo, R. Lima, F. Paulino, and A. Lima, "Estudo fenológico de algumas espécies lenhosas e herbaceas da caatinga," *Ciência Agronômica*, vol. 20, no. 1-2, pp. 11–20, 1989.

[173] E. Neves, L. Funch, and B. Viana, "Comportamento fenológico de três espécies de Jatropha (Euphorbiaceae) da Caatinga, semi-árido do Brasil," *Revista Brasileira de Botânica*, vol. 33, no. 1, pp. 155–166, 2010.

[174] Y. Nunes, M. Fagundes, H. Almeida, and M. Veloso, "Aspectos ecologicos da aroeira (Myracrodruon urundeuva Allemão-Anacardiaceae): fenologia e germinação de sementes," *Revista Árvore*, vol. 32, no. 2, pp. 233–243, 2008.

[175] E. Lima, E. Araújo, E. Sampaio, E. Ferraz, and K. Silva, "Fenologia e Dinâmica de duas populações herbaceas da caatinga," *Revista de Geografia*, vol. 24, no. 1, pp. 120–136, 2007.

[176] L. D. A. de Araújo, Z. G. M. Quirino, and I. C. Machado, "Fenologia reprodutiva, biologia floral e polinização de Allamanda blanchetii, uma apocynaceae endêmica da Caatinga," *Revista Brasileira de Botânica*, vol. 34, no. 2, pp. 211–222, 2011.

[177] E. Bezerra, A. Lopes, and I. Machado, "Biologia reprodutiva de Byrsonima gardnerana A Juss, (Malpighiaceae) e interações com abelhas centris (Centridini) no nordeste do Brasil," *Revista Brasileira de Botânica*, vol. 32, pp. 95–108, 2009.

[178] R. B. S. Fonseca, L. S. Funch, and E. L. Borba, "Reproductive phenology of Melocactus (Cactaceae) species from Chapada Diamantina, Bahia, Brazil," *Revista Brasileira de Botânica*, vol. 31, no. 2, pp. 237–244, 2008.

[179] L. M. S. Griz and I. C. S. Machado, "Fruiting phenology and seed dispersal syndromes in caatinga, a tropical dry forest in the northeast of Brazil," *Journal of Tropical Ecology*, vol. 17, no. 2, pp. 303–321, 2001.

[180] T. Nadia, I. Machado, and A. Lopes, "Polinização de Spondias tuberosa Arruda (Anacardiaceae) e analise da partilha de polinizadores com Ziziphus joazeiro Mart-(Rhamnaceae), espécies frutiferas e endemicas da caatinga," *Revista Brasileira de Botânica*, vol. 30, no. 1, pp. 89–100, 2007.

[181] D. Barbosa, M. Barbosa, and L. Lima, "Fenologia de espécies lenhosas da Caatinga," in *Ecologia e Conservação da Caatinga*, I. R. Leal, M. Tabarelli, and J. M. C. Silva, Eds., pp. 657–693, Universidade Federal de Pernambuco (UFPE), Recife, Brazil, 2003.

[182] A. Lima, *Tipos funcionais fenológicos em espécies lenhosas da caatinga, Nordeste do Brasil [M.S. thesis]*, Universidade Federal de Pernambuco (UFPE), Recife, Brazil, 2010.

[183] R. Borchert, S. Renner, Z. Calle, D. Vavarrete, and A. Tye, "Photoperiodic induction of synchronous flowering near the Equador," *Nature*, vol. 433, pp. 627–629, 2005.

[184] R. Borchert and G. Rivera, "Photoperiodic control of seasonal development and dormancy in tropical stem-succulent trees," *Tree Physiology*, vol. 21, no. 4, pp. 213–221, 2001.

[185] Z. Calle, B. O. Schlumpberger, L. Piedrahita et al., "Seasonal variation in daily insolation induces synchronous bud break and flowering in the tropics," *Trees-Structure and Function*, vol. 24, no. 5, pp. 865–877, 2010.

[186] C. Parmesan, "Influences of species, latitudes and methodologies on estimates of phenological response to global warming," *Global Change Biology*, vol. 13, no. 9, pp. 1860–1872, 2007.

[187] G. Rivera, S. Elliott, L. Caldas, G. Nicolssi, and V. Coradin, "Increasing day-length induces spring flushing of tropical dry forest trees in the absence of rain," *Trends in Ecology and Evolution*, vol. 16, no. 7, pp. 445–456, 2002.

[188] M. Westoby, "A leaf-height-seed (LHS) plant ecology strategy scheme," *Plant and Soil*, vol. 199, no. 2, pp. 213–227, 1998.

[189] P. B. Reich, M. B. Walters, and D. S. Ellsworth, "Leaf life-span in relation to leaf, plant, and stand characteristics among diverse ecosystems," *Ecological Monographs*, vol. 62, no. 3, pp. 365–392, 1992.

[190] P. B. Reich, I. J. Wright, J. Cavender-Bares et al., "The evolution of plant functional variation: Traits, spectra, and strategies," *International Journal of Plant Sciences*, vol. 164, no. 3, pp. S143–S164, 2003.

[191] H. Shugart, "Plant and ecosystem functional types," in *Plant Functional Types: Their Relevance to Ecosystem Properties and Global Change*, T. M. Smith, H. H. Shugart, and F. I. Woodward, Eds., International Geosphere-Biosphere Programme Book Series, pp. 20–43, Cambridge University Press, Cambridge, UK, 1997.

[192] F. Woodward and C. Kelly, "Plant functional types: towards a definition by environmental constraints," in *Plant Functional Types: Their Relevance to Ecosystem Properties and Global Change*, T. M. Smith, H. H. Shugart, and F. I. Woodward, Eds., International Geosphere-Biosphere Programme Book Series, pp. 47–65, Cambridge University Press, Cambridge, UK, 1997.

[193] V. Pillar, "How can we define optimal plant functional types?" in *Proceedings of the 43th IAVS Symposium*, pp. 352–356, 2000.

[194] L. E. Newstrom, G. W. Frankie, and H. G. Baker, "A new classification for plant phenology based on flowering patterns in lowland tropical rain forest trees at La Selva, Costa Rica," *Biotropica*, vol. 26, no. 2, pp. 141–159, 1994.

[195] R. Aerts, J. H. C. Cornelissen, and E. Dorrepaal, "Plant performance in a warmer world: General responses of plants from cold, northern biomes and the importance of winter and spring events," *Plant Ecology*, vol. 182, no. 1-2, pp. 65–77, 2006.

[196] D. Barbosa, "Crescimento e estabelecimento de plantas," in *Pesquisa Botanica Nordestina—Progresso e Perspectivas*, E. V. S. B. Sampaio, S. J. Mayo, and M. R. V. Barbosa, Eds., pp. 173–177, Sociedade Botanica do Brasil, Recife, Brazil, 1996.

[197] D. Barbosa, "Estrategias de germinação e crescimento de espécies lenhosas da caatinga com germinação rapida," in *Biodiversidade, Conservação e Uso Sustentavel da flora do Brasil*, E. L. Araújo, A. N. Moura, E. V. S. B. Sampaio, L. M. S. Gestinari, and J. M. T. Carneiro, Eds., pp. 172–174, Imprensa Universitaria, Recife, Brazil, 2002.

[198] S. Prazeres, "Germinação e propagacao vegetativa," in *Pesquisa Botanica nordestina: Progresso e Perspectivas*, E. V. S. B. Sampaio, S. J. Mayo, and M. R. Barbosa, Eds., pp. 179–189, Sociedade Botanica do Brasil, Recife, 1996.

[199] E. Araújo, V. Barretto, F. Leite, V. Lima, and N. Canuto, "Germinação e protocolos de quebra de dormencia de plantas do semi-árido," in *Recursos Geneticos do semi-árido*, A. M. Giulietti and L. P. Queiroz, Eds., vol. 5, pp. 73–110, APNE/ Instituto do Milenio do semi-árido, Recife, Brazil, 2006.

[200] E. Araújo, R. Nogueira, S. Silva et al., "Ecofisiologia de plantas da caatinga e implicações na Dinâmica das populações e do ecossistema," in *Biodiversidade, Potencial Economico e Processos Eco-Fisiologicos em Ecossistemas Nordestinos*, A. N. Moura, E. L. Araújo, and U. P. Albuquerque, Eds., vol. 1, pp. 329–361, Comunigraf, Recife, Brazil, 1st edition, 2008.

[201] F. Araújo, M. Rodal, M. Barbosa, and F. Martins, "Repartição da flora lenhosa no dominio da caatinga," in *Analise das Variações da Biodiversidade do Bioma Caatinga*, F. S. de Araújo, M. J. Nogueira, and R. M. R. de Barbosa, Eds., vol. 1, pp. 15–33, Ministirio do Maio Ambiente, Brasília, 1st edition, 2005.

[202] R. Nogueira, "Mecanismos de superação de estresse hidrico em plantas da caatinga," in *Biodiversidade, Conservação e Uso Sustentavel da Flora do Brasil*, E. L. Araújo, A. N. Moura, E. V. S. B. Sampaio, L. M. S. Gestinari, and J. M. T. Carneiro, Eds., vol. 1, pp. 162–164, Editora Universitaria UFPE, Recife, Brazil, 1st edition, 2002.

[203] I. Silva and D. Barbosa, "Crescimento e sobrevivencia de Anadenanthera macrocarpa (Benth) Brenan (Leguminosae), em uma área de caatinga, Alagoinha, PE," *Acta Botanica Brasilica*, vol. 14, no. 3, pp. 251–261, 2000.

[204] D. Barbosa, P. Silva, and M. Barbosa, "Tipos de frutos e síndromes de dispersão de espécies lenhosas da caatinga de Pernambuco," in *Diagnóstico da Biodiversidade do Estado de Pernambuco*, J. M. Silva and M. Tabarelli, Eds., pp. 609–621, SECTMA, Recife, Brazil, 2002.

[205] R. Costa and F. Araújo, "Densidade, germinação e flora do banco de sementes no solo, no final da estação seca, em uma área de caatinga, Quixada, CE," *Acta Botanica Brasilica*, vol. 17, no. 2, pp. 259–264, 2003.

[206] L. Griz, I. Machado, and M. Tabarelli, "Ecologia de dispersao de sementes: progressos e perspectivas," in *Diagnostico da Biodiversidade do Estado de Pernambuco*, J. M. Silva and M. Tabarelli, Eds., pp. 596–608, SECTMA, Recife, Brazil, 2002.

[207] A. Vicente, A. Santos, and M. Tabarelli, "Variação no modo de dispersão de espécies lenhosas em um gradiente de

precipitação entre floresta seca e amida no nordeste do Brasil," in *Ecologia e Conservação da Caatinga*, I. R. Leal, M. Tabarelli, and J. M. C. Silva, Eds., Universidade Federal de Pernambuco (UFPE), Recife, Brazil, 2003.

[208] J. Andrade, J. Santos, E. Lima et al., "Estudo populacional de Panicum trichoides Swart. (Poaceae) em uma área de caatinga em Caruaru, Pernambuco," *Revista Brasileira de Biociências*, vol. 5, pp. 858–860, 2007.

[209] E. Araújo, K. Silva, E. Ferraz, E. Sampaio, and S. Silva, "Diversidade de herbáceas em microhabitats rochoso, plano e ciliar em uma área de caatinga, Caruaru, PE, Brasil," *Acta Botanica Brasilica*, vol. 19, pp. 285–294, 2005.

[210] J. Santos, K. Silva, E. Lima et al., "Dinâmica de duas populações herbaceas de uma área de caatinga, Pernambuco, Brasil," *Revista de Geografia*, vol. 26, no. 2, pp. 142–160, 2009.

[211] E. Araújo, "Estresses abioticos e bioticos como forças modeladoras da Dinâmica de populações vegetais da caatinga," in *Estresses Ambientais-Danos ou Benefecifios em Plantas*, R. J. M. C. Nogueira, E. L. Araújo, L. G. Willadino, and U. M. T. Cavalcante, Eds., vol. 1, pp. 50–64, MXM Grafica e Editora, Recife, Brazil, 1st edition, 2005.

[212] J. Santos, J. Andrade, E. Lima, K. Silva, and E. Araújo, "Dinâmica populacional de uma espécie herbácea em uma área de floresta tropical seca no Nordeste do Brasil," *Revista Brasileira de Biociências*, vol. 5, pp. 855–857, 2007.

[213] E. Araújo and E. Ferraz, "Processos ecológicos mantenedores da diversidade vegetal na caatinga: estado atual do conhecimento," in *Ecossistemas Brasileiros-Manejo e Conservação*, V. Claudino-Sales, Ed., vol. 1, pp. 115–128, Expressão Gráfica, Fortaleza, Brazil, 1st edition, 2003.

[214] A. D. S. Freitas, E. V. S. B. Sampaio, C. E. R. S. Santos, and A. R. Fernandes, "Biological nitrogen fixation in tree legumes of the Brazilian semi-arid caatinga," *Journal of Arid Environments*, vol. 74, no. 3, pp. 344–349, 2010.

[215] I. Salcedo and E. Sampaio, "Dinâmica da matéria orgânica no bioma caatinga," in *Fundamentos da Matéria Orgânica do Solo*, G. A. Santos, L. S. Silva, L. P. Canellas, and F. A. O. Camargo, Eds., pp. 419–441, Metropole, Porto Alegre, Brazil, 2nd edition, 2008.

[216] P. Santos, J. Souza, J. Santos, D. Santos, and E. Araújo, "Diferenças Sazonais no aporte de Serrapilheira em uma área de Caatinga em Pernambuco," *Revista Caatinga*, vol. 24, pp. 94–101, 2011.

[217] H. Tiessen, C. Feller, E. V. S. B. Sampaio, and P. Garin, "Carbon sequestration and turnover in semiarid savannas and dry forest," *Climatic Change*, vol. 40, no. 1, pp. 105–117, 1998.

[218] H. Castelletti, J. Silva, M. Tabarelli, and A. Santos, "Quando ainda resta da Caatinga? uma estimativa preliminar," in *Biodiversidade da Caatinga: áreas e ações Prioritárias para a Conservação*, J. M. C. da Silva, M. Tabarelli, M. T. Fonseca, and L. V. Lins, Eds., pp. 91–100, Ministerio do Meio Ambiente e Universidade Federal de Pernambuco, Brasília, Brazil, 2004.

[219] J. Silva, M. Tabarelli, and M. Fonseca, "Áreas e ações Prioritárias para a Conservação da biodiversidade da Caatinga," in *Biodiversidade da Caatinga: Áreas e ações Prioritárias para a Conservação*, J. M. C. Silva, M. Tabarelli, M. T. Fonseca, and L. V. Lins, Eds., pp. 349–374, Ministerio do Meio Ambiente e Universidade Federal de Pernambuco, Brasília, Brazil, 2004.

[220] J. Figueroa, F. Pareyn, E. Araújo, C. Silva, and V. Santos, "Effects of cutting regimes in the dry and wet season on survival and sprouting of woody species from the semi-arid caatinga of northeasth Brazil," *Forest Ecology and Management*, vol. 229, no. 1–3, pp. 294–303, 2006.

[221] M. Pagano and F. Araújo, "Semi-arid vegetation in Brazil: biodiversity, impacts and management," in *Semi-arid environments: agriculture, water supply and vegetation*, K. M. Degenovine, Ed., pp. 99–114, Nova Publishers, New York, NY, USA, 2011.

[222] E. Sampaio, E. Araújo, I. Salcedo, and H. Tiessen, "Regeneração da vegetação da caatinga apos corte e queima, em Serra Talhada, PE," *Pesquisa Agropecuária Brasileira*, vol. 33, no. 5, pp. 621–632, 1998.

[223] E. Sampaio, "Usos das plantas da caatinga," in *Virginio J, Rojas CFLG (Org.) vegetação e Flora da Caatinga*, E. V. S. B. Sampaio and A. M. Giulietti, Eds., vol. 1, pp. 49–90, APNE-CNIP, Recife, Brazil, 2002.

[224] E. Sampaio and C. Rojas, "Uso das plantas em pernambuco," in *Diagnostico da Biodiversidade de Pernambuco, Secretaria de Ciênciga*, M. Tabarelli and J. M. C. Silva, Eds., vol. 2, pp. 633–660, Tecnologia e Meio Ambiente, Recife, Brazil, 2002.

[225] M. Pereira, L. Andrade, J. Costa, and J. Dias, "2001 Regeneração natural em um remanescente de caatinga sob diferentes niveis de perturbação, no agreste paraibano," *Acta Botanica Brasilica*, vol. 15, no. 3, pp. 413–426.

[226] R. R. N. Alves, L. E. T. Mendonça, M. V. A. Confessor, W. L. S. Vieira, and L. C. S. Lopez, "Hunting strategies used in the semi-arid region of northeastern Brazil," *Journal of Ethnobiology and Ethnomedicine*, vol. 5, p. 12, 2009.

[227] R. N. N. Alves, J. A. A. Barbosa, S. L. D. X. Santos, W. M. S. Souto, and R. R. D. Barboza, "Animal-based remedies as complementary medicines in the semi-arid region of northeastern Brazil," *Evidence-based Complementary and Alternative Medicine*, vol. 2011, pp. 1–15, 2011.

[228] R. R. N. Alves, L. E. T. Mendonça, M. V. A. Confessor, W. L. S. Vieira, and K. S. Vieira, "Caça no semi-árido paraibano-uma abordagem etnozoologica," in *A Etnozoologia no Brasil-Importancia, Status Atual e Perspectivas*, F. N. Alves, W. M. S. Souto, and J. S. Mouro, Eds., vol. 7, pp. 347–378, NUPEEA, Recife, Brazil, 1st edition, 2010.

[229] R. R. N. Alves, E. E. G. Nogueira, H. F. P. Araújo, and S. E. Brooks, "Bird-keeping in the Caatinga, NE Brazil," *Human Ecology*, vol. 38, no. 1, pp. 147–156, 2010.

[230] R. R. N. Alves, K. S. Vieira, G. G. Santana et al., "A review on human attitudes towards reptiles in Brazil," *Environmental Monitoring and Assessment*. In press.

[231] J. A. A. Barbosa, V. A. Nobrega, and R. R. N. Alves, "Hunting practices in the semiarid region of Brazil," *Indian Journal of Traditional Knowledge*, vol. 10, no. 3, pp. 486–490, 2011.

[232] D. M. M. Bezerra, H. F. P. Araújo, and R. R. N. Alves, "The use of wild birds by rural communities in the semi-arid region of Rio Grande do Norte State, Brazil," *Bioremediation, Biodiversity & Bioavailability*, vol. 5, pp. 117–120, 2011.

[233] D. Santos-Fita, E. Costa-Neto, and A. Schiavetti, "'Offensive' snakes: cultural beliefs and practices related to snakebites in a Brazilian rural settlement," *Journal of Ethnobiology and Ethnomedicine*, vol. 6, no. 1, pp. 1–13, 2010.

[234] R. R. D Barboza, J. S. Mourão, W. M. S. Souto, and R. R. N. Alves, "Knowledge and strategies of armadillo (Dasypus novemcinctus L, 1758 and Euphractus sexcinctus L, 1758) hunters in the "Sertão Paraibano", paraiba State, NE Brazil," *Bioremediation, Biodiversity and Bioavailability*, vol. 5, pp. 1–7, 2011.

[235] P. Dantas-Aguiar, R. Barreto, D. Santos-Fita, and E. Santos, "Hunting activities and wild fauna use: a profile of queixo d'antas community, campo formoso, Bahia, Brazil," *Bioremediation, Biodiversity and Bioavailability*, vol. 5, pp. 1–10, 2011.

[236] H. Fernandes-Ferreira, S. V. Mendonça, C. Albano, F. S. Ferreira, and R. R. N. Alves, "Hunting, use and conservation of birds in Northeast Brazil," *Biodiversity and Conservation*, vol. 21, pp. 221–244, 2012.

[237] M. S. P. Rocha, P. C. M. Cavalcanti, R. L. Sousa, and R. R. N. Alves, "Aspectos da comercialização ilegal de aves nas feiras livres de Campina Grande, Paraiba, Brasil," *Revista de Biologia e Ciências da Terra*, vol. 6, no. 2, pp. 204–221, 2006.

[238] J. Marques and W. Guerreiro, "Répteis em uma Feira nordestina (Feira de Santana, Bahia)-contextualização Progressiva e Analise Conexivo-Tipologica," *Sitientibus Série Ciências Biológicas*, vol. 7, no. 3, pp. 289–295, 2007.

[239] R. R. N. Alves, M. G. G. Oliveira, R. R. D. Barboza, R. Singh, and L. C. S. Lopez, "Medicinal Animals as therapeutic alternative in a semi-arid region of Northeastern Brazil," *Research in Complementary Medicine*, vol. 16, no. 5, pp. 305–312, 2009.

[240] M. V. A. Confessor, L. E. T. Mendonça, J. S. Mourao, and R. R. N. Alves, "Animals to heal animals: ethnoveterinary practices in semi-arid region, Northeastern Brazil," *Journal of Ethnobiology and Ethnomedicine*, vol. 5, p. 37, 2009.

[241] F. S. Ferreira, S. V. Brito, S. C. Ribeiro, W. O. Almeida, and R. R. N. Alves, "Zootherapeutics utilized by residents of the community Poço Dantas, Crato-CE, Brazil," *Journal of Ethnobiology and Ethnomedicine*, vol. 5, p. 21, 2009.

[242] F. S. Ferreira, A. V. Brito, S. C. Ribeiro, A. A. F. Saraiva, W. O. Almeida, and R. R. N. Alves, "Animal-based folk remedies sold in public markets in Crato and Juazeiro do Norte, Ceará, Brazil," *BMC Complementary and Alternative Medicine*, vol. 9, p. 17, 2009.

[243] W. M. S. Souto, J. S. Mour;ão, R. R. D. Barboza, and R. R. N. Alves, "Parallels between zootherapeutic practices in ethnoveterinary and human complementary medicine in NE Brazil," *Journal of Ethnopharmacology*, vol. 134, no. 3, pp. 753–767, 2011.

[244] F. Oliveira, U. Albuquerque, and V. Fonseca-Kruel, "Avanços nas pesquisas etnobotânicas no Brasil," *Acta Botanica Brasilica*, vol. 23, no. 2, pp. 590–605, 2009.

[245] T. A. de Sousa Araújo, N. L. Alencar, E. L. C. de Amorim, and U. P. de Albuquerque, "A new approach to study medicinal plants with tannins and flavonoids contents from the local knowledge," *Journal of Ethnopharmacology*, vol. 120, no. 1, pp. 72–80, 2008.

[246] M. F. Agra, G. S. Baracho, K. Nurit, I. J. L. D. Basílio, and V. P. M. Coelho, "Medicinal and poisonous diversity of the flora of "Cariri Paraibano", Brazil," *Journal of Ethnopharmacology*, vol. 111, no. 2, pp. 383–395, 2007.

[247] S. L. Cartaxo, M. M. de Almeida Souza, and U. P. de Albuquerque, "Medicinal plants with bioprospecting potential used in semi-arid northeastern Brazil," *Journal of Ethnopharmacology*, vol. 131, no. 2, pp. 326–342, 2010.

[248] M. C. S. Cruz, P. O. Santos, A. M. Barbosa et al., "Antifungal activity of Brazilian medicinal plants involved in popular treatment of mycoses," *Journal of Ethnopharmacology*, vol. 111, no. 2, pp. 409–412, 2007.

[249] C. D. F. C. B. R. De Almeida, M. A. Ramos, E. L. C. de Amorim, and U. P. de Albuquerque, "A comparison of knowledge about medicinal plants for three rural communities in the semi-arid region of northeast of Brazil," *Journal of Ethnopharmacology*, vol. 127, no. 3, pp. 674–684, 2010.

[250] N. L. G. Silva, F. S. Ferreira, H. D. M. Coutinho, and R. R. N. Alves, "Zooterapicos utilizados em comunidades rurais do municipio de Sume, semiarido da Paraiba, Nordeste do Brasil," in *Zooterapia: Os Animais na Medicina Popular Brasileira*, E. M. Costa-Neto and R. R. N. Alves, Eds., vol. 2, pp. 243–267, NUPEEA, Recife, Brazil, 1st edition, 2010.

[251] V. T. Nascimento, L. G. Sousa, A. G. C. Alves, E. L. Araújo, and U. P. Albuquerque, "Rural fences in agricultural landscapes and their conservation role in an área of caatinga (dryland vegetation) in northeast Brazil," *Environment, Development and Sustainability*, vol. 11, no. 5, pp. 1005–1029, 2009.

[252] M. Ramos, P. Medeiros, A. Almeida, A. Feliciano, and U. Albuquerque, "Can wood quality justify local preferences for rewood in an área of caatinga (dryland) vegetation?" *Biomass and Bioenergy*, vol. 32, no. 6, pp. 503–509, 2008.

[253] U. P. de Albuquerque, P. M. de Medeiros, A. L. S. de Almeida et al., "Medicinal plants of the caatinga (semi-arid) vegetation of NE Brazil: A quantitative approach," *Journal of Ethnopharmacology*, vol. 114, no. 3, pp. 325–354, 2007.

[254] R. Oliveira, N. E. Lins, E. Araújo, and U. Albuquerque, "Conservation priorities and population structure of woody medicinal plants in an área of caatinga vegetation (Pernambuco State, NE Brazil)," *Environmental Monitoring and Assessment*, vol. 132, no. 1–3, pp. 189–206, 2007.

[255] U. P. De Albuquerque, T. A. De Sousa Araújo, M. A. Ramos et al., "How ethnobotany can aid biodiversity conservation: Reflections on investigations in the semi-arid region of NE Brazil," *Biodiversity and Conservation*, vol. 18, no. 1, pp. 127–150, 2009.

[256] R. Lucena, U. Albuquerque, J. Monteiro et al., "Useful plants of the semi-arid Northeastern region of Brazil—a look at their conservation and sustainable use," *Environmental Monitoring and Assessment*, vol. 125, no. 1–3, pp. 281–290, 2007.

[257] R. F. P. De Lucena, E. D. L. Araújo, and U. P. De Albuquerque, "Does the local availability of woody Caatinga plants (Northeastern Brazil) explain their use value?" *Economic Botany*, vol. 61, no. 4, pp. 347–361, 2007.

[258] G. T. Soldati and U. P. De Albuquerque, "Impact assessment of the harvest of a medicinal plant (Anadenanthera colubrina (Vell.) Brenan) by a rural semi-arid community (Pernambuco), northeastern Brazil," *International Journal of Biodiversity Science, Ecosystems Services and Management*, vol. 6, no. 3-4, pp. 106–118, 2010.

[259] N. L. Alencar, T. A. de Sousa Araújo, E. L. C. de Amorim, and U. P. de Albuquerque, "The inclusion and selection of medicinal plants in traditional pharmacopoeias-evidence in support of the diversification hypothesis," *Economic Botany*, vol. 64, no. 1, pp. 68–79, 2010.

Application of Scenario Analysis and Multiagent Technique in Land-Use Planning: A Case Study on Sanjiang Wetlands

Huan Yu,[1] Shi-Jun Ni,[1] Bo Kong,[2] Zheng-Wei He,[1] Cheng-Jiang Zhang,[1] Shu-Qing Zhang,[3] Xin Pan,[3] Chao-Xu Xia,[1] and Xuan-Qiong Li[1]

[1] College of Earth Sciences, Chengdu University of Technology, Chengdu 610059, China
[2] Institute of Mountain Hazards and Environment, Chinese Academy of Sciences, Chengdu 610041, China
[3] Northeast Institute of Geography and Agroecology, Chinese Academy of Sciences, Changchun 130012, China

Correspondence should be addressed to Huan Yu; yuhuan0622@126.com

Academic Editors: J. Bai, H. Cao, B. Cui, A. Li, and Y. J. Xu

Land-use planning has triggered debates on social and environmental values, in which two key questions will be faced: one is how to see different planning simulation results instantaneously and apply the results back to interactively assist planning work; the other is how to ensure that the planning simulation result is scientific and accurate. To answer these questions, the objective of this paper is to analyze whether and how a bridge can be built between qualitative and quantitative approaches for land-use planning work and to find out a way to overcome the gap that exists between the ability to construct computer simulation models to aid integrated land-use plan making and the demand for them by planning professionals. The study presented a theoretical framework of land-use planning based on scenario analysis (SA) method and multiagent system (MAS) simulation integration and selected freshwater wetlands in the Sanjiang Plain of China as a case study area. Study results showed that MAS simulation technique emphasizing quantitative process effectively compensated for the SA method emphasizing qualitative process, which realized the organic combination of qualitative and quantitative land-use planning work, and then provided a new idea and method for the land-use planning and sustainable managements of land resources.

1. Introduction

As a consequence of global increase of economic and societal prosperity, ecosystems and natural resources have been substantially exploited, degraded, and destroyed in the last century [1–3]. Land is one of the most valuable natural resources because of its close relation with human daily lives, and it is suffering high strength of landscape transformation activities such as mine exploitation, infrastructure construction, and agriculture cultivation, which have an important influence on the composition and quality of land resources [4]. The sustainable management of land resource has become the broadly accepted backdrop for policy and management decisions in most parts of the world [5–8]. Described as an activity that envisages future land arrangements [9], land-use planning has been recognized as a key instrument for identifying and ensuring sustainable land resource uses, improving the livelihoods of rural communities, and thereby achieving sustainable development [10].

Land use/cover change (LUCC) is the result of diverse interactions between society and the environment [11–13]. As such, land-use planning has triggered debates on social and environmental values and on the need for participatory processes to address individual differences in these values [14–19]. Over the past years, a number of efforts were undertaken for land-use planning with the consideration of individual participatory processes. For example, Ishii et al. proposed a new needs analysis method for the conceptual land-use planning of contaminated sites and illustrated this method with a case study of an illegal dumping site for hazardous waste [20]. Helbron et al. presented the use of indicators in a site-specific assessment method for strategic environmental assessment in regional land-use planning [21].

Koschke et al. presented a multicriteria assessment framework for the qualitative estimation of regional potentials to provide ecosystem services as a prerequisite to support regional development planning [2]. Lestrelin et al. examined the extent to which the evolution of Laos' village land-use planning has resulted in increased local participation and improved livelihoods [17]. Fitzsimons et al. aimed to create a quantitative, community-engaged basis for the evolution of multiple land uses [22]. Lagabrielle et al. considered participatory modelling to integrate biodiversity conservation into land-use planning and to facilitate the incorporation of ecological knowledge into public decision making for spatial planning [23]. Magigi and Drescher analyzed local communities' involvement in land-use planning to regulate land use change and customary land tenure challenges in a rapidly expanding city in Tanzania [24]. Those researches show that collaborative planning has become an increasingly popular approach in land-use decision making, particularly in situations where there are multiple actors with conflicting interests.

Over the last couple of decades, scenario analysis (SA) has become a broadly used tool to provide support and advice to policy makers [25]. In decision-making processes, scenarios can help the decision makers to anticipate possible or potential strategies according to different plausible scenarios, which is usually designed to identify a set of possible futures, where the occurrence of each is plausible, although not assured and not necessarily probable [26]. In this way, SA can be seen as a process of understanding, analyzing, and describing the behavior of complex systems consistently and completely. This kind of systematic analysis is crucial in collaborative planning, and it is widely used in land-use planning [27–33]. These approaches are mainly based on the elicitation of information from a set of people, or a panel of experts or stakeholders, and they are therefore characterized by a high level of subjectivity. Indeed, the quality and performance of SA as a basis for decision support become critically dependent on the quality and performance of the assessments expressed throughout the entire land-use planning process [34]. This represents the primary limitation of such qualitative approaches, particularly when the dynamic complexity of coupled systems is not well understood. Then different planners may get completely diverse planning results, and their scientific creditability has frequently been questioned. Hence, a land-use planning methodology based on a systems approach involving realistic computational modeling and metaheuristic optimization is still lacking in many current practices [35]. Through the above analysis, two key questions will be faced during land-use planning process: one is how to see different planning simulation results instantaneously and apply the results back to assist planning work interactively; the other is how to ensure that the planning simulation result is scientific and accurate.

Landscape digital reconstruction and spatial-temporal distribution simulation based on multiperiod regional land cover data can help to understand the mechanisms and laws of land use succession, recognize the relationship between human activities and land-use changes, predict the future trend of the land use, and ultimately provide strategies to

decision maker for land-use planning. A multiagent system (MAS) can be defined as a set of agents that interact in a common environment, able to modify their attributes and their environment [36]. MAS may increase understanding of complex coupled social-ecological systems [37], more particularly in the context of land-use planning [38–41]. The MAS technique can simulate the different planning results based on a systems approach involving realistic computational modeling, so as to provide reference for planning work. More importantly, the data mining technique will be used in the process of multiagent simulation, which ensures the planning process can conform to various regions and get a more accurate, more scientific planning result.

The objective of this paper is to analyze whether and how a bridge can be built between qualitative and quantitative approaches for land-use planning work and to find out a way to overcome the gap that exists between the ability to construct computer simulation models to aid integrated land-use plan making and the demand for them by planning professionals. The specific topic is the integration of scenario analysis with the multiagent system technique, with its application in land-use planning process.

2. Theoretical Framework

MAS simulation and SA method both emphasize the role of human factors in the regional land use change and outstand human intervention in the simulation and planning process. This provides the theoretical basis for the combination of two methods. SA method aims to separate uncertain factors and establish system variables, which focuses on qualitative or qualitative and quantitative combination analysis process. The MAS simulation emphasizes the laws of landscape spatial-temporal distribution under influences of various geographic, economic, and other factors, which focuses on quantitative or quantitative and qualitative combination analysis process. This simulation process is based on actual multiperiod land cover data to obtain the transformation rules, which can reflect the regional actual change situations more effectively.

Through the analysis of MAS simulation and SA method basic principles, combination of them is mainly reflected in the qualitative analysis and design for different scenarios using SA, and actual simulation model and process are completed by MAS quantitatively. Furthermore, factors analysis in scenario design process will be affected by the data mining results of landscape dynamic knowledge database in the MAS simulation process. This makes the scenario analysis process more consistent with the actual situation of regional development and gain a strong geographical significance. The study presents a theoretical framework of land-use planning based on SA method and MAS simulation integration (Figure 1).

3. Case Implementation and Results

3.1. Study Area and Data

3.1.1. Study Area. Wetlands are integral parts of the global ecosystem as they can prevent or reduce the severity offloods,

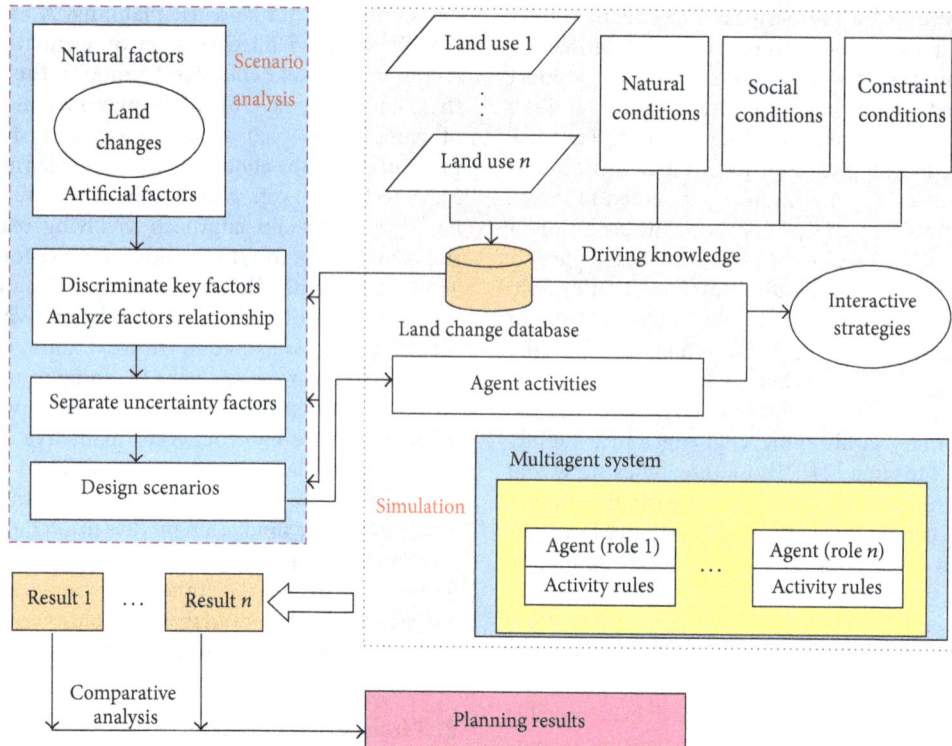

FIGURE 1: The theoretical framework of landscape planning based on MAS and SA integration.

feed ground water, and provide unique habitats for flora and fauna [42, 43]. The Sanjiang Plain, located in the Northeastern region of China, is one of the largest freshwater wetland in the country (Figure 2). Since the end of the 1950s, large-scale development in the Sanjiang Plain marsh land has occurred [44]. By 2003, about 80% of natural wetlands had been converted to farm land and the progressive loss of wetland is continuing [45]. With a local population of 7.8 million in this region, of which 53.4% is engaged in farming the Sanjiang Plain has become an important grain and bean production region for China [46]. The regional climate is mild humid to subhumid continental monsoon feature. The average temperatures range from −18°C in January to 21-22°C in July with a frost-free period of 120–140 days. Annual precipitation is somewhere around 500–650 mm with 80% occurring from May to September. Most of the rivers at the area have riparian wetlands supporting meadow and marsh vegetation. Sedge (*Carex* spp.) is the dominant plants with *Phragmites* spp. scattered across some parts of the landscape [47].

The study area is limited within 47°21′42″–48°15′9″ north altitude and 133°25′52″–134°33′37″ east longitude in the Northeast of Sanjiang Plain at (Figure 2). Several factors had been taken into consideration when this region was chosen to start this study. Firstly, the Sanjiang Plain is one of the largest marsh distribution region. Secondly, it is a typical representation in the global temperate wetland ecosystems. Thirdly, due to the relative cold weather, deep surface waters, large marsh patches, and sparse population, reclamation of marsh lands in this region is relatively late. Fourthly, study

area contains two national nature reserves and three major river systems: Honghe Reserve, Sanjiang Reserve, and Yalu River, Dongjiang River, Bielahong River. They make the study area possess natural original scenery relatively. In addition, during the process of development and utilization in recent decades, the conflict between people and land is a constant game of war. Wetland degradation process under the disturbance from human activities is representative, which makes it suitable for carrying out simulation of wetland landscape spatial-temporal evolution.

3.1.2. Data. To complete the simulation using MAS, land-use data were collected during three-year period (1995, 2000, and 2006). The 1995 and 2000 datasets are used in decision ruling on transformation, while 2006 dataset is used to verify predicted results. Each land use data set contains 5 types of covers such as water, farmlands, resident area, forest, and wetlands. The data of soil, topography, terrain, location, and other thematic parameters are sorted to formulate the transformation probability under the influence of many geographical conditions. The soil data represent 22 different types; topography data contains 14 types of landforms; River distance is a grid file that reflects the distance to rivers and road distance reflects the distance to road. The units of river distance, road distance, and digital elevation model (DEM) data are meters, and the slope is degree. In addition, the existing data collections previously include planning, feasibility reports, scientific research reports, maps and documentation of Honghe National Natural Reserve and Sanjiang National Natural Reserve, and meteorology, hydrology, groundwater

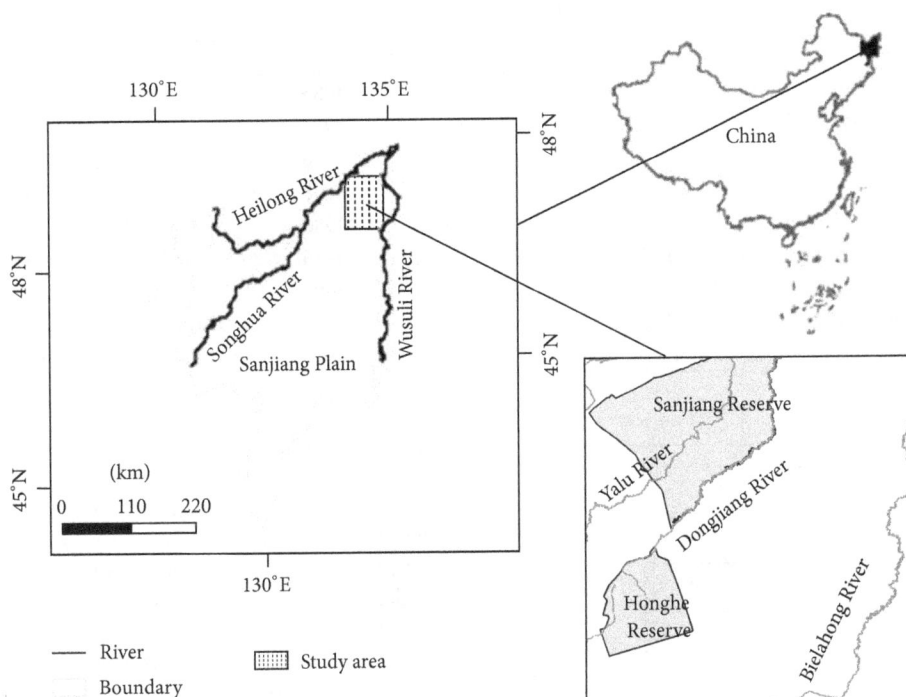

FIGURE 2: Location of the study area in Sanjiang Plain, China.

observations and other statistical records are available for reference in this research. The detail of each dataset is listed in Table 1. All of the data were coregistered and formatted as GRID format under ArcGIS 9.3.

3.2. Scenario Analysis and Design

3.2.1. The Key Variables and Their Interactions.
There are mainly two kinds of factors influencing land use [48, 49]. The first one is direct, and it consists of various forms of activities including conservation and development. The second one is indirect, and it relies on the legal instruments of public policy to influence the behavior of landowners [50].

Considering the actual situation of study area, one direct factor is mainly performed by protectors who are the staff of national nature reserves. They prevent wetlands from developing to other landscape directly. Secondly, the statistic results of land use/cover change cells between the years of 1995 and 2006 show that total area changing from wetland to other cover types is 2402.68 km^2, among which 2215.75 km^2 is from wetlands to farmlands. The farmlands count about 92.22% of total altered area, and this indicates that the cultivation is the main factor that results in the wetland shrink. Then, another direct factor is mainly performed by farmers who change the land cover through reclaiming wetland. Thirdly, because local governments plan for and decide current and future land use, their role in land use/cover change is crucial. The behavior of the governments includes the government's macromanagement and policy establishment, which are indirectly influence regional landscape change. At last, three main variables that cause regional land use change can be simplified as protectors, farmers, and governments according to study area actual situations.

Although one variable has certain functions, however, relying on a single variable cannot always describe and solve complex large-scale problems in reality. Therefore, an application system often includes multiple variables. Each variable is not isolated but an interactive part of the group. Those variables can follow some kind of specific agreement and possess multilinguistic communication skill to complete a specific task. According to the actual situation of study area, logical interaction rules among governments, protectors, and farmers variables are designed as shown in Figure 3.

During the land-use change process, land cover status of certain position is determined by governments, farmers, and protectors variable jointly. First, farmers variable determine whether to reclaim wetlands under various environmental conditions. If the farmers wish to do so, a small part of them will illegally reclaim wetlands, and when this part of farmers goes around obstacles from protectors, the land cover status will be changed. On the contrary, if they are hampered successfully by protectors, then the land cover remains unchanged. The rest of them will apply to the governments variable to reclaim wetlands. Two total diverse consequences will result depending on the government approval to their petition: if the government approved and farmers avoided obstacles from protectors, the land cover status would be changed; if not, the land cover status will be unaffected and unchanged.

TABLE 1: List of data description.

Name	Content	Resolution	Time	Source	Size
Soil	Spatial distribution of soil types	30 m	1985	Digitizing	8.41 MB
Landform	Spatial distribution of geomorphologic types	30 m	1985	Digitizing	8.41 MB
River distance	Distance to rivers	30 m	1998	Euclidean distance calculation	33.66 MB
Road distance	Distance to roads	30 m	1998	Euclidean distance calculation	33.66 MB
DEM	Digital elevation model	30 m	1986	Digitizing	33.66 MB
Slope	Spatial distribution of slope	30 m	1986	Calculated from DEM	33.66 MB
Land use	Spatial distribution of land cover types	30 m	1995	TM image classification	16.82 MB
Land use	Spatial distribution of land cover types	30 m	2000	TM image classification	16.82 MB
Land use	Spatial distribution of land cover types	30 m	2006	TM image classification	16.82 MB

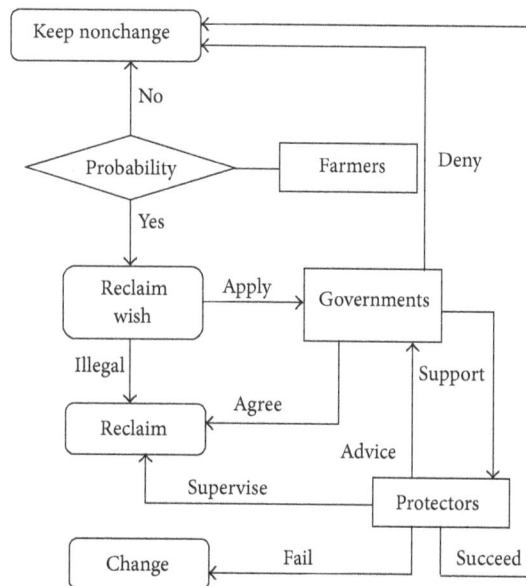

FIGURE 3: Interaction logistics of variables.

3.2.2. The Scenario Design. After finishing the key factors discrimination and interaction relationship analysis, study defined 3 specific planning scenarios in order to verify the availability of theoretical framework presented by this paper.

Undisturbed Scenario. According to actual land-use change rules of study region from 1995 to 2000, the land cover of year 2006 will be predicted based on MAS simulation technique.

Ecotype Scenario. In this scene, governments' criterions of wetland development approval become strict, and farmers and protectors' awareness of ecological environment protection is strengthened, which cause the reclaim desire to reduce and protection scrutiny to increase. Under these conditions, damage degree of wetland landscape will be degraded. However, it is also very likely to cause regional economy development to be slowed down, and then the income of farmers and governments may drop.

Economy Scenario. Governments encourage farmland development for the needs of economic construction, which affects the speed and manner of the entire regional land-use change. Being driven by economic interests, farmers also strongly destroy wetlands for increasing farmland quantity. Protectors abandon wetland conservation efforts and even join in the wetland destruction and agriculture development in action. Under these conditions, the regional land use subordinates the economic construction and ignores the protection of ecological environment. It is a nonsustainable development mode, but a certain degree of economic achievements may arise in this case.

Research realizes different planning scenarios through modifying decision-making behavior of governments, farmers, and protectors variables. It can clearly explain the specific reasons for the differences of diverse planning scenario simulation results. In detail, undisturbed scene is realized through MAS simulation based on the transformation rules that were gained by data mining technique. Governments, protectors and farmers variables are only used to reflect actual land-use change process. An ecotype scenario is realized through governments auditing standards or reducing the rate of approvals, protectors reinforcing supervision to prevent wetlands being destroyed, and famers reducing their cultivation will. On the other side, an economy scenario is realized through which governments lowering standards or increasing the rate of approvals, protectors reducing supervision to increase rate of development, and famers increasing willingness to reclaim wetlands. After finishing the discrimination of key factors, the analysis of their interactions, and the design of scenarios, the next step is how to quantitatively descript them based on MAS computer simulation models.

3.3. Model Construction of MAS

3.3.1. Construction of Environmental Factors Layer. Environmental factor layer in the model is the natural and social environment of MAS, the database for land cover spatial-temporal evolution simulation, and a key element of the model [51, 52]. In this model, environmental factor layer is defined as an integral body including the status of initial land cover, elevation, slope, soil, topography, distance to road, distance to river, and other environmental factors.

3.3.2. Definition of Roles and Conduct Rules. A key issue of multiagent model construction is how to abstract and descript agents properly [53]. Analysis of study area land cover reveals the driving force of regional landscape changes that are caused by human activities. Therefore, the simulation of landscape spatial-temporal evolution using MAS is to link up human activities and agents based on multi-agent characteristics. According to scenario analysis and design results, three agent types are defined as, farmer agent, protector agent and government agent.

Protector Agent. Protectors who are the staff of national nature reserves in this research prevent farmers from agricultural developing. Protector agent becomes the main driving force slowing down wetland landscape degradation with the support from government agent. The protective efforts of the protector agent are directly reflected on wetland area changes in a specific period of time. The wetland area reduces enormously and quickly, and it is indicative of a poor effect of protection and a small effect on wetland protection. Thus, an equation assessing the protective effect can be quantitatively expressed as

$$P_{\text{omit}} = \frac{A_{\text{marsh}}^{t1} - A_{\text{marsh}}^{t2}}{A_{\text{marsh}}} \times 100\%, \tag{1}$$

where P_{omit} is the probability that omits hindering agriculture development, A_{marsh}^{t1} is the wetland area at time $t1$, A_{marsh}^{t2} is the wetland area at time $t2$, and A_{marsh} is the reserve total area.

Formula (1) will be used to assess the protect effect on Honghe and Sanjiang National Nature Reserves separately. When land cover change position (i, j) is within those reserves, the omit probability will be calculated using this equation. The omit probability will be one hundred percent if the position is out of the reserves. In such case, protector's activity can be expressed as

$$P_{\text{pro}}(i, j) = \begin{cases} P_{\text{omit}}(i, j) & \text{(within reserves)}, \\ 1 & \text{(out of reserves)}. \end{cases} \tag{2}$$

Farmer Agent. The behaviors of farmer agent can be classified into two categories: development and undevelopment. This behavior will cause two types of possible results: one is negative behavior that can reduce the wetland area, and the other is a positive activity that does not change the land cover. In reality, farmers perform such activities under the approval of government. In this model, one part of the farmers' behavior is carried out directly (illegal development), while the other part applies for government agent approval selectively. The whole behavior is also affected by protector agent. It must get the approval of government and avoid the hindering effect from protector agent. As a result, the land cover status can be changed eventually.

When the model starts running, the farmers' probability of reclaim wetlands will be calculated by the formula below

$$P_{\text{far}}(i, j) = \left(w_1 E_{\text{eleva}}, w_2 E_{\text{slop}}, w_3 E_{\text{soil}}, \right. \tag{3}$$
$$\left. w_4 E_{\text{landf}}, w_5 E_{\text{rivd}}, w_6 E_{\text{road}}, w_7 R_{\text{disb}} \right),$$

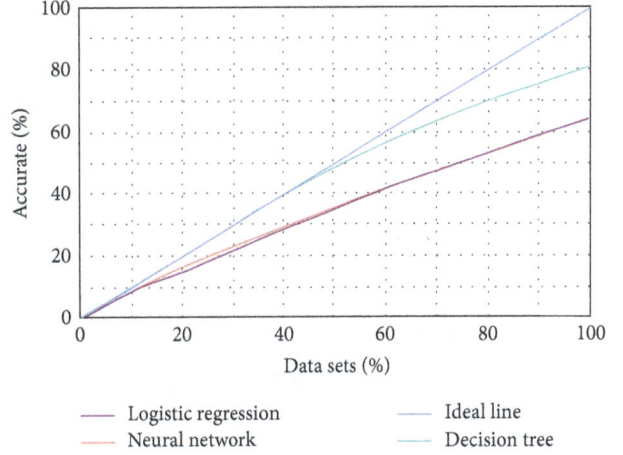

FIGURE 4: The lift map of data mining.

where $P_{\text{far}}(i, j)$ is the change probability for position (i, j); E_{eleva}, E_{slop}, E_{soil}, E_{landf}, E_{rivd}, E_{road}, and R_{disb} are the environment factors of elevation, slope, soil, landform, distance to river, distance to road, and random disturbance. These indicators were selected because they are representative of the most critical environmental issues of the study area, and they are easy to understand and communicate. $w_1, \ldots, w_6$ is the influence weight for each factor, which is calculated through data mining method. This instance completed the calculation of farmland development probability using Microsoft SQL Server 2008 software, data mining process used neural network, decision tree, and logistic regression mathematical model, respectively, and results of them described by lift chart (Figure 4). A lift chart is used for comparing the accuracy of each prediction model. The x-axis represents the percentage of the test data set for prediction, and the y-axis indicates the percentage of accurate prediction. An ideal line is a diagonal line, which means 50 percent of the data accurately predicted 50 percent of the cases (the expected maximum). Then we found that decision tree method was the best, and then it was used to calculate the probability of land use change under the influence of various geographical conditions.

The random disturbance is expressed as

$$R_{\text{disb}} = 1 + (-\ln \theta)^{\alpha}, \tag{4}$$

where θ is the random number between 0 and 1; α is the parameter that controls the size of random disturbance; weight w_7 for it is set as 1 [54].

Government Agent. Government achieves its own wish through the planning actions, while residents affect the probability of land cover conversion through the cooperation with the government [55]. The behavior of the government agent at the area includes the government's macromanagement, action planning, and decision making in response to farmer agent's application [56]. When farmers apply to convert wetlands for agricultural purpose, government agent will make decisions based on current land cover status and future planning of utilization. And degree of support from government also indirectly determines the specific action of

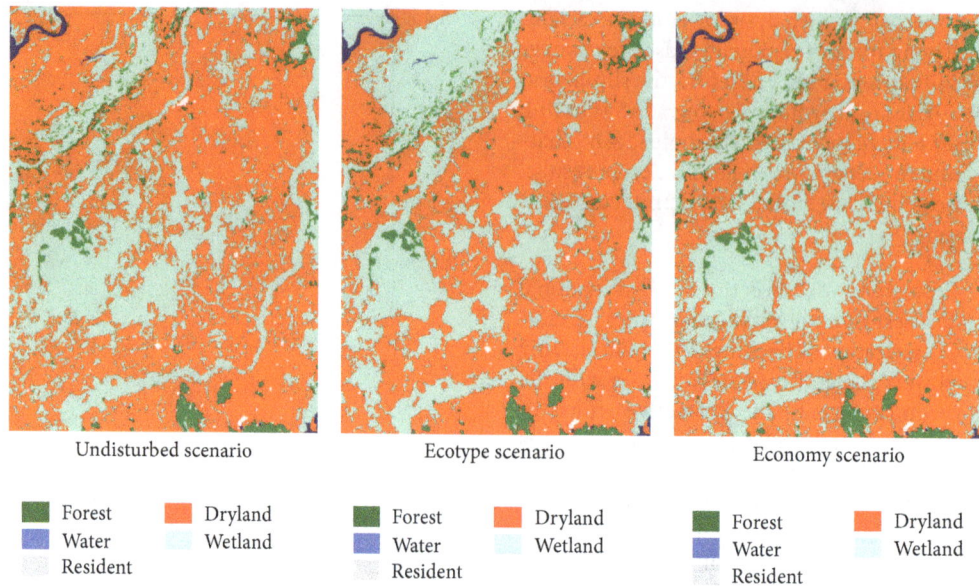

Undisturbed scenario

Ecotype scenario

Economy scenario

Forest	Dryland
Water	Wetland
Resident	

FIGURE 5: The results of scenarios planning based on multi-agent adjustment.

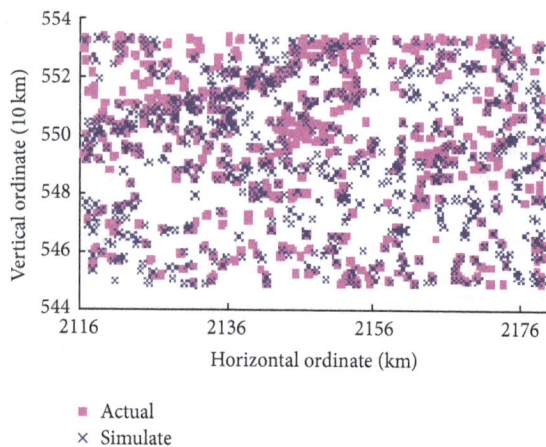

- Actual
- × Simulate

FIGURE 6: The spatial distribution of centroids.

protector. A strongly support will obtain an excellent performance of protection and vice versa. Then the behaviors of the government agent can be expressed through the activities of farmer and protector indirectly as

$$P_{\text{gov}}(i, j) = \left| 1 - \left(a P_{\text{pro}}(i, j) + b P_{\text{far}}(i, j) \right) \right|, \qquad (5)$$

where $P_{\text{gov}}(i, j)$ is the probability that government approves to change land cover for position (i, j); a is the adjustment coefficient for protector between 0 and 1; b is the adjustment coefficient for famers between 0 and 1; coefficients are varied with the government policy changing.

Different scenarios are realized through revising 3 kinds of agents, and the details are shown in Table 2.

3.4. Simulation Results. Based on the characteristics of three planning scenarios, combined with the framework of

interaction relationships between governments, protectors, and farmers variables (Figure 4), MAS model is used to generate the different planning scenarios (Figure 5).

3.5. Assessment of Accuracy. In order to facilitate observation of the differences between simulation results and reality situations, centroids of each landscape patch including undisturbed scenario simulation results and reality situations of year 2006 are calculated. The spatial distribution of them is plotted, which is shown in Figure 6.

The centroids distribution plots show that simulation results of undisturbed scenario have a high degree of consistency with 2006 year's actual situation. For the quantitative evaluation of the simulation accuracy, study overlaps the simulation result map of undisturbed scenario and the actual land cover map of 2006 together to gain the simulation accuracy of points to points. The overall simulation accuracy of undisturbed scenario reaches 85.12%, and it validates the feasibility and effectiveness of land cover change simulation using MAS. At the same time, it assures the scientific creditability of the other two planning scenarios' simulation results based on SA method and MAS technique integration.

4. Discussions and Conclusions

Simulation results show that three different scenarios show vast differences, especially that the eco-type scenario is significantly different from the other two scenarios, which means that the current ecological conditions of the study area are not ideal, and that adjustment and optimization work should promptly be carried out to protect wetlands. The differences between economy scenario and undisturbed scenario are relatively small, but obvious differences still exist on closer inspection. Wetland patches distribution is loose in

TABLE 2: Realization of scenarios through agent parameter adjusting.

Scenarios	Protectors	Famers	Governments	Illustrations
Undisturbed	P_{pro}	P_{far}	P_{gov}	All the parameters keep unchanged
Eco-type	$1/3 * P_{pro}$	$1/3 * P_{far}$	$1/3 * P_{gov}$	Governments reduce the rate of approvals, protectors reinforce supervision, and famers reduce their cultivation will.
Economy	$3 * P_{pro}$	$3 * P_{far}$	$3 * P_{gov}$	Governments increase rate of approvals, protectors reduce supervision, and famers increase willingness to reclaim.

economy scenario, and many ecologically significant wetland patches are disappeared.

According to the simulation results of different scenarios, depending on the development goals, different planning strategies can be gained. Furthermore, the similarities and differences between actual situations and simulation scenarios can be used to assist land use optimization problem, even to provide reference for landscape reconfiguration including wetland sustainable development, returning farmland to wetland, and so forth.

To sum up, a theoretical framework was proposed for the land-use planning based on the integration SA method with MAS simulation. Taking Sanjiang Plain inland freshwater wetland areas as an example, the study verified the availability of this framework. Results showed that MAS simulation technique emphasizing quantitative process effectively compensated for the SA method emphasizing qualitative process, which realized the organic combination of qualitative and quantitative land-use planning work and then provided a new idea and method for the land-use planning and sustainable managements of land resources.

Applying SA method originated in enterprise management and MAS technique from the field of artificial intelligence into the complex geographical system problem, there is still a lot of refinement work that needs to be further completed. How to establish criteria for the classification of different planning scenarios in SA more scientifically based on regional characteristics and how to define the simulation variables and their interaction relationships based on land-use planning objective more accurately still need to be explored in further practical work.

Acknowledgments

This study was supported by the National Natural Science Foundation of China (Grant no. 41101174), the Strategic Priority Research Program Climate Change: Carbon Budget, and Relevant Issues of the Chinese Academy of Sciences (Grant no. XDA05050105), Remote Sensing Investigation and Assessment of National Ecological Environment Change for Ten Years (2000–2010) (Grant no. STSN-01-04-02), Landslide Hazard Information Extraction based on Multisources Data (Grant no. SKLGP2011Z005). The authors also thank Honghe and Sanjiang National Nature Reserves for providing relative data.

References

[1] MA, "Ecosystems and human well-being: synthesis," *A report of the millennium ecosystem assessment*, Island Press, Washington, DC, USA, 2005.

[2] L. Koschke, C. Fürst, S. Frank, and F. Makeschin, "A multi-criteria approach for an integrated land-cover-based assessment of ecosystem services provision to support landscape planning," *Ecological Indicators*, vol. 21, pp. 54–66, 2012.

[3] J. H. Bai, B. S. Cui, B. S. Chen et al., "Spatial distribution and ecological risk assessment of heavy metals in surface sediments from a typical plateau lake wetland, China," *Ecological Modelling*, vol. 222, no. 2, pp. 301–306, 2011.

[4] H. Yu, B. Kong, S. Q. Zhang, and X. Pan, "Wetlands spatial-temporal distribution multi-scale simulation using multi-agent system," *International Journal of Intelligent Systems and Applications*, vol. 4, no. 9, pp. 29–38, 2012.

[5] M. G. Zhang, Z. K. Zhou, W. Y. Chen, J. W. Ferry Slik, C. H. Cannon, and N. Raes, "Using species distribution modeling to improve conservation and land use planning of Yunnan, China," *Biological Conservation*, vol. 153, pp. 257–264, 2012.

[6] D. E. Orenstein, L. Jiang, and S. P. Hamburg, "An elephant in the planning room: political demography and its influence on sustainable land-use planning in drylands," *Journal of Arid Environments*, vol. 75, no. 6, pp. 596–611, 2011.

[7] J. Bourgoin, "Sharpening the understanding of socio-ecological landscapes in participatory land-use planning. A case study in Lao PDR," *Applied Geography*, vol. 34, pp. 99–110, 2012.

[8] J. G. Underwood, J. Francis, and L. R. Gerber, "Incorporating biodiversity conservation and recreational wildlife values into smart growth land use planning," *Landscape and Urban Planning*, vol. 100, no. 1-2, pp. 136–143, 2011.

[9] FAO, Guidelines for land-use planning, Rome: Food and Agriculture Organization of the United Nations, 1993.

[10] J. Bourgoin, J. C. Castella, D. Pullar, G. Lestrelin, and B. Bouahom, "Toward a land zoning negotiation support platform: "Tips and tricks" for participatory land use planning in Laos," *Landscape and Urban Planning*, vol. 104, no. 2, pp. 270–278, 2012.

[11] P. H. Verburg, D. B. van Berkel, A. M. van Doorn, M. van Eupen, and H. A. R. M. van den Heiligenberg, "Trajectories of land use change in Europe: a model-based exploration of rural futures," *Landscape Ecology*, vol. 25, no. 2, pp. 217–232, 2010.

[12] G. I. Díaz, L. Nahuelhual, C. Echeverría, and S. Marín, "Drivers of land abandonment in Southern Chile and implications for landscape planning," *Landscape and Urban Planning*, vol. 99, no. 3-4, pp. 207–217, 2011.

[13] J. H. Bai, Q. Q. Lu, J. J. Wang et al., "Landscape pattern evolution processes of alpine wetlands and their driving factors in

the Zoige plateau of China," *Journal of Mountain Science*, vol. 10, no. 1, pp. 54–67, 2013.

[14] J. Hillier, "Habitat's habitus: nature as sense of place in land use planning decision-making," *Urban Policy and Research*, vol. 17, no. 3, pp. 191–204, 1999.

[15] S. Owens, "Land, limits and sustainability: a conceptual framework and some dilemmas for the planning system," *Transactions of the Institute of British Geographers*, vol. 19, no. 4, pp. 439–456, 1994.

[16] Y. Rydin, "Sustainable development and the role of land use planning," *Area*, vol. 27, no. 4, pp. 369–377, 1995.

[17] G. Lestrelin, J. Bourgoin, B. Bouahom, and J. Castella, "Measuring participation: case studies on village land use planning in Northern Lao PDR," *Applied Geography*, vol. 31, no. 3, pp. 950–958, 2011.

[18] Z. F. Yao, F. Yang, and X. T. Liu, "Quantitative assessment of impacts of climate and economic-technical factors on grain yield in Jilin Province from 1980 to 2008," *Chinese Geographical Science*, vol. 21, no. 5, pp. 543–553, 2011.

[19] A. Durán, H. Morrás, and G. Studdert, "Distribution, properties, land use and management of mollisols in South America," *Chinese Geographical Science*, vol. 21, no. 5, pp. 511–530, 2011.

[20] K. Ishii, T. Furuichi, and Y. Nagao, "A needs analysis method for land-use planning of illegal dumping sites: a case study in Aomori-Iwate, Japan," *Waste Management*, vol. 33, no. 2, pp. 445–455, 2013.

[21] H. Helbron, M. Schmidt, J. Glasson, and N. Downes, "Indicators for strategic environmental assessment in regional land use planning to assess conflicts with adaptation to global climate change," *Ecological Indicators*, vol. 11, no. 1, pp. 90–95, 2011.

[22] J. Fitzsimons, C. J. Pearson, C. Lawson, and M. J. Hill, "Evaluation of land-use planning in greenbelts based on intrinsic characteristics and stakeholder values," *Landscape and Urban Planning*, vol. 106, no. 1, pp. 23–34, 2012.

[23] E. Lagabrielle, A. Botta, W. Daré, D. David, S. Aubert, and C. Fabricius, "Modelling with stakeholders to integrate biodiversity into land-use planning—lessons learned in Réunion Island (Western Indian Ocean)," *Environmental Modelling & Software*, vol. 25, no. 11, pp. 1413–1427, 2010.

[24] W. Magigi and A. W. Drescher, "The dynamics of land use change and tenure systems in Sub-Saharan Africa cities; learning from Himo community protest, conflict and interest in urban planning practice in Tanzania," *Habitat International*, vol. 34, no. 2, pp. 154–164, 2010.

[25] R. Meyer, "Comparison of scenarios on futures of European food chains," *Trends in Food Science & Technology*, vol. 18, no. 11, pp. 540–545, 2007.

[26] S. P. Schnaars, "How to develop and use scenarios," *Long Range Planning*, vol. 20, no. 1, pp. 105–114, 1987.

[27] Y. Z. Wu, X. L. Zhang, and L. Y. Shen, "The impact of urbanization policy on land use change: a scenario analysis," *Cities*, vol. 28, no. 2, pp. 147–159, 2011.

[28] R. C. Estoque and Y. Murayama, "Examining the potential impact of land use/cover changes on the ecosystem services of Baguio city, the Philippines: a scenario-based analysis," *Applied Geography*, vol. 35, no. 1-2, pp. 316–326, 2012.

[29] T. L. Sohl, B. M. Sleeter, K. L. Sayler et al., "Spatially explicit land-use and land-cover scenarios for the Great Plains of the United States," *Agriculture, Ecosystems and Environment*, vol. 153, pp. 1–15, 2012.

[30] L. M. Normana, M. Feller, and M. L. Villarreal, "Developing spatially explicit footprints of plausible land-use scenarios in the Santa Cruz Watershed, Arizona and Sonora," *Landscape and Urban Planning*, vol. 107, no. 3, pp. 225–235, 2012.

[31] P. H. Verburg, A. Tabeau, and E. Hatna, "Assessing spatial uncertainties of land allocation using a scenario approach and sensitivity analysis: a study for land use in Europe," *Journal of Environmental Management*, pp. 1–13, 2012.

[32] B. M. Sleeter, T. L. Sohl, M. A. Bouchard et al., "Scenarios of land use and land cover change in the conterminous United States: utilizing the special report on emission scenarios at ecoregional scales," *Global Environmental Change*, vol. 22, no. 4, pp. 896–914, 2012.

[33] J. Alcamo, R. Schaldach, J. Koch, C. Kölking, D. Lapola, and J. Priess, "Evaluation of an integrated land use change model including a scenario analysis of land use change for continental Africa," *Environmental Modelling & Software*, vol. 26, no. 8, pp. 1017–1027, 2011.

[34] D. Gambelli, D. Vairo, and R. Zanoli, "Exploiting qualitative information for decision support in scenario analysis," *Journal of Decision Systems*, vol. 19, no. 4, pp. 407–422, 2010.

[35] H. Qi and M. S. Altinakar, "A conceptual framework of agricultural land use planning with BMP for integrated watershed management," *Journal of Environmental Management*, vol. 92, no. 1, pp. 149–155, 2011.

[36] N. Ferrand, *Modèles Multi-Agents pour l'aide à la décision et la négociation en aménagement du territoire [Ph.D. thesis]*, Université Joseph Fourier, Grenoble, France, 1997.

[37] M. F. Acevedo, J. B. Callicott, M. Monticino et al., "Models of natural and human dynamics in forest landscapes: cross-site and cross-cultural synthesis," *Geoforum*, vol. 39, no. 2, pp. 846–866, 2008.

[38] D. C. Parker, S. M. Manson, M. A. Janssen, M. J. Hoffmann, and P. Deadman, "Multi-agent systems for the simulation of land-use and land-cover change: a review," *Annals of the Association of American Geographers*, vol. 93, no. 2, pp. 314–337, 2003.

[39] M. Etienne, C. Le Page, and M. Cohen, "A step-by-step approach to building land management scenarios based on multiple viewpoints on multi-agent system simulations," *Journal of Artificial Societies and Social Simulation*, vol. 6, no. 2, 2003, http://jasss.soc.surrey.ac.uk/6/2/2.html.

[40] F. Bousquet and C. Le Page, "Multi-agent simulations and ecosystem management: a review," *Ecological Modelling*, vol. 176, no. 3-4, pp. 313–332, 2004.

[41] P. Schreinemachers and T. Berger, "Land-use decisions in developing countries and their representation in multi-agent systems," *Journal of Land-Use Science*, vol. 1, no. 1, pp. 29–44, 2006.

[42] J. H. Bai, H. F. Gao, and R. Xiao, "A review of soil nitrogen mineralization as affected by water and salt in Coastal Wetlands: issues and methods," *Clean—Soil, Air, Water*, vol. 40, no. 10, pp. 1099–1105, 2012.

[43] J. H. Bai, H. Ouyang, R. Xiao et al., "Spatial variability of soil carbon, nitrogen, and phosphorus content and storage in an alpine wetland in the Qinghai-Tibet Plateau, China," *Australian Journal of Soil Research*, vol. 48, no. 8, pp. 730–736, 2010.

[44] X. T. Liu, "Marsh resource and its sustainable utility in the Songnen-Sanjiang Plain," *Scientia Geographica Sinica*, vol. 16, pp. 451–460, 1997.

[45] H. Y. Liu, S. K. Zhang, and X. G. Lu, "Wetland landscape structure and the spatial-temporal changes in 50 years in

the Sanjiang Plain," *Acta Geographica Sinica*, vol. 59, no. 3, pp. 391–400, 2004.

[46] G. Q. Chen and X. H. Ma, "A study on the underground and its water balance change after development in the Sanjiang Plain," *Scientia Geographica Sinica*, vol. 16, pp. 427–433, 1997.

[47] Y. Y. Chen, *Study of Wetlands in China*, Jilin Sciences and Technology Press, Changchun, China, 1995.

[48] R. H. Nelson, *Zoning and Property Rights: An Analysis of the American System of Land-Use Regulation*, MIT Press, Cambridge, Mass, USA, 1977.

[49] H. Doremus, "A policy portfolio approach to biodiversity protection on private lands," *Environmental Science and Policy*, vol. 6, no. 3, pp. 217–232, 2003.

[50] J. D. Gerber, "The difficulty of integrating land trusts in land use planning," *Landscape and Urban Planning*, vol. 104, no. 2, pp. 289–298, 2012.

[51] J. Guo, *Dynamic simulation of vegetation spatial pattern based on multi-agent [M.S. thesis]*, Beijing Forestry University, Beijing, China, 2009.

[52] X. P. Liu, X. Li, and B. Ai, "Multi-agent systems for simulating and planning land use development," *Acta Geographica Sinica*, vol. 61, no. 10, pp. 1101–1112, 2006.

[53] J. M. Zhang, B. Wu, and T. Y. Shen, "Research on dynamic simulation of Beijing land covering & changing by applying agent modeling," *Journal of East China Institute of Technology*, vol. 27, pp. 80–83, 2004.

[54] R. White and G. Engelen, "Cellular automata and fractal urban form: a cellular modelling approach to the evolution of urban land-use patterns," *Environment & Planning A*, vol. 25, no. 8, pp. 1175–1199, 1993.

[55] X. L. Huang, *Research on urban ecological land evolution based on cellular automata and multi-agent [M.S. thesis]*, Central South University, Changsha, China, 2008.

[56] X. Li, J. A. Ye, and X. P. Liu, *Geographical Simulation Systems: Cellular Automata and Multi-Agent System*, Science Press, Beijing, China, 2007.

8

A Review of Surface Water Quality Models

Qinggai Wang, Shibei Li, Peng Jia, Changjun Qi, and Feng Ding

Appraisal Center for Environment and Engineering, Ministry of Environmental Protection, Beijing 100012, China

Correspondence should be addressed to Qinggai Wang; qinggaiwang@126.com

Academic Editors: J. Bai, H. Cao, B. Cui, A. Li, and B. Zhang

Surface water quality models can be useful tools to simulate and predict the levels, distributions, and risks of chemical pollutants in a given water body. The modeling results from these models under different pollution scenarios are very important components of environmental impact assessment and can provide a basis and technique support for environmental management agencies to make right decisions. Whether the model results are right or not can impact the reasonability and scientificity of the authorized construct projects and the availability of pollution control measures. We reviewed the development of surface water quality models at three stages and analyzed the suitability, precisions, and methods among different models. Standardization of water quality models can help environmental management agencies guarantee the consistency in application of water quality models for regulatory purposes. We concluded the status of standardization of these models in developed countries and put forward available measures for the standardization of these surface water quality models, especially in developing countries.

1. Signature of Water Quality Models

Water quality models can be effective tools to simulate and predict pollutant transport in water environment [1–3], which can contribute to saving the cost of labors and materials for a large number of chemical experiments to some degree. Moreover, it is inaccessible for on-site experiments in some cases due to special environmental pollution issues. Therefore, water quality models become an important tool to identify water environmental pollution and the final fate and behaviors of pollutants in water environment [3]. These construction projects such as petrochemical, hydrological, and paper-making projects can bring serious effects on aquatic environment after enforcement [4, 5]. Therefore, these environmental effects have to be simulated, predicted, and assessed using numerical models before these construction projects are implemented. These modeling results under different pollution scenarios using water quality models are very important components of environmental impact assessment. Moreover, they are also the important basis for environmental management decisions as they not only provide data assistance for environmental management agencies to authorize the construction projects but also provide technical supports for water environmental protection agencies [6, 7]. Whether these model results are right or not can

greatly impact the reasonability and scientific significance of the authorized construction projects and the availability of pollution control measures.

With the development of model theory and the fast-updating computer technique [8], more and more water quality models have been developed with various model algorithms [3, 4]. Up to date, tens of types of water quality models including hundreds of model softwares have been developed for different topography, water bodies, and pollutants at different space and time scales [3, 9]. However, there are often big differences between these modeling results due to different theories and algorithms of these models, which can lead to the insistency of the predicted results using different models, and thus bringing different environment management decisions as these modeling results cannot be referred or compared to each other [10].

The uniform model standardization system has not been established yet in most developing countries [9, 11], which limits the wide applications of these models to environmental management due to no references and comparisons among different modeling results. Therefore, it is very necessary for most developing countries to better understand the availability and precisions of different water quality models and their methods of calculation and calibration and progress in the model standardization in order to apply effectively these models and form a good model regulation system [11, 12]. In particular, this work can contribute to making better environmental management policies and authorizing reasonable construction projects.

2. Development of Surface Water Quality Models

Surface water quality models have undergone a long period of development since Streeter and Phelps built the first water quality model (S-P model) to control river pollution in Ohio state of the US [13]. Surface water quality models have made a big progress from single factor of water quality to multifactors of water quality, from steady-state model to dynamic model, from point source model to the coupling model of point and nonpoint sources, and from zero-dimensional mode to one-dimensional, two-dimensional, and three-dimensional models [31, 32]. More than 100 surface water quality models have been developed up to now. Cao and Zhang [11] classified these models based on water body types, model-establishing methods, water quality coefficient, water quality components, model property, spatial dimension, and reaction kinetics. However, each surface water quality model has its own constraint conditions [33]. Therefore, water quality models still need to be further studied to overcome the shortcomings of these current models. Generally, the surface water quality models have undergone three important stages since 1925 to now.

2.1. The Primary Sage (1925–1965). Water quality of water bodies has been paid much more attention to at this stage. The water quality models focused on the interactions among different components of water quality in river systems as affected by living and industrial point source pollution [9, 11, 34]. Like hydrodynamic transmission, sediment oxygen demand and algal photosynthesis and respiration were considered as external inputs, whereas the nonpoint source pollution was just taken into account as the background load [35, 36].

At the beginning of this stage (from 1925 to 1965), the simple BOD-DO bilinear system model was developed and achieved a success in water quality prediction, and the one-dimensional model was applied to solve pollution issues in rivers and estuaries [33]. After that, most researchers modified and further developed the Streeter-Phelps models (S-P models). For example, Thomas Jr. [14] believed that BOD could be reduced without oxygen consumption due to sediment deposition and flocculation, and the reduction rate was proportional to the number of remained BOD; thus, the flocculation coefficient was introduced in the steady-state S-P model to distinguish the two BOD removal pathways. O'Connor [15] divided BOD parameter into carbonized BOD and nitrified BOD and added the effects of dispersion based on the equation. Dobbins-Camp [16, 17] added two coefficients, including the changing rate of BOD caused by sediment release and surface runoff as well as the changing rate of DO controlled by algal photosynthesis and respiration, to Thomas's equation.

2.2. The Improving Stage (1965–1995). From 1965 to 1970, water quality models were classified as six linear systems and made a rapid progress based on further studies on multidimensional coefficient estimation of BOD-DO models. The one-dimensional model was updated to a two-dimensional one which was applied to water quality simulation of lakes and gulfs [37, 38]. Nonlinear system models were developed during the period from 1970 to 1975 [39]. These models included the N and P cycling system, phytoplankton and zooplankton system and focused on the relationships between biological growing rate and nutrients, sunlight and temperature, and phytoplankton and the growing rate of zooplankton [35, 37, 39]. The finite difference method and finite element method were applied to these water quality models due to the previous nonlinear relationships and they were simulated using one- or two-dimensional models.

After 1975, the number of state variables in the models increased greatly, and the three-dimensional models were developed at this stage, and the hydrodynamic mode and the influences of sediments were introduced to water quality models [40, 41]. Meanwhile, water quality models were combined with watershed models to consider nonpoint source pollution input as a variable [42, 43]. The effects of sediments were coped with inner interaction processes of the models [43]; so, the sediment fluxes could vary accordingly under different input conditions. Therefore, the water quality management policies were greatly improved due to more constraint conditions and nonpoint source pollution simulation at watershed scale. The typical models including QUAL models [18, 19], MIKE11 model [22], and WASP models [23, 44] were developed and used at this stage. Meanwhile, the one-dimensional OTIS model developed by USGS was also applied to water quality simulation [45, 46].

2.3. The Deepening Stage (after 1995). Nonpoint source pollution has been reduced due to strong control in developed countries. However, the dry and wet atmospheric deposition such as organic compounds, heavy metals, and nitrogen compounds showed increasing effects on water quality of rivers [47–49]. Although nutrients and toxic chemical materials depositing to water surface have been included in model framework, these materials not only deposited directly on water surface but also they can be deposited on the land surface of a watershed and sequentially transferred to water body [20, 50], which has been an important pollutant source. From the viewpoint of management demands, an air pollution model has to be developed to introduce this proceed in the model, indicating that the static or dynamic atmospheric deposition should be related to a given watershed [51]. Therefore, at this stage, some air pollution models were integrated to water quality models to evaluate directly the contribution of atmospheric pollutant deposition [20].

With the exception of the typical models such as QUAL 2K model [52], WASP 6 model [24], QUASAR model [25, 53], SWAT model [21], and MIKE 21 [26] and MIKE 31 models [27] (Table 1), other water quality models have also been developed to simulate complicated water environmental conditions. For example, Whitehead et al. [54] developed

TABLE 1: Main surface water quality models and their versions and characteristics.

Models	Model version	Characteristics
Streeter-Phelps models	S-P model [13]; Thomas BOD-DO model [14]; O'Connor BOD-DO model [15]; Dobbins-Camp BOD-DO model [16, 17]	Streeter and Phelps established the first S-P model in 1925. S-P models focus on oxygen balance and one-order decay of BOD and they are one-dimensional steady-state models.
QUAL models	QUAL I [11]; QUAL II [18]; QUAL2E [19]; QUAL2E UNCAS [19]; QUAL 2K [20, 21]	The USEPA developed QUAL I in 1970. QUAL models are suitable for dendritic river and non-point source pollution, including one-dimensional steady-state or dynamic models.
WASP models	WASP1-7 models [22, 23]	The USEPA developed WASP model in 1983. WASP models are suitable for water quality simulation in rivers, lakes, estuaries, coastal wetlands, and reservoirs, including one-, two-, or three-dimensional models.
QUASAR model	QUASAR model [11, 24, 25]	Whitehead established this model in 1997. QUASAR model is suitable for dissolved oxygen simulation in larger rivers, and it is a one-dimensional dynamic model including PC_QUA SAR, HERMES, and QUESTOR modes.
MIKE models	MIKE11 [22]; MIKE 21 [26]; MIKE 31 [27]	Denmark Hydrology Institute developed these MIKE models, which are suitable for water quality simulation in rivers, estuaries, and tidal wetlands, including one-, two-, or three dimensional models.
BASINS models	BASINS 1 [11, 28]; BASINS 2 [11, 28]; BASINS 3 [11, 28]; BASINS 4 [28]	The USEPA developed these models in 1996. BASINS models are multipurpose environmental analysis systems, and they integrate point and nonpoint source pollution. BASINS models are suitable for water quality analysis at watershed scale.
EFDC model	EFDC model [29, 30]	Virginia Institute of Marine Science developed this model. The USEPA has listed the EFDC model as a tool for water quality management in 1997. EFDC model is suitable for water quality simulation in rivers, lakes, reservoirs, estuaries, and wetlands, including one-, two-, or three-dimensional models.

a semidistributed integrated nitrogen model (INCA) based on the effects of atmospheric and soil N inputs, land uses, and hydrology. More recently, Fan et al. [55] integrated QUAL 2K water quality model and HEC-RAS model to simulate the impact of tidal effects on water quality simulation. For the integration of point and nonpoint sources, the US Environmental Protection Agency (USEPA) developed a multipurpose environmental analysis system (BASINS), which makes it possible to assess quickly large amounts of point and nonpoint source [28]. Meanwhile, the USEPA also listed the EFDC model as a tool for water quality management.

Among the previously mentioned surface water quality models, these models including the Streeter-Phelps model, QUASAR model, QUAL model, WASP model, CE-QUAL-W 2 model, BASINS model, MIKE model, and EFDC model were widely applied worldwide [56, 57]. Recently, Kannel et al. [58] concluded that these public domain models (e.g., QUAL2EU, WASP7, and QUASAR) are the most suitable for simulating dissolved oxygen along rivers and streams. Generally, most developed countries (especially the US or European countries) have developed better and advanced surface water quality models [22, 27, 28, 30]. Some surface water quality models have also been established in some universities or institutes of China over the past years [11], but these models were still not widely utilized like MIKE models, EFDC model, and WASP models [59, 60].

3. Standardization of Surface Water Quality Models

Water quality models should be more available, standardized, and reliable when they are utilized to aid the important and valid reports (e.g., environmental impact assessment report). Therefore, it is very necessary for environmental management agencies to mandate or list some water quality models in order to guarantee the consistency of water quality models for regulatory purposes [61]. The models can be regulated and standardized through these pathways such as the establishment of the national model assessment indicator and validating system, published articles, workshops, or setting up local workgroup [62]. For example, The USEPA holds regular academic conferences on water quality models to identify and update regulatory models [62]. The European Union organizes regular workshops on the consistency of water quality models to evaluate the regulatory models. Moreover, the standardized models should be able to be downloaded free and have open origin codes.

Special research institutes of water quality models have been built to do a lot of researches on the regulation and standardization of water quality models in some regions or countries [62, 63]. They recommended some prediction models based on the requirements of environmental management. Compared to other countries, most water environmental models have been standardized in the US. The Water Science

Center belonging to the USEPA focuses on the following studies regarding water resources management and conservation, the theory and methods applied in water environments, numerical models, calculating tools, and databanks. Meanwhile, the USEPA also provides foundations for some universities, institutes, or companies to develop and compare related models and finish a series of research reports. In 2002, the USEPA mandated the Guidance for Quality Assurance Project Plans for Models, and some advices and guidance principles were given for the applications of water quality models in this guidance [64]. Additionally, the USEPA also authorized Tetra Tech Inc. to do the project of TMDL Model Evaluation and Research Needs, through which the modeling capacity, availability, and scopes of more than 60 models have been evaluated and compared using detailed appraisal forms [65]. Based on the above researches, the USEPA finally published the Guidance on the Development, Evaluation, and Application of Environmental Models in 2009. This guidance introduces concisely the characteristics and appropriate environment process modeling of these surface water environment models such as HSPF model, WASP model, and QUAL2E model and also gave the website links for more details of these models. The best practices for model evaluation are also appended to this guidance, which describes the methods, objectives, and procedures of model evaluation in detail [66]. Besides the guidance, the Council for Regulatory Environmental Modeling of the USEPA provides the model banks on its website. The United States Geological Survey, Federal Emergency Management Agency, and the United States Army Corps of Engineers also have similar model banks and detailed introduction for different types of models. The USEPA recommended its own developed models and those models developed by other research institutes or companies, but an announcement has been provided in the recommendation report that the recommended models do not denote that they have been authenticated by the USEPA [66]. The USEPA only suggested how to select appropriate models under different environmental conditions as each model has its own appropriate scope and scale. However, Kannel et al. [58] pointed out that the choice of a model depends upon availability of time, financial cost, and a specific application.

Similarly special research institute of model development and evaluation has been set up by the United Kingdom Environment Agency (UKEA). This institute helped the UKEA finish the Framework for Assessing the Impact of Contaminated Land on Groundwater and Surface Waterand put forward the procedure, method, and prediction models of surface water environmental impact assessment of potential pollution sources, which can assess the influencing degree of pollution sources on water environment. The Her Majesty's Inspectorate of Pollution (HMIP) recommended 54 surface water quality models and limiting conditions for rivers, lakes, reservoirs, estuaries, and sea pollution assessment. Aspinwall and Company Limited recommended 11 models for different conditions including 1 one-dimensional model, 4 two-dimensional models, and 6 three-dimensional models, of which 11 models for steady-state simulation and 10 models for dynamic simulation [67]. In Korea, the Ministry

of Environment made a general plan for water environmental management in 2006, which described 6 water quality prediction models in detail and recommended a series of numerical models including widely-used Qual2E model and EFDC model [68]. The MIKE models and Tuflow model were widely applied to predict surface water quality in Australia. MIKE models were adapted in Denmark to solve some issues in these fields such as ecology, environmental chemistry, water resources, hydraulic engineering, and hydrological dynamics. In China, the Delft 3D hydrological dynamic-water quality model has been used to simulate water environmental quality in Hong kong since 1970s and now become the standard model of Hong kong Environment Agency. Taiwan Environmental Protection Bureau issued the guidance on methods of water quality assessment of rivers and environmental impact assessment and provided a water quality model list for different conditions in this guidance. The Ministry of Environmental Protection of China formally published the Technical Guidelines for Environmental Impact Assessment (Surface water Environment) in 1993 and recommended some numerical models for rivers, lakes, estuaries, and marine environment under different conditions [69]. However, the standardized numerical models in China are still not provided yet up to date. Most models such as MIKE models, EFDC model, and Delft 3D model have been applied to simulate water environmental quality in most institutes of environmental impact assessment [70, 71]. However, little information is available on the differences in model results from different models and the suitability and parameter sensitivity of these models. Moreover, it is also an urgent task to standardize some numerical models to compare the modeling results among different regions efficiently. Additionally, Moriasi et al. [72] suggested to develop the consistent framework of model calibration and validation guidelines, as it is difficult to compare modeling results from different studies with different calibration and validation methods.

4. Measurements for the Standardization of Surface Water Quality Models

The appraisal techniques of the standardization of water quality models and their authentication system can provide an important scientific basis for the development of software informatization for water quality models and environmental impact assessment [68]. To improve the standardization of surface water quality models, the best way is to understand fully the status, progress, frame structure, assessing indicators, and authentication system of the standardization system of surface water quality models in developed countries, especially in some European or North American countries. Based on the previously mentioned, it is necessary for environmental management agencies of those countries without standardization models of water environmental quality to develop their own construction and frame structure of standardized model system of surface water quality, screen assessing indicators, procedures, and methods to establish their own authentication and standardization system for surface water quality models.

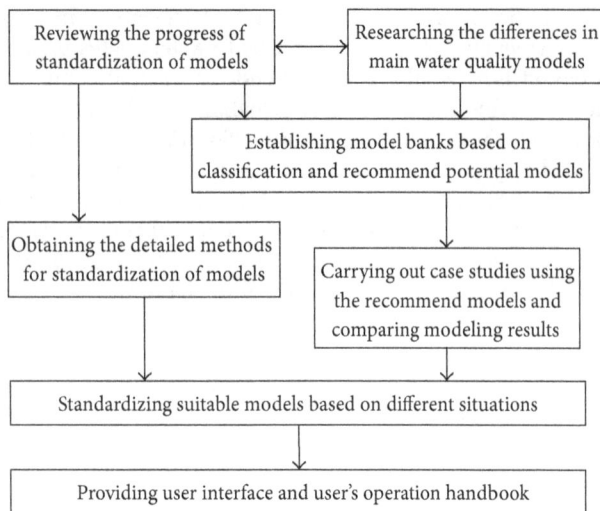

FIGURE 1: Flow chart of the standardization of water quality models.

The specific measures for the standardization of surface water quality models are given as follows (Figure 1).

(1) To research the water quality models which are widely used in the fields of surface water environmental impact assessment to know well the model mechanisms, suitable conditions, appropriate scopes, model parameters, stability, and the differences in modeling results.

(2) To develop case bank and data bank for surface water quality models through indoor experiments, case collection, and field monitoring.

(3) To compare the modeling results among different models and to conclude and analyze the input and output files, equations, theories, frames, and calculating methods of water quality models based on some case studies.

(4) To provide the screening indicators and appraisal methods for water quality models to establish the appraisal authentication system of these models and standardize the standard interfaces of input and output data for these models. To standardize some water quality models and list the standardized models for environmental impact assessment based on each country's actual conditions.

(5) To give the parameter calibration and validation methods and the access, sources, and recommended values of these parameters and put forward some standard proposals for typical model parameters considering the actual conditions of each country.

(6) To provide user interface of model graphs in native language and publish detailed model operation handbook including model inputs (data access, data processing), model structure, model calibration, model validation, parameter assessment, and model outputs.

5. Conclusions

Water quality models are very important to predict the changes in surface water quality for environmental management in the world. Worldwide, hundreds of surface water quality models have been developed. Moreover, some developed countries have mandated the guidance on water environmental quality assessment and provided some regulated models for surface water quality simulation. Therefore, it is very necessary for most developing countries to standardize some widely used water quality models for efficient environmental impact assessment. However, it is also a big challenge to standardize these models based on their own countries' actual conditions as a lot of investigations and researchers are still needed.

Acknowledgments

This work was financially supported by the Project of Environmental Protection Public Welfare Scientific Research Project, Ministry of Environmental Protection of the People's Republic of China (nos. 201309062 and 201309003).

References

[1] L. B. Huang, J. H. Bai, R. Xiao, H. F. Gao, and P. P. Liu, "Spatial distribution of Fe, Cu, Mn in the surface water system and their effects on wetland vegetation in the Pearl River Estuary of China," *CLEAN—Soil, Air, Water*, vol. 40, no. 10, pp. 1085–1092, 2012.

[2] J. H. Bai, B. S. Cui, B. Chen et al., "Spatial distribution and ecological risk assessment of heavy metals in surface sediments from a typical plateau lake wetland, China," *Ecological Modelling*, vol. 222, no. 2, pp. 301–306, 2011.

[3] Q. G. Wang, W. N. Dai, X. H. Zhao, F. Ding, S. B. Li, and Y. Zhao, "Numerical model of thermal discharge from Laibin power plant based on Mike 21," *Research of Environmental Sciences*, vol. 22, no. 3, pp. 332–336, 2009 (Russian).

[4] S.-M. Liou, S.-L. Lo, and C.-Y. Hu, "Application of two-stage fuzzy set theory to river quality evaluation in Taiwan," *Water Research*, vol. 37, no. 6, pp. 1406–1416, 2003.

[5] R. Xiao, J. H. Bai, H. F. Gao, J. J. Wang, L. B. Huang, and P. P. Liu, "Distribution and contamination assessment of heavy metals in water and soils from the college town in the Pearl River Delta, China," *CLEAN—Soil, Air, Water*, vol. 40, no. 10, pp. 1167–1173, 2012.

[6] J. H. Bai, H. F. Gao, R. Xiao, J. J. Wang, and C. Huang, "A review of soil nitrogen mineralization in coastal wetlands: issues and methods," *CLEAN—Soil, Air, Water*, vol. 40, no. 10, pp. 1099–1105, 2012.

[7] J. H. Bai, R. Xiao, K. J. Zhang, and H. F. Gao, "Arsenic and heavy metal pollution in wetland soils from tidal freshwater and salt marshes before and after the flow-sediment regulation regime in the Yellow River Delta, China," *Journal of Hydrology*, vol. 450-451, pp. 244–253, 2012.

[8] M. A. Ashraf, M. J. Maah, and I. Yusoff, "Morphology, geology and water quality assessment of former tinmining catchment," *The Scientific World Journal*, vol. 2012, Article ID 369206, 15 pages, 2012.

[9] J. Q. Wang, Z. Zhong, and J. Wu, "Steam water quality models and its development trend," *Journal of Anhui Normal University (Natural Science)*, vol. 27, no. 3, pp. 243–247, 2004.

[10] C. C. Obropta, M. Niazi, and J. S. Kardos, "Application of an environmental decision support system to a water quality trading program affected by surface water diversions," *Environmental Management*, vol. 42, no. 6, pp. 946–956, 2008.

[11] X. J. Cao and H. Zhang, "Commentary on study of surface water quality model," *Journal of Water Resources and Architectural Engineering*, vol. 4, no. 4, pp. 18–21, 2006 (Russian).

[12] M. Politano, M. M. Haque, and L. J. Weber, "A numerical study of the temperature dynamics at McNary Dam," *Ecological Modelling*, vol. 212, no. 3-4, pp. 408–421, 2008.

[13] H. W. Streeter and E. B. Phelps, *A Study of the Pollution and Natural Purification of the Ohio River*, United States Public Health Service, U.S. Department of Health, Education and Welfare, 1925.

[14] H. A. Thomas Jr., "The pollution load capacity of streams," *Water Sewage Works*, vol. 95, pp. 409–413, 1948.

[15] D. J. O'Connor, "The temporal and spatial distribution of dissolved oxygen in streams," *Water Resource Research*, vol. 3, no. 1, pp. 65–79, 1967.

[16] W. E. Dobbins, "BOD and oxygen relationships in streams," *Sanitary Engineering Division, American Society of Civil Engineers*, vol. 90, no. 3, pp. 53–78, 1964.

[17] T. R. Camp, *Water and Its Impurities*, Reinhold, New York, NY, USA, 1963.

[18] W. J. Grenney, M. C. Teuscher, and L. S. Dixon, "Characteristics of the solution algorithms for the QUAL II river model," *Journal of the Water Pollution Control Federation*, vol. 50, no. 1, pp. 151–157, 1978.

[19] L. C. Brown and T. O. Barnwell Jr., *The Enhanced Stream Water Quality Models QUAL2E and QUAL2E—UNCAD: Documentation and User Manual*, US Environmental Protection Agency, Environmental Research Laboratory, Athens, Ga, USA, 1987.

[20] S. R. Esterby, "Review of methods for the detection and estimation of trends with emphasis on water quality applications," *Hydrological Processes*, vol. 10, no. 2, pp. 127–149, 1996.

[21] B. Grizzetti, F. Bouraoui, K. Granlund, S. Rekolainen, and G. Bidoglio, "Modelling diffuse emission and retention of nutrients in the Vantaanjoki watershed (Finland) using the SWAT model," *Ecological Modelling*, vol. 169, no. 1, pp. 25–38, 2003.

[22] Danish Hydraulics Institute, *MIKE11, User Guide & Reference Manual*, Danish Hydraulics Institute, Horsholm, Denmark, 1993.

[23] R. B. Ambrose, T. A. Wool, and J. P. Connolly, *WASP4, A Hydrodynamic and Water Quality Model-Model Theory, User's Manual and Programmer's Guide*, US Environmental Protection Agency, Athens, Ga, USA, 1988.

[24] Y. Artioli, G. Bendoricchio, and L. Palmeri, "Defining and modelling the coastal zone affected by the Po river (Italy)," *Ecological Modelling*, vol. 184, no. 1, pp. 55–68, 2005.

[25] P. G. Whitehead, R. J. Williams, and D. R. Lewis, "Quality simulation along river systems (QUASAR): model theory and development," *Science of the Total Environment*, vol. 194-195, pp. 447–456, 1997.

[26] Danish Hydraulic Institute, *MIKE21: User Guide and Reference Manual*, Danish Hydraulic Institute, Horsholm, Denmark, 1996.

[27] Danish Hydraulic Institute, *MIKE 3 Eutrophication Module, User Guide and Reference Manual, Release 2. 7*, Danish Hydraulic Institute, Horsholm, Denmark, 1996.

[28] "The US Environmental Protection Agency," http://water.epa .gov/scitech/datait/models/basins/fs-basins4.cfm.

[29] The U.S. Environmental Protection Agency, "Compendium of tools for watershed assessment and TMDL development," Tech. Rep. EPA 841-B-97-006, The U.S. Environmental Protection Agency, Washington, DC, USA, 1997.

[30] The U.S. Environmental Protection Agency, "Review of potential modeling tools and approaches to support the BEACH Program," "Rep. No. EPA 823-R-99-002, The U.S. Environmental Protection Agency, Washington, DC, USA, 1999.

[31] Q. G. Wang, X. H. Zhao, M. S. Yang, Y. Zhao, K. Liu, and Q. Ma, "Water quality model establishment for middle and lower reaches of Hanshui river, China," *Chinese Geographical Sciences*, vol. 21, no. 6, pp. 647–655, 2011.

[32] Z.-X. Xu and S.-Q. Lu, "Research on hydrodynamic and water quality model for tidal river networks," *Journal of Hydrodynamics*, vol. 15, no. 2, pp. 64–70, 2003.

[33] D. H. Burn and E. A. McBean, "Optimization modeling of water quality in an uncertain environment," *Water Resources Research*, vol. 21, no. 7, pp. 934–940, 1985.

[34] S. Rinaldi and R. Soncini-Sessa, "Sensitivity analysis of generalized Streeter-Phelps models," *Advances in Water Resources*, vol. 1, no. 3, pp. 141–146, 1978.

[35] R. Riffat, *Fundamentals of Wastewater Treatment and Engineering*, CRC Press, Boca Raton, Fla, USA, 2012.

[36] P. P. Mujumdar and V. R. S. Vemula, "Fuzzy waste load allocation model: simulation-optimization approach," *Journal of Computing in Civil Engineering*, vol. 18, no. 2, pp. 120–131, 2004.

[37] D. I. Gough, "Incremental stress under a two-dimensional artificial lake," *Canadian Journal of Earth Sciences*, vol. 6, no. 5, pp. 1067–1075, 1969.

[38] P. Welander, "Wind-driven circulation in one-and two-layer oceans of variable depth," *Tellus*, vol. 29, pp. 1–16, 1968.

[39] S.-M. Yih and B. Davidson, "Identification in nonlinear, distributed parameter water quality models," *Water Resources Research*, vol. 11, no. 5, pp. 693–704, 1975.

[40] E. Wolanski, Y. Mazda, and P. Ridd, "Mangrove hydrodynamics," *Coastal and Estuarine Studies*, vol. 41, pp. 43–62, 1992.

[41] M. J. Zheleznyak, R. I. Demchenko, S. L. Khursin, Y. I. Kuzmenko, P. V. Tkalich, and N. Y. Vitiuk, "Mathematical modeling of radionuclide dispersion in the Pripyat-Dnieper aquatic system after the Chernobyl accident," *Science of the Total Environment*, vol. 112, no. 1, pp. 89–114, 1992.

[42] C. T. Hunsaker and D. A. Levine, "Hierarchical approaches to the study of water quality in rivers—spatial scale and terrestrial processes are important in developing models to translate research results to management practices," *BioScience*, vol. 45, no. 3, pp. 193–203, 1995.

[43] U. S. Tim and R. Jolly, "Evaluating agricultural nonpoint-source pollution using integrated geographic information systems and hydrologic/water quality model," *Journal of Environmental Quality*, vol. 23, no. 1, pp. 25–35, 1994.

[44] R. B. Ambrose, T. A. Wool, and J. L. Martin, *WASP5 X, A Hydrodynamic and Water Quality Model-Model Theory, User's Manual and Programmer's Guide*, Environmental Research Laboratory, US Environmental Protection Agency, Washington, DC, USA, 1993.

[45] K. E. Bencala and R. A. Walters, "Simulation of solute transport in a mountain pool-and-riffle stream: a transient storage model," *Water Resources Research*, vol. 19, no. 3, pp. 718–724, 1983.

[46] H. M. Valett, J. A. Morrice, C. N. Dahm, and M. E. Campana, "Parent lithology, surface-groundwater exchange, and nitrate retention in headwater streams," *Limnology and Oceanography*, vol. 41, no. 2, pp. 333–345, 1996.

[47] N. Poor, R. Pribble, and H. Greening, "Direct wet and dry deposition of ammonia, nitric acid, ammonium and nitrate to the Tampa Bay Estuary, FL, USA," *Atmospheric Environment*, vol. 35, no. 23, pp. 3947–3955, 2001.

[48] D. Golomb, D. Ryan, J. Underhill, T. Wade, and S. Zemba, "Atmospheric deposition of toxics onto Massachusetts Bay - II. Polycyclic aromatic hydrocarbons," *Atmospheric Environment*, vol. 31, no. 9, pp. 1361–1368, 1997.

[49] L. Morselli, P. Olivieri, B. Brusori, and F. Passarini, "Soluble and insoluble fractions of heavy metals in wet and dry atmospheric depositions in Bologna, Italy," *Environmental Pollution*, vol. 124, no. 3, pp. 457–469, 2003.

[50] R. P. Mason, N. M. Lawson, and K. A. Sullivan, "Atmospheric deposition to the Chesapeake Bay watershed—Regional and local sources," *Atmospheric Environment*, vol. 31, no. 21, pp. 3531–3540, 1997.

[51] J. H. Bai, R. Xiao, B. C. Cui et al., "Assessment of heavy metal pollution in wetland soils from the young and old reclaimed regions in the Pearl River Estuary, South China," *Environmental Pollution*, vol. 159, no. 3, pp. 817–824, 2011.

[52] X. Fang, J. Zhang, Y. Chen, and X. Xu, "QUAL2K model used in the water quality assessment of Qiantang River, China," *Water Environment Research*, vol. 80, no. 11, pp. 2125–2133, 2008.

[53] A. M. Sincock and M. J. Lees, "Extension of the QUASAR river-water quality model to unsteady flow conditions," *Journal of the Chartered Institution of Water and Environmental Management*, vol. 16, no. 1, pp. 12–17, 2002.

[54] P. G. Whitehead, E. J. Wilson, and D. Butterfield, "A semi-distributed Integrated Nitrogen model for multiple source assessment in Catchments (INCA): part I—model structure and process equations," *Science of the Total Environment*, vol. 210-211, pp. 547–558, 1998.

[55] C. Fan, C.-H. Ko, and W.-S. Wang, "An innovative modeling approach using Qual2K and HEC-RAS integration to assess the impact of tidal effect on River Water quality simulation," *Journal of Environmental Management*, vol. 90, no. 5, pp. 1824–1832, 2009.

[56] S. F. Fan, M. Q. Feng, and Z. Liu, "Simulation of water temperature distribution in Fenhe reservoir," *Water Science and Engineering*, vol. 2, no. 2, pp. 32–42, 2009.

[57] N. J. Morley, "Anthropogenic effects of reservoir construction on the parasite fauna of aquatic wildlife," *EcoHealth*, vol. 4, no. 4, pp. 374–383, 2007.

[58] P. R. Kannel, S. R. Kanel, S. Lee, Y.-S. Lee, and T. Y. Gan, "A review of public domain water quality models for simulating dissolved oxygen in rivers and streams," *Environmental Modeling and Assessment*, vol. 16, no. 2, pp. 183–204, 2011.

[59] S.-J. Zhang and W.-Q. Peng, "Water temperature structure and influencing factors in Ertan Reservoir," *Shuili Xuebao/Journal of Hydraulic Engineering*, vol. 40, no. 10, pp. 1254–1258, 2009.

[60] G. Wang, L. X. Han, and W. T. Chang, "Modeling water temperature distribution in reservoirs with 2D laterally averaged flow-temperature coupled model," *Water Resources Protection*, vol. 25, no. 2, pp. 59–63, 2009.

[61] FR 21506, "Requirements for Preparation, Adoption, and Submittal of State Implementation Plans (Guideline on Air Quality Models)," 40 CFR Part 51, April 2000.

[62] 68 FR 18440, "Revision to the Guideline on Air Quality Models: Adoption of a Preferred Long Range Transport Model and Other Revisions," April 2003.

[63] Project # 387-2006, "Evaluation of Potential Standardization Models for Canadian Water Quality Guidelines," January 2007.

[64] The US Environmental Protection Agency, "Guidance for quality assurance project plans for modeling," Tech. Rep. EPA QA/G-5M, 2002.

[65] The US Environmental Protection Agency, "TMDL model evaluation and research needs," Tech. Rep. EPA/600/R-05/149, November 2005.

[66] The US Environmental Protection Agency, "Guidance on the Development, Evaluation, and Application of Environmental Models," Tech. Rep. EPA/100/K-09/003, March 2009.

[67] Aspinwall and Company, "A framework for assessing the impact of contaminated land on groundwater and surface water. Volumes 1 & 2," CLR Report, 1994.

[68] B. K. Lee, "Water environment management master plan outline (2006–2015)—clean water, Eco River 2015," *Korea Environmental Policy Bulletin*, vol. 4, no. 3, pp. 1–10, 2006.

[69] HJ/T2.3-93, *Technical Guidelines for Environmental Impact Assessment (Surfacewater Environment)*, Ministry of Environmental Protection of the People's Republic of China, 1993.

[70] S.-J. Zhang and W.-Q. Peng, "Water temperature structure and influencing factors in Ertan Reservoir," *Shuili Xuebao/Journal of Hydraulic Engineering*, vol. 40, no. 10, pp. 1254–1258, 2009.

[71] Q. G. Wang, "Prediction of water temperature as affected by a pre-constructed reservoir project based on MIKE11," *CLEAN—Soil, Air, Water*, 2013.

[72] D. N. Moriasi, B. N. Wilson, K. R. Douglas-Mankin, J. G. Arnold, and P. H. Gowda, "Hydrologic and water quality models: use, calibration, and validation," *Transactions of the ASABE*, vol. 55, no. 4, pp. 1241–1247, 2012.

9

N : P Stoichiometry in a Forested Runoff during Storm Events: Comparisons with Regions and Vegetation Types

Lanlan Guo,[1,2] Yi Chen,[1] Zhao Zhang,[1] and Takehiko Fukushima[3]

[1] State Key Laboratory of Earth Surface Processes and Resource Ecology, Academy of Disaster Reduction and Emergency Management, Beijing Normal University, Beijing 100875, China
[2] Key Laboratory of Environmental Change and Natural Disaster, MOE, Beijing Normal University, Beijing 100875, China
[3] Graduate School of Life and Environmental Science, University of Tsukuba, Tsukuba 305-8572, Japan

Correspondence should be addressed to Zhao Zhang, zhangzhao@bnu.edu.cn

Academic Editor: Joao Torres

Nitrogen and phosphorus are considered the most important limiting elements in terrestrial and aquatic ecosystems. however, very few studies have focused on which is from forested streams, a bridge between these two systems. To fill this gap, we examined the concentrations of dissolved N and P in storm waters from forested watersheds of five regions in Japan, to characterize nutrient limitation and its potential controlling factors. First, dissolved N and P concentrations and the N : P ratio on forested streams were higher during storm events relative to baseflow conditions. Second, significantly higher dissolved inorganic N concentrations were found in storm waters from evergreen coniferous forest streams than those from deciduous broadleaf forest streams in Aichi, Kochi, Mie, Nagano, and with the exception of Tokyo. Finally, almost all the N : P ratios in the storm water were generally higher than 34, implying that the storm water should be P-limited, especially for Tokyo.

1. Introduction

Nitrogen (N) and phosphorus (P) are considered the most important limiting elements for vegetation in terrestrial ecosystems, especially for the forested headwater watersheds, where there is no direct application of fertilizer and the soils are commonly considered to be infertile [1, 2]. At longer time scales of ecosystem development, the predominant source of P is from rock weathering while N is of atmospheric origin. Accordingly, plants growing on young soils tend to be N-limited while vegetation on older, highly weathered soils is often P-limited [3–5]. Similarly, P is commonly believed to be the limiting nutrient in the freshwater whereas N as a limiting factor in estuarine or marine waters [6–8]. These broad-scale trends in the aquatic environment, soil features, and climatic conditions, coupled with smaller-scale heterogeneity in environment-vegetation interactions, produce a spectrum of nutrient availabilities and patterns of nutrient limitation [8–13]. As for nutrient-poor forested ecosystems, many researchers have suggested that N alone or N and P together may be limited elements based on the investigations such as

decomposition, nutrient mineralization, trace gas emissions, and leaching losses [13–15]. However, most observations have only focused on the foliage, root dynamics, litter, ground vegetation, and the soil, even only a few have been on stream water with the exception of studies on forested streams [15–17].

The N : P ratio has gained worldwide acceptance as an indicator of biological growth and nutrient cycling and has been successfully used in several studies of both aquatic and terrestrial areas [9, 10, 17]. As an important part of forest ecosystems, stream water serves an export function, and creates a bridge between terrestrial and aquatic ecosystems. Accordingly, the N : P ratio in stream runoff from forested watersheds has been used as an index to diagnose the nutrient status of both terrestrial forests and the water body downstream [2, 18, 19]. As the interface between terrestrial and aquatic systems, runoff from a forested watershed is a critical, however, complex process, since the two types of ecosystem have different dynamics which interact together [18, 20]. Additionally, viewed as a space-time matrix of possible outcomes, researchers have assumed that spatial factors predominate in terrestrial systems, and time factors are more critical in aquatic systems [7, 21]. The interface combines extreme hydrologic events (both varying in magnitude and time) with biogeochemical environments, which makes it critical for elucidating nutrient dynamics in the forested ecosystems [22, 23]. Most previous studies have shown the significant role of storm runoff in nutrient cycling relative to baseflow conditions, which is generally considered as an equilibrium state by many ecologists. However, studies on nutrient dynamics and nutrient limitation during the nonequilibrium situations, for example, extreme storms, have been generally ignored [18, 24].

In 2003, Turner et al. [25] found seasonal differences in nutrient limitations by investigating their temporal dynamics in soil and a stream, in which marked P limitation was expected throughout the year, with the likelihood of some communities becoming N-limited during the spring. More-over, in a previous study on runoff from forested watersheds during baseflow conditions, we found that the N : P ratio showed clear regional characteristics, and a uniformly higher N : P ratio than from an evergreen conifer (EC) than decid-uous broadleaf (DB) forest among five Japanese forests [2]. But water chemistry in small forest watersheds is very sen-sitive to precipitation/discharge events [26, 27], particularly in response to extreme ones such as flooding. Considering the vital role of storm runoff in regulating the nutrient cycles in a forest ecosystem [2], it would be beneficial to characterize the nutrient concentration and N : P ratio in streams during storm events. Therefore, our objectives in this study were the following: (1) to characterize the N : P ratios in storm runoff at both regional and ecosystem (forest) scales; (2) to determine the potential factors controlling their spatial variability; (3) to understand what the potential limiting nutrients in storm runoff waters are from forested watersheds. Our study aimed to provide evidence to fill the knowledge gap between terrestrial and aquatic systems, with a special focus on storm events, a nonequilibrium situation which generally receives little attention.

2. Methods

2.1. Study Sites.
We chose a set of headwater streams located throughout five regions in Japan (Aichi, Kochi, Mie, Nagano, and Tokyo), which were characterized by different climatic, geological, and topographical conditions, matching as closely as possible the following watershed criteria: (1) the presence of first- or second-order streams, (2) the watersheds were entirely forested, and (3) a lack of obvious lakes or wetlands in the watershed. All of the streams were fast flowing and the streambeds were composed of boulders and large cobbles with little accumulation of fine sediment. Within each region we selected two to six streams, for which the vegetation (including evergreen conifer (EC) plantations, and natural deciduous broadleaf forest (DB)) was representative of the region and shared similar geological environments. The topographic, meteorological, and vegetation details of each sampling site are summarized in supplementary material see Table 1 and Figure 1 in Supplementary Material available online at doi: 10.1100/2012/257392).

The Aichi sites were located in the Aichi Research Forest of The University of Tokyo, east of Inuyama in Aichi Prefecture. The watershed was composed mainly of Neogene sediments and had an undulating topography. The Kochi sites were located in the Tsuzuragawa watershed, which is part of the east tributary of the Shimanto River, southeast of Tashouchou in Kochi Prefecture. The terrain comprised somewhat steep and incised hillsides. Sandstone and pelitic rocks were dominant in these areas. The Mie sites were located in Taikichiou, Mie Prefecture, and were characterized by generally steep slopes. The prevalent rock was gneiss, with a typical brown forest soil cover. The Nagano sites were located in the Terasawayama Education and Research Forest of Shinshu University in Ina, Nagano Prefecture. The Tanazawagawa, a small tributary of the Tenryu River, discharges from this area. As in Mie, steep slopes are a common feature of this area. The watersheds were underlain by granite. The Tokyo sites were located in the Joubanzawa watershed, a tributary of the Arakawa River, and near the headwaters of the Narikigawa in Tokyo. Steep, incised hillsides are also common in this area. The watershed is underlain by sandstone, pelitic rocks, and chert.

2.2. Field Sampling and Observations.
We performed our sampling regime from June 2004 to July 2005 during 10 storm events (not every storm event occurred in every region, and ensured that samples were taken from both the EC and DB streams during the same event and in the same region). Samples were collected in 500 mL clean polyethylene bottles during rainfall storm events using autosamplers (Teledyne Isco Inc., USA, model 6712) which were activated automatically when the stream water level increased. Discrete samples were usually collected half-hourly, or sometimes hourly or two-hourly during the falling limb of the hydro-graph. All samples were shipped by refrigerated express to our laboratory and placed in a cool store until analysis. Next, the samples were filtered through precombusted (at $450°C$ for 3 hours) and preweighed glass fiber filters (0.45μm Whatman GF/F, Chicago, USA). The filters were dried at $90°C$ for 24 hours then reweighed. The filtrates were retained at $1°C$ for further analysis. A series of analyses were carried out on the filtered subsamples. The following were measured using an Auto-analyzer (Traacs 800, Bran Luebbe Co., New York, USA): $NH_4-N+NO_3+NO_2-N$, as dissolved inorganic N (DIN), the molybdate-reactive fraction of P (DIP), total dissolved N (DTN: persulfate oxidation method), and the total dissolved P (DTP: persulfate oxidation method). Dissolved organic N (DON) and dissolved organic P (DOP) were determined as the difference between DTN and DIN and that between DTP and DIP, respectively. The N : P ratio was calculated from the molar ratio of DTN and DTP.

2.3. Data Analysis.
Descriptive statistics were first conducted to investigate the nutrient concentrations for the complete dataset, and to make comparisons with baseflow conditions. Then we investigated the correlation between the water quality indicators. We grouped the data by region, and performed a one-way analysis of variance (ANOVA) to examine the regional distribution of N and P constituents

and the N : P ratio. The Turkey multiple comparison and F-test were used to identify significant differences. We also conducted ANOVA analyses to investigate the differences between the two vegetation types, that is, the EC and DB forests and the factors influencing the concentration of N and P and the N : P ratio. Finally, nutrient limitations and during storm flow conditions were discussed.

3. Results

3.1. Characterization of Water Quality in Storm Runoff. Table 1 showed that higher concentrations of dissolved N, and P occurred in storm flow than during baseflow conditions with the exception of DIP. The means for DTN, DTP, DIN, and DOP during storm flows were about double those during the baseflow conditions. Moreover, the mean DON reached a peak 4.8-times greater than that during baseflow conditions. As for the N : P ratio, the mean in stormflow water flow was about 1.4-times greater than at baseflow. Furthermore, we found that DIN was the dominant N component, being 28.7 and 17.9 μmol L^{-1} higher than DON concentrations during baseflow, and stormflow conditions, respectively. Of the inorganic N, NO$_3$–N was the dominant constituent, comprising 55 to 95% in storm runoff for each site (data not shown). However, DOP concentrations were the dominant portion of DTP (accounting for 62–75% of the total) in storm runoff, and the concentrations of DOP were also higher on a weekly basis than during baseflow conditions.

The correlation analysis revealed that there were strong positive correlations between any two parameters, with the exception of the correlation between DON and DOP ($r =$ 0.081, $P =$ 0.062; Table 2). The significant correlations (all $P < 0.001$) between both the inorganic components and others indicated a potential similarity in factors controlling the nutrient runoff from forested watersheds during storm events. However, the lack of a correlation between two organic components (DOP and DON) implied the presence of a complex mechanism controlling their runoff during extreme storm events.

3.2. Regional Characteristics of Nutrient Concentrations and N : P Ratios during Storm Events. Since many environmental factors (including climate, topography, and soil) interact on a regional scale [28], we summarized factors controlling water quality for each site, and grouped them into five regions (Table 3). There were regional differences in nutrient concentrations among the five regions during storm events. The concentrations of DIN in Tokyo were significantly higher than those from the other regions ($P < 0.01$), with an average of 142 μmol L^{-1}, followed by Nagano (at 37 μmol L^{-1}), Kochi (at 21 μmol L^{-1}), Mie (at 11 μmol L^{-1}), and Aichi (at 10 μmol L^{-1}). The spatial patterns of DIN during storm events were consistent with our previous study during baseflow conditions [2]. The regional pattern of DTN was similar to DIN, but the highest concentration of DON existed in Mie and with a completely different regional pattern to that during baseflow conditions: Mie (44 μmol L^{-1}) > Nagano

(13 μmol L^{-1}) > Aichi (7 μmol L^{-1}) > Tokyo (6 μmol L^{-1}) > Kochi (3.5 μmol L^{-1}), implying that there were different controlling factors for DIN and DON in forested ecosystems, especially during storm events [27].

The concentrations of all the forms of P in the storm runoff in Nagano were the highest amongst the five regions with averages of 0.77, 0.37, and 0.40 μmol L^{-1} for DTP, DIP, and DOP, respectively (Table 3). Additionally, a consistent regional pattern was found for the concentration of DTP, DIP, and DOP in storm runoff, being Nagano > Tokyo > Kochi > Mie > Aichi. However, there was an inverse order between Nagano and Tokyo when compared with the regional pattern of DIN during both storm and baseflow conditions. The dissolved inorganic phosphorus and total dissolved P in Nagano and Tokyo were significantly ($P < 0.01$) greater than those in other three regions. Although there was great diversity in climate, forest type, and geological conditions among the regions, taking account into the similar spatial patterns of the different dissolved P constituents (Table 3) and the significantly positive correlation between them (Table 2), we were convinced that the same potential factors controlling the dissolved P runoff during storm events were operating on a large spatial scale rather than just in each watershed.

As for the N : P ratio in the five regions, a distinct regional pattern was shown, with significantly lower N : P ratios in Kochi than those in the four other sites ($P < 0.05$; Table 3). Interestingly, different spatial patterns relating to the N : P ratio of storm flow (Kochi < Nagano < Mie < Aichi < Tokyo) were found when comparing them with baseflow conditions (Kochi < Mie < Aichi < Tokyo < Nagano) [2]. These results indicated that the "N-saturation" region, that is, Tokyo, would be more affected by being P-limited than the other four regions during storm events when N inputs from atmospheric sources increased.

3.3. Different Responses to Nutrient Concentrations and the N : P Ratio in Evergreen Conifer and Deciduous Broadleaf Forests. We found no significant differences in the DIP concentration between the DB and EC in the four regions including Aichi, Kochi, and Mie while significantly higher DIP was shown in the DB than that of the EC in Nagano and Tokyo ($P < 0.001$; top and left in Figure 1). Concentrations of DTP in the DB were significantly greater than those of the EC in Nagano ($P < 0.001$) and Tokyo ($P < 0.001$; top and right in Figure 1). However, the differences in N concentrations between the two forests were more significant when comparing the differing P constituents. Concentrations of DIN (mainly in the form of NO$_3$–N) in the EC were significantly higher than those in the DB for Aichi ($P < 0.001$), Kochi ($P < 0.001$), Mie ($P < 0.001$), and Nagano ($P < 0.001$) while there was an inverse trend for Tokyo ($P < 0.001$; bottom and left in Figure 1). The similar site responses of DTN in storm flow to DIN are shown in each region (bottom and right in Figure 1). The different responses of the two types of vegetation to the differing nutrient concentrations as a result of storm flow was a consistently higher N : P ratio (DTN : DTP) in the EC than in the DB in each region (right part of Figure 1),

TABLE 1: Summary of forested runoff nutrient concentrations (μmol L^{-1}) and the molar N : P ratio in storm and baseflow states for the complete dataset (all sites).

Condition	Value	DTP	DTN	DIP	DIN	DOP	DON	N : P Ratio
Storm	Min	0.02	1.36	0.02	0.18	0.02	0.43	7
	Max	2.29	226.36	2.26	210.72	1.45	416	2567
	Mean	0.33	51.62	0.12	45.95	0.21	17.23	236
	SD	0.34	56.32	0.21	55.91	0.21	32.53	279
Baseflow*	Min	0.03	0.75	0.02	0.60	0.02	0.18	7
	Max	1.23	112	1.03	106	0.41	14.1	1858
	Mean	0.16	24.6	0.13	21.5	0.09	3.59	165
	SD	0.21	25.2	0.17	23.6	0.07	2.52	198

N : P ratios are defined as DTN : DTP molar ratios; molar N: $P \approx 0.45$ N : P (by mass); *data from Zhang et al. [2].

TABLE 2: Relationships between the different forms of N and P.

r	DTP	DTN	DIP	DIN	DOP	DON
			P			
1-7 DTP		<0.001	<0.001	<0.001	<0.001	<0.001
DTN	0.466**		<0.001	<0.001	<0.001	<0.001
DIP	0.798**	0.374**		<0.001	<0.001	<0.001
DIN	0.427**	0.992**	0.360**		0.062	<0.001
DOP	0.804**	0.371**	0.284**	0.321**		<0.001
DON	0.136**	0.151**	0.135**	0.190**	0.081	

** Correlation is significant at the $P < 0.01$ level.

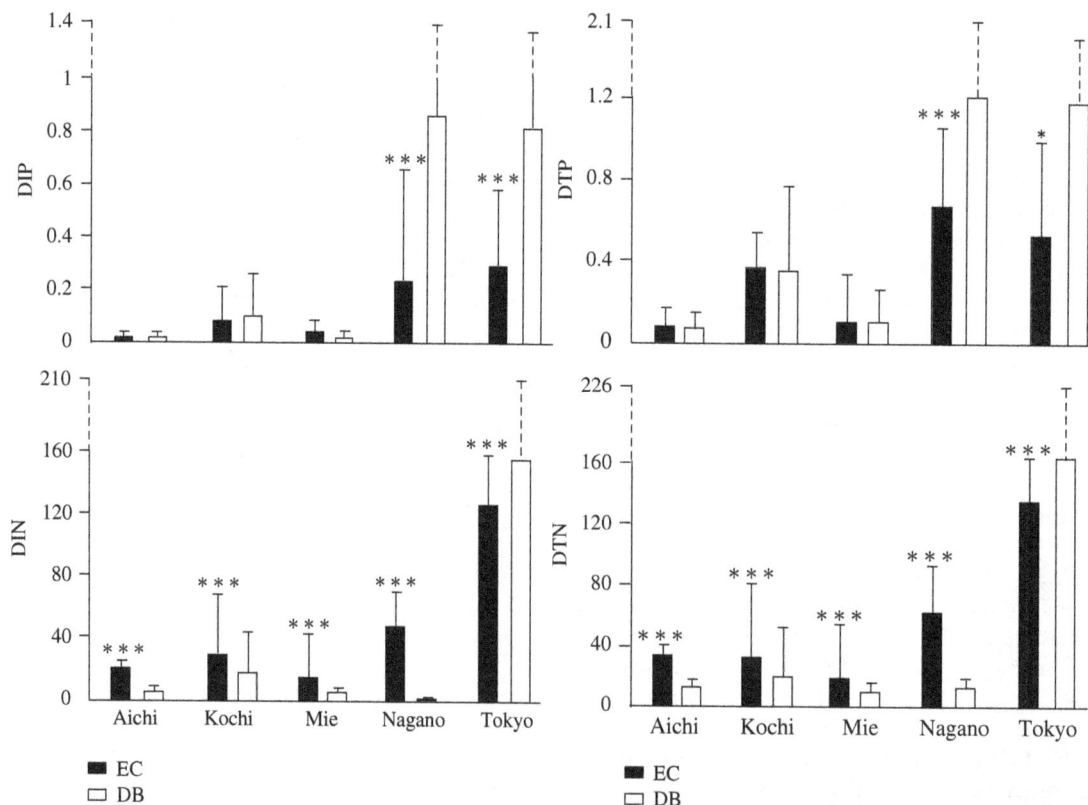

FIGURE 1: Forested stream nutrient concentrations for the different vegetation types in the five regions during storm events. The symbol on the top of columns shows a significant difference at either *** $P \leq 0.001$, ** $0.001 \leq P \leq 0.01$, or * $0.01 \leq P \leq 0.05$.

TABLE 3: Means for nutrient concentrations (μmol L^{-1}) and the molar N : P ratio at the 20 sites across the five study regions in Japan (standard deviations are in parentheses).

Site	Number of samples	DTP	DTN	DIP	DIN	DOP	DON	N : P ratio
EC-A3	19	0.07 (0.06)	31.73 (4.71)	0.02 (0.01)	21.38 (7.56)	0.06 (0.05)	10.35 (10.11)	638 (296.37)
DB-A4	53	0.06 (0.04)	11.95 (3.92)	0.02 (0.01)	6.16 (3.57)	0.04 (0.03)	5.79 (1.44)	241 (129.75)
Aichi	**72**	**0.06 (0.04)**	**17.17 (9.69)**	**0.02 (0.01)**	**10.18 (8.33)**	**0.05 (0.04)**	**7.00 (5.62)**	**346 (256.25)**
DB-K2	24	0.48 (0.29)	14.86 (3.22)	0.14 (0.16)	12.75 (2.77)	0.34 (0.14)	2.11 (0.94)	40 (23.19)
DB-K3	64	0.30 (0.16)	20.90 (14.27)	0.09 (0.12)	17.24 (12.48)	0.22 (0.12)	3.66 (2.88)	79 (47.65)
EC-K4	4	0.06 (0.03)	1.74 (0.27)	0.20 (0.00)	0.26 (0.16)	0.04 (0.23)	1.47 (0.30)	40 (23.36)
EC-K5	5	0.41 (0.16)	37.67 (2.86)	0.04 (0.05)	34.58 (2.53)	0.37 (0.18)	3.09 (0.79)	107 (48.05)
EC-K6	21	0.42 (0.11)	18.94 (1.65)	0.05 (0.05)	16.56 (1.51)	0.37 (0.12)	2.38 (0.69)	49 (15.97)
EC-K7	19	0.35 (0.11)	54.53 (15.14)	0.16 (0.99)	47.90 (11.98)	0.19 (0.15)	6.62 (5.04)	169 (70.19)
Kochi	**137**	**0.35 (0.19)**	**24.26 (17.37)**	**0.10 (0.12)**	**20.74 (15.34)**	**0.26 (0.15)**	**3.52 (3.08)**	**80 (59.28)**
EC-M1	29	0.06 (0.08)	15.17 (11.63)	0.02 (0.00)	12.01 (8.47)	0.05 (0.08)	53.97 (50.00)	352 (221.19)
EC-M2	40	0.13 (0.10)	13.87 (11.89)	0.04 (0.02)	10.01 (9.86)	0.10 (0.10)	53.77 (39.13)	117 (76.00)
EC-M3	32	0.08 (0.09)	16.92 (8.67)	0.03 (0.04)	13.18 (7.43)	0.06 (0.08)	52.36 (31.13)	370 (248.20)
EC-M4	4	0.10 (0.08)	20.56 (10.33)	0.02 (0.00)	9.31 (0.74)	0.08 (0.07)	157.4 (150.69)	433 (471.02)
EC-M5	10	0.12 (0.13)	37.93 (12.19)	0.02 (0.00)	34.46 (11.77)	0.11 (0.13)	48.70 (26.98)	670 (493.82)
DB-M8	39	0.09 (0.09)	9.46 (4.75)	0.03 (0.18)	5.44 (2.69)	0.07 (0.08)	4.01 (3.61)	182 (107.83)
Mie	**154**	**0.10 (0.10)**	**15.40 (11.68)**	**0.03 (0.02)**	**11.44 (10.18)**	**0.07 (0.09)**	**43.95 (50.03)**	**275 (261.32)**
EC-N2	24	0.56 (0.50)	68.36 (6.18)	0.14 (0.46)	61.91 (8.01)	0.42 (0.31)	6.45 (4.79)	190 (112.23)
EC-N4	9	1.01 (0.22)	53.34 (14.74)	0.24 (0.07)	5.05 (2.86)	0.78 (0.17)	48.29 (12.11)	52.67 (7.04)
EC-N5	20	0.61 (0.24)	56.75 (13.60)	0.34 (0.06)	50.20 (13.68)	0.27 (0.25)	6.55 (2.13)	100 (32.62)
DB-N6	15	1.20 (0.29)	11.50 (4.32)	0.87 (0.20)	0.77 (0.83)	0.33 (0.10)	10.73 (3.66)	9 (2.67)
Nagano	**68**	**0.77 (0.44)**	**50.41 (23.73)**	**0.37 (0.40)**	**37.45 (27.95)**	**0.40 (0.29)**	**12.96 (15.02)**	**106 (98.13)**
EC-T5	57	0.51 (0.33)	133.92 (28.21)	0.30 (0.32)	126.93 (30.28)	0.22 (0.16)	6.98 (4.83)	483 (464.92)
DB-T6	60	0.70 (0.46)	161.54 (31.80)	0.29 (0.27)	156.38 (32.97)	0.41 (0.25)	5.16 (5.38)	390 (297.56)
Tokyo	**117**	**0.61 (0.41)**	**148.08 (33.02)**	**0.29 (0.30)**	**142.03 (34.8)**	**0.32 (0.23)**	**6.05 (5.18)**	**435 (389.30)**

which is also in accordance with those during baseflow conditions [2]. Therefore, it could be implied that the water bodies downstream of the EC will be more affected by being P-limited than the DB streams considering the increasing amounts of N inputs from atmospheric sources [24, 29] and the natural low fertility of the forest soil [30].

4. Discussion

4.1. Factors Controlling the Spatial Pattern of Nutrient Concentrations and the N : P Ratio in Storm Runoff. To compare the relative impacts of the two main variables, that is, region and vegetation (DB, EC), we calculated the weighting of the two factors with respect to the nutrient concentration and the N : P ratio using the whole dataset from the five regions. All F values for the regional factors were higher than those for the differing vegetation types, suggesting that region plays a more important role in controlling the nutrient concentrations and the N : P ratio under storm conditions.

4.2. Potential Limiting Nutrients in Storm Runoff Waters from Forested Watersheds. To quantitatively describe the limiting

nutrients, we assumed that phototrophic bacteria are likely to be limited by N when the ambient N : P ratio is less than 20, and limited by P when the N : P ratio is greater than 34 [25, 31]. Trends in DTP and DTN concentration by vegetation type for each region are shown in Figure 2. Three patterns could generally be determined according the size of the N : P ratios. Values located above the 34 : 1 line indicate P-limited conditions; between two lines for nonnutrient limited and on the 20 : 1 line as being N-limited. We found almost all the N : P ratios were greater than 34, except the values for the DB in Nagano, 26% and 4% of the values for the DB and EC, respectively, in Kochi, 11% and 4% of the values for the DB and EC in Mie, the ratios being mostly located near the 34 : 1 line. From the plots in Aichi and Tokyo, it could be concluded that storm water was generally P-limited from both the DB and EC forested streams. However, in Nagano, all storm water from the EC-forested streams was P-limited while storm water from the DB streams was N-limited. In Kochi, 96 and 74% of the storm water from EC and DB streams in storm events would be categorized as being P-limited while 4 and 26% of the storm water from the EC and DB forested streams in storm events would be N-limited. In Mie, for the EC and DB about 96 and 89% of the storm water

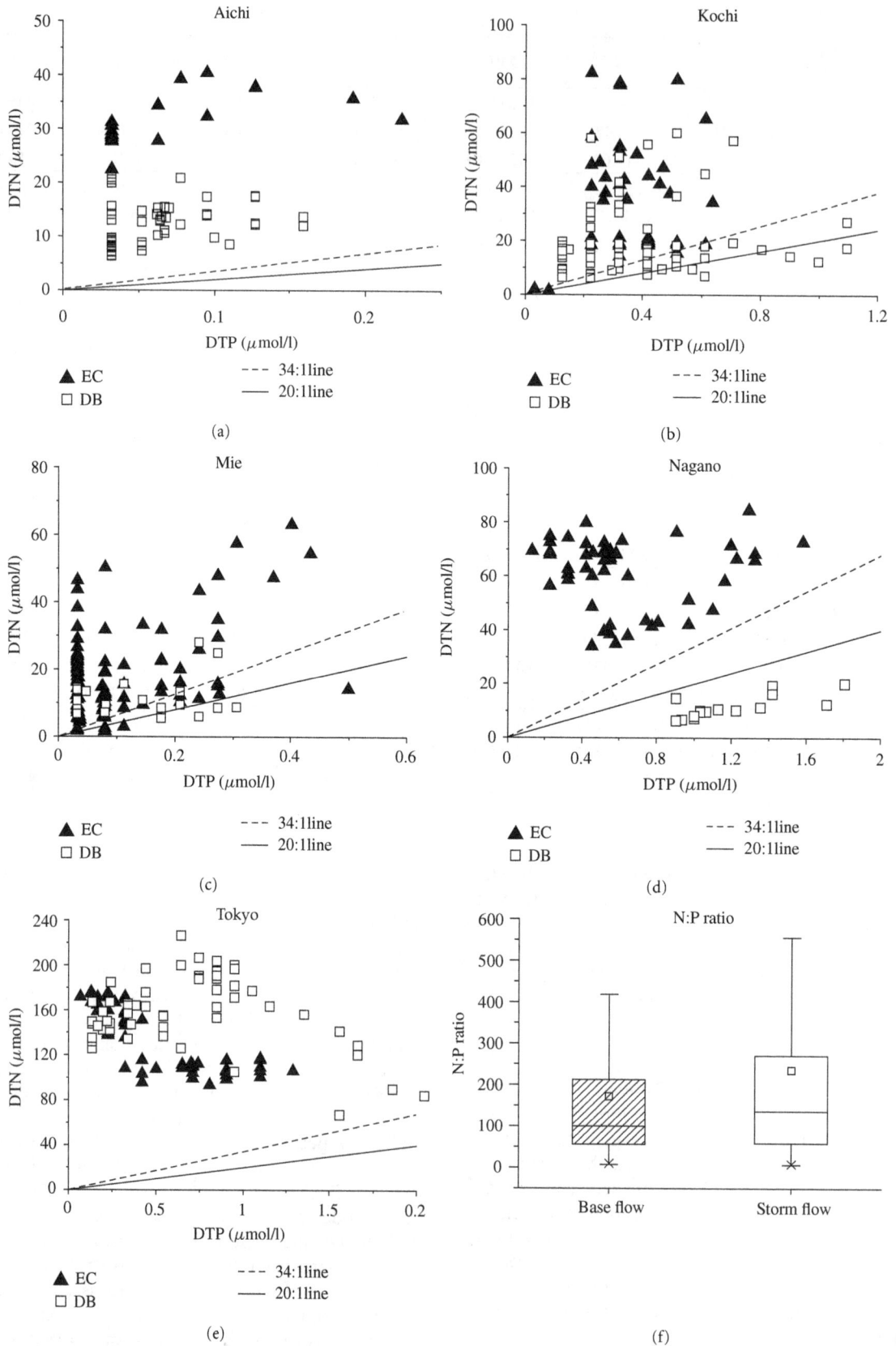

FIGURE 2: Relationships between the runoff concentrations of DTN and DTP in the EC and DB forests during storm events in each region.

were P-limited, respectively, and likewise 4 and 11% of the storm water from the EC and DB forested streams were N-limited. Generally, the N : P ratios of the EC sites were located to the left of the 34 : 1 line than those for the DB forests. These results suggest that the storm water from the EC streams should be more P-limited than that from the DB-streams, which is consistent with our previous study [2]. Moreover, the N : P ratios in the storm water were higher than under baseflow condition (bottom and right in Figure 2), implying that the storm water should be more sensitive to being P-limited relative to baseflow conditions.

5. Conclusions

In this study, the concentrations of dissolved N and P in storm water from forested watersheds in five regions in Japan during storm events were analyzed and compared. First, the concentrations of DIN, DON, DOP, DTN, DTP, and the N : P ratio were higher during storm events relative to baseflow conditions at all the sites, and the same spatial pattern for DIN in storm water was found as for baseflow conditions. Additionally, the spatial patterns across the five regions for all P constituents including DIP, DOP, and DTP followed the same order as being the greatest in Nagano > Tokyo > Kochi > Mie > Aichi, a little different from that of the trends in DIN during both storm and baseflow conditions, in which there was an inverse order for Nagano and Tokyo. The differences in concentrations and spatial patterns of nutrients between stormflow and baseflow conditions across the five regions verified the different mechanisms controlling the nutrient runoff during these two stream states.

Second, significantly higher concentrations of DIN were found in storm water from the EC sites than from the DB with the exception of Tokyo. However, the DTP concentrations from the EC were significantly lower than those from the DB in Nagano and Tokyo, with no significant differences evident between the two types of vegetation in the other three regions. Interestingly, a consistently higher N : P ratio in the storm water from the EC than those of the DB among each region were indicated, which is also in accordance with those measured during baseflow conditions.

Finally, almost all the N : P ratios in the storm water were generally higher than 34 with the exception of those from the DB in Nagano and a very small fraction of the sites in Kochi and Mie, implying that forested stream water in storm events is P-limited, especially in Tokyo, and at the EC sites in Nagano. When comparing the N : P ratios during baseflow conditions, there should be a greater sensitivity to P limitation in the storm water from forested watersheds to increasing trends in atmospheric N deposition and extreme storm events, more of which are possible in the future.

Acknowledgments

This paper was funded by the Chinese National Science Foundation projects (no. 41101074), the Fundamental Research Funds for the Central Universities, the National Basic Research Program of China (2012CB955404), State Key Laboratory of Earth Surface Processes and Resource Ecology (2010-ZY-10, 2011-KF-06), and International Cooperation Project, the Ministry of Science and Technology of China (40821140354, 2010DFB20880). This paper was also funded by the Japan Science and Technology Agency, Core Research for Evolutional Science and Technology Research for Evolutional Science and Technology.

References

[1] R. W. Dana, S. B. Emily, O. H. J. Robert, and E. L. Gene, "Forest age, wood and nutrient dynamics in headwater streams of the Hubbard Brook Experimental Forest, NH," *Earth Surface Processes and Landforms*, vol. 32, no. 8, pp. 1154–1163, 2007.

[2] Z. Zhang, T. Fukushima, P. Shi et al., "Baseflow concentrations of nitrogen and phosphorus in forested headwaters in Japan," *Science of the Total Environment*, vol. 402, no. 1, pp. 113–122, 2008.

[3] E. G. Jobbágy and R. B. Jackson, "The distribution of soil nutrients with depth: global patterns and the imprint of plants," *Biogeochemistry*, vol. 53, no. 1, pp. 51–77, 2001.

[4] P. B. Reich and J. Oleksyn, "Global patterns of plant leaf N and P in relation to temperature and latitude," *Proceedings of the National Academy of Sciences of the United States of America*, vol. 101, no. 30, pp. 11001–11006, 2004.

[5] R. J. Raison and K. K. Partap, "Possible impacts of climate change on forest soil health," *Soil Health and Climate Change*, vol. 29, no. 3, pp. 257–285, 2011.

[6] P. M. Vitousek and R. W. Howarth, "Nitrogen limitation on land and in the sea: how can it occur?" *Biogeochemistry*, vol. 13, no. 2, pp. 87–115, 1991.

[7] C. A. Klausmeler, E. Litchman, T. Daufreshna, and S. A. Levin, "Optimal nitrogen-to-phosphorus stoichiometry of phytoplankton," *Nature*, vol. 429, no. 6988, pp. 171–174, 2004.

[8] J. J. Elser, M. E. S. Bracken, E. E. Cleland et al., "Global analysis of nitrogen and phosphorus limitation of primary producers in freshwater, marine and terrestrial ecosystems," *Ecology Letters*, vol. 10, no. 12, pp. 1135–1142, 2007.

[9] J. J. Elser, M. M. Elser, N. A. Mackay, and S. R. Carpenter, "Zooplankton-mediated transitions between N- and P-limited algal growth," *Limnology & Oceanography*, vol. 33, no. 1, pp. 1–14, 1988.

[10] A. R. Townsend, C. C. Cleveland, G. P. Asner, and M. M. C. Bustamante, "Controls over foliar N:P ratios in tropical rain forests," *Ecology*, vol. 88, no. 1, pp. 107–118, 2007.

[11] L. S. Robert, L. L. Christian, N. W. Michael et al., "Stoichiometry of soil enzyme activity at global scale," *Ecology Letters*, vol. 11, no. 11, pp. 1252–1264, 2008.

[12] Z. Y. Yuan and H. Y. H. Chen, "Global-scale patterns of nutrient resorption associated with latitude, temperature and precipitation," *Global Ecology and Biogeography*, vol. 18, no. 1, pp. 11–18, 2009.

[13] Z. Y. Yuan and H. Y. H. Chen, "Global trends in senesced-leaf nitrogen and phosphorus," *Global Ecology and Biogeography*, vol. 18, no. 5, pp. 532–542, 2009.

[14] C. N. Jason, E. H. Sarah, and M. V. Peter, "Nutrient and mineralogical control on dissolved organic C, N and P fluxes and stoichiometry in Hawaiian soils," *Biogeochemistry*, vol. 51, no. 3, pp. 283–302, 2000.

[15] M. Stefand, A. T. John, B. J. Robert, and P. Amilcare, "Stoichiometric controls on carbon, nitrogen, and phosphorus dynamics in decomposing litter," *Ecological Monographs*, vol. 80, no. 1, pp. 89–106, 2010.

[16] R. B. Jackson, H. A. Mooney, and E. D. Schulze, "A global bud-
     get for fine root biomass, surface area, and nutrient contents,"
     *Proceedings of the National Academy of Sciences of the United
     States of America*, vol. 94, no. 14, pp. 7362–7366, 1997.

[17] E. M. Megan, D. Tanguy, and O. H. Lars, "Scaling of C:N:P
     stoichiometry in forests worldwide: implications of terrestrial
     redfield-type ratios," *Ecology*, vol. 85, no. 9, pp. 2390–2401,
     2004.

[18] P. C. Christopher and J. M. Jeffrey, "Linking the hydrologic and
     biogeochemical controls of nitrogen transport in near-stream
     zones of temperate-forested catchments: a review," *Journal of
     Hydrology*, vol. 199, no. 1-2, pp. 88–120, 1997.

[19] J. M. Myron, "Linkages of nitrate losses in watersheds to
     hydrological processes," *Hydrological Processes*, vol. 15, pp.
     3305–3307, 2001.

[20] S. Hideaki, S. Osamu, T. Hisano et al., "Nitrogen dynamics in
     the hyporheic zone of a forested stream during a small storm,
     Hokkaido, Japan," *Biogeochemistry*, vol. 69, no. 1, pp. 83–103,
     2004.

[21] D. S. John, F. E. Javier, A. K. Christopher, E. M. Megan,
     A. T. Steven, and L. X. Zhang, "A conceptual framework
     for ecosystem stoichiometry: balancing resource supply and
     demand," *Oikos*, vol. 109, no. 1, pp. 40–51, 2005.

[22] P. J. Mulholland, "Regulation of nutrient concentrations in
     a temperate forest stream: roles of upland, riparian, and
     instream processes," *Limnology & Oceanography*, vol. 37, no.
     7, pp. 1512–1526, 1992.

[23] V. A. Brown, J. J. McDonnell, D. A. Burns, and C. Kendall,
     "The role of event water, a rapid shallow flow component, and
     catchment size in summer stormflow," *Journal of Hydrology*,
     vol. 217, no. 3-4, pp. 171–190, 1999.

[24] C. D. Roderick, F. Oskar, M. Annikki, E.M. Ross, and T. V.
     Harry, "Optimal function explains forest responses to global
     change," *BioScience*, vol. 59, no. 2, pp. 127–139, 2009.

[25] B. L. Turner, R. Baxter, and B. A. Whitton, "Nitrogen
     and phosphorus in soil solutions and drainage streams in
     Upper Teesdale, northern England: implications of organic
     compounds for biological nutrient limitation," *Science of the
     Total Environment*, vol. 314–316, pp. 153–170, 2003.

[26] N. Ohte, N. Tokuchi, M. Katsuyama, S. Hobara, Y. Asano,
     and K. Koba, "Episodic increases in nitrate concentrations in
     streamwater due to the partial dieback of a pine forest in Japan:
     runoff generation processes control seasonality," *Hydrological
     Processes*, vol. 17, no. 2, pp. 237–249, 2003.

[27] Z. Zhang, T. Fukushima, Y. Onda et al., "Nutrient runoff from
     forested watersheds in central Japan during typhoon storms:
     implications for understanding runoff mechanisms during
     storm events," *Hydrological Processes*, vol. 21, no. 9, pp. 1167–
     1178, 2007.

[28] J. M. Paruelo, I. C. Burke, and W. K. Lauenroth, "Land-use
     impact on ecosystem functioning in eastern Colorado, USA,"
     Global Change Biology, vol. 7, no. 6, pp. 631–639, 2001.

[29] S. P. Steven and O. H. Lars, "Nitrogen loss from unpolluted
     South American forests mainly via dissolved organic com-
     pounds," *Nature*, vol. 415, no. 6870, pp. 416–419, 2002.

[30] R. Bobbink, M. Hornung, and J. G. M. Roelofs, "The effects
     of air-borne nitrogen pollutants on species diversity in natural
     and semi-natural European vegetation," *Journal of Ecology*, vol.
     86, no. 5, pp. 717–738, 1998.

[31] M. Sakamoto, "Primary production by phytoplankton com-
     munity in some Japanese lakes and its dependence on lake
     depth," *Archiv für Hydrobiologie*, vol. 62, pp. 1–28, 1966.

10

Diurnal Characteristics of Ecosystem Respiration of Alpine Meadow on the Qinghai-Tibetan Plateau: Implications for Carbon Budget Estimation

Yu Qin and Shuhua Yi

State Key Laboratory of Cryospheric Sciences, Cold and Arid Regions Environmental and Engineering Research Institute, Chinese Academy of Sciences, 320 Donggang West Road, Lanzhou 730000, China

Correspondence should be addressed to Shuhua Yi; yis@lzb.ac.cn

Academic Editors: D. M. Lloyd and L. Olsvig-Whittaker

Accurately estimating daily mean ecosystem respiration rate (Re) is important for understanding how ecosystem carbon budgets will respond to climate change. Usually, daily mean Re is represented by measurement using static chamber on alpine meadow ecosystems from 9:00 to 11:00 h a.m. local time directly. In the present study, however, we found that the calculated daily mean Re from 9:00 to 11:00 h a.m. local time was significantly higher than that from 0:00 to 23:30 h local time in an alpine meadow site, which might be caused by special climate condition on the Qinghai-Tibetan Plateau. Our results indicated that the calculated daily mean Re from 9:00 to 11:00 h a.m. local time cannot be used to represent daily mean Re directly.

1. Introduction

Great concerns over global warming and climate change have been proposed for improving the accuracy in estimating carbon flux in terrestrial ecosystems [1–3]. Static chamber method has widely been applied to measure ecosystem respiration in different ecosystems [4–6]. This method requires manual operation, therefore, it is a common practice to measure respirations of a period in a day to represent daily mean value, for example, 06:00–11:00 h a.m. [7], 9:00–11:00 h a.m. [8], and 14:00–16:00 h p.m [9]. Re measured between 9:00 and 11:00 h a.m. local time has been used to represent the daily mean value for the alpine meadow ecosystems on the Qinghai-Tibetan Plateau [10–15] and other ecosystems [8, 16, 17]. However, it is unknown whether this method is valid on the Qinghai-Tibetan Plateau, which has unique climate condition [18]. Therefore, in this study, we carried out round-the-clock field observation of ecosystem respiration per half-hour using an automated soil CO_2 flux system to test this method.

The field measurement was carried out in an alpine meadow at 3,887 m a.s.l. in Shule River Basin at the southeast 45 km far away Suli county ($98°18'33.2''$ E, $38°25''13.5''$ N),

the western part of Qilian Mountain, which is on the northeast edge of the Qinghai-Tibetan Plateau, Qinghai Province, China. The climate belongs to continental climate and is mainly controlled by westerly winds, with annual average precipitation being 200–400 mm, of which nearly 90% falls in the growing season (May–September), and the annual mean temperature ranged from -4.0 to $-19.4°C$ [19]. Soils are classified as felty soils [20]. The study site, $\sim 100 \times 100\,m^2$, has been fenced in 2010 to exclude the grazing activities of sheep and yaks. The dominant vegetations are *Kobresia capillifolia* and *Carex moorcroftii*. The permafrost type is transition according to the classified method by Cheng and Wang [21]. Three $2 \times 2\,m^2$ plots were set up randomly for the measurement of ecosystem respiration, and all the selected plots were expected to be less in spatial heterogeneity by visual inspection of the vegetation. Half-hour Re values were measured every 3 to 15 days depending on weather conditions during the whole growing season (May–September) in 2012 using an automated soil CO_2 flux system (LI-8150, LI-COR Biosciences, Lincoln, NE, USA) equipped with LI-COR-8100-104 long-term chamber. Three polyvinyl chloride collars 20 cm in diameter and 12 cm in height were used for measurements. Collars were inserted into soil at 8-9 cm. To reduce

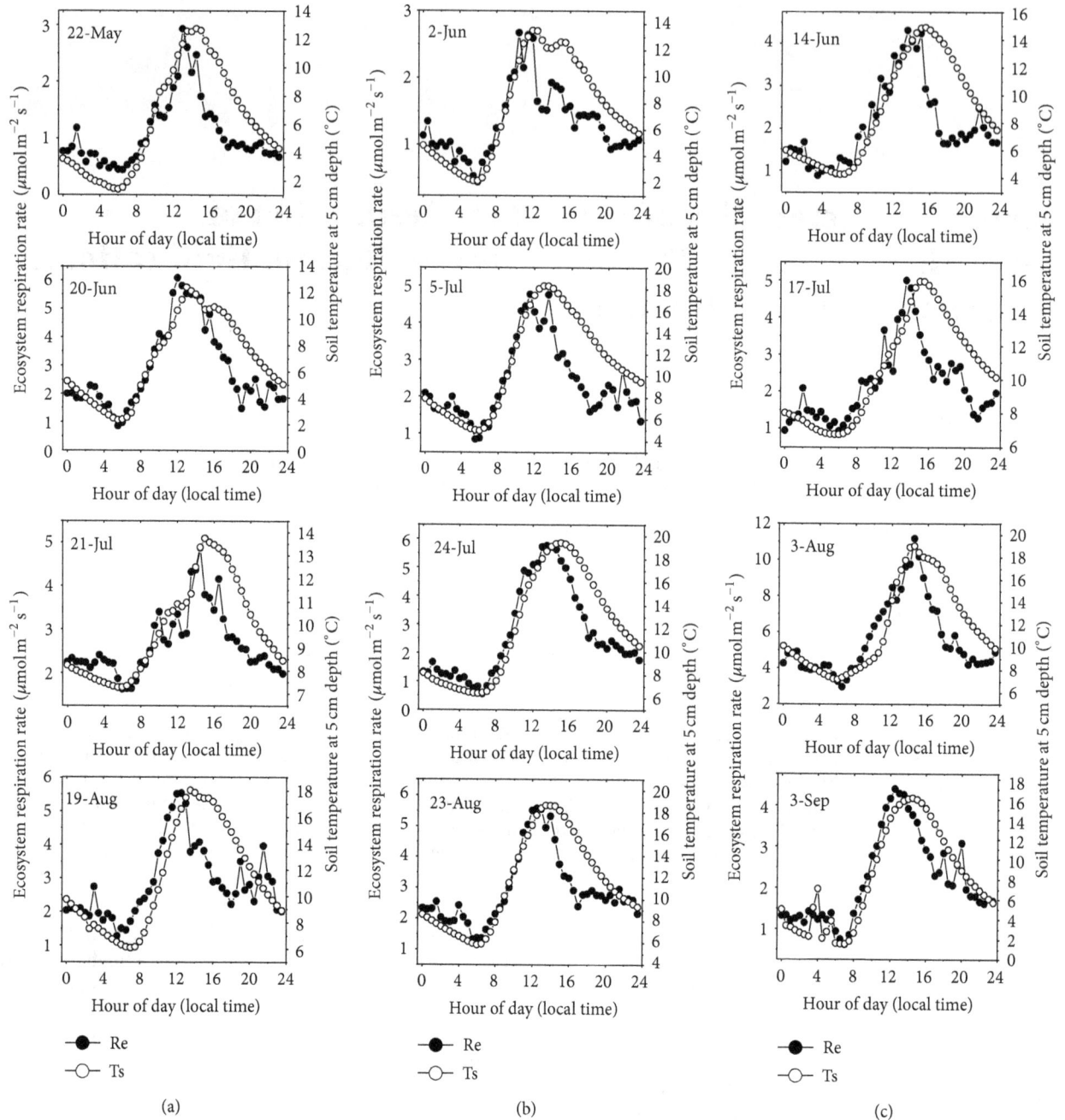

FIGURE 1: Diurnal variations of ecosystem respiration rate (μmoL m^{-2} s^{-1}) and soil temperature at 5 cm depth (°C) during the grown season from May to September.

a disturbance-induced CO_2 efflux, all collars were installed 24 h prior to the first measurement. Soil temperature at 5 cm and moisture at 7 cm below soil surface were measured at each chamber simultaneously while Re was measured. To eliminate artifacts due to the Venturi effect [22], we excluded measured values of Re when wind speed exceeded 7.5 m s^{-1} according to Xu et al. [23]. Data of the half-hour wind speed was derived from the meteorological stations in our study site for the period of May to September, 2012. However, our exclusion can maintain the reliability of values because field observation supplied enough data for estimating daily mean Re.

The maximum and minimum Re values of ecosystem respiration occurred at 12:00 to 16:00 h p.m. and 4:00 to 8:00 h a.m. local time, respectively (Figure 1), which corresponded well with the diurnal pattern of soil temperatures. Although the daily mean Re from 9:00 to 11:00 h a.m. local time was strongly correlated with that from 0:00 to 23:30 h (Figure 2), the calculated daily mean values of Re from 9:00 to 11:00 h a.m. local time were significantly higher than those from 0:00 to 23:00 h for both conditions with and without exclusion of wind effects (Table 1). Compared with daily mean values from 0:00 to 23:00 h, the Re values from 9:00 to 11:00 h a.m. local time were 23.90%/24.08% greater than the daily mean values

Diurnal Characteristics of Ecosystem Respiration of Alpine Meadow on the Qinghai-Tibetan Plateau: Implications
for Carbon Budget Estimation

121

TABLE 1: Daily mean values of Re calculated from 9:00 to 11:00 h a.m. local time and from 0:00 to 23:00 h before and after exclusion of the effect of wind.

Date	Before exclusion of the effect of wind			After exclusion of the effect of wind		
	0:00–23:30	9:00–11:00	Overestimate (%)	0:00–23:30	9:00–11:00	Overestimation (%)
22-May	1.09	1.32	21.13	1.15	1.25	9.09
02-Jun	1.33	2.09	57.69	1.32	2.09	58.13
14-Jun	2.09	2.54	21.94	2.11	2.54	20.71
20-Jun	2.86	3.72	30.23	2.86	3.72	30.23
05-Jul	2.39	3.65	53.05	2.46	3.65	48.55
17-Jul	2.19	2.52	15.07	2.19	2.52	15.07
21-Jul	2.65	2.88	8.42	2.65	2.88	8.42
24-Jul	2.68	3.46	28.96	2.72	3.46	27.09
03-Aug	5.63	6.20	10.18	5.35	6.20	15.85
19-Aug	2.87	3.63	26.58	2.86	3.63	27.14
23-Aug	2.93	3.56	21.54	2.93	3.56	21.54
03-Sep	2.21	2.71	22.91	2.21	2.71	22.91
Mean	2.58	3.19	23.90	2.57	3.19	24.08

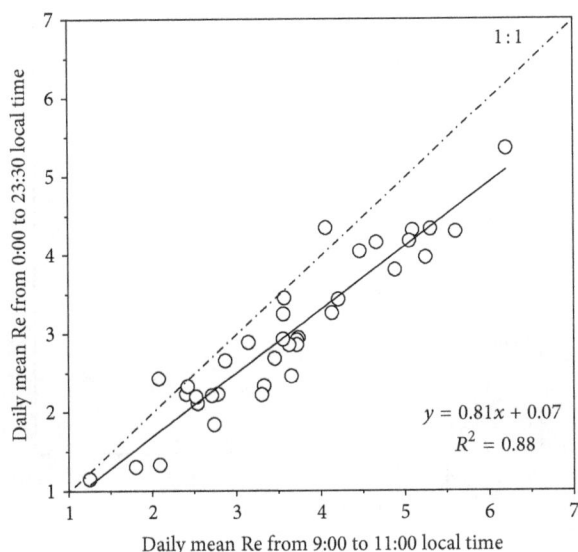

FIGURE 2: Relationships between the daily mean values of Re (μmoL m^{-2} s^{-1}) calculated from 0:00–23:00 h local time and the Re estimated by 9:00–11:00 h a.m. local time; all the measurements for three chambers for all the sampling dates were pooled together, $n = 36$.

before/after exclusion the effects of wind during the whole growing season (Table 1).

Our results are consistent with Cao et al. [24] and Zhang et al. [25], whose results showed that diurnal variations of soil respiration on the alpine meadow presented significant single peak dynamics with the maximum value in 12:00–16:00 h p.m. local time and the minimum value in 4:00–8:00 h a.m. local time by using static chamber. However, our results are inconsistent with Tang et al. [26] and Li et al. [27], who have demonstrated that CO_2 fluxes measured at 9:00–11:00 h a.m. local time were close to daily means in subtropical forest ecosystem in southern China and in cropland ecosystem on the Loess Plateau in northern China. Those correlations that existed in forest and cropland ecosystems were associated

with relatively low range of soil temperature and flat daily curves of CO_2 fluxes during daytime and nighttime [28]. In the present study, daily mean Re calculated from 9:00 to 11:00 h a.m. local time was significantly higher than 0:00 to 23:30 h local time, which might be caused by special climate conditions on the Qinghai-Tibetan Plateau. The solar radiation can penetrate thin atmosphere easily to heat land surface, the soil temperature increases quickly after sunrise; the upward land surface longwave radiation can also dissipate quickly due to thin atmosphere, and surface soil temperature decreases quickly in the afternoon (Figure 3). Stronger diurnal variation of Re was apparent in this alpine meadow ecosystem, it is possible that daily mean Re will be overestimated if it is represented by Re of any period during daytime. To improve the accuracy in estimation of carbon budget, hence, automated continuous all-day field observation for Re initiated here will provide more definitive studies [29, 30]. Compared with static chamber, automated soil CO_2 flux system can measure wide temporal variability in ecosystem respiration with the range from half-hour to seasonal and even to interannual [31]. An alternative way is to calculate the ratios of instant Re from 9:00–11:00 a.m. local time to the daily mean Re and then to use these ratios to calculate daily mean Re [32].

Authors' Contribution

Shuhua Yi contributed to experiment design and paper revision. Yu Qin contributed to field data collection, data analysis, and paper writing.

Acknowledgments

The authors acknowledge Shilong Ren, Jianjun Chen, and Xiaoyun Wang for for their help with the fieldwork and Professors Tianding Han, Jinkui Wu, and Hao Wu for providing meteorological data. This study was jointly supported through Grants of the Major State Basic Research

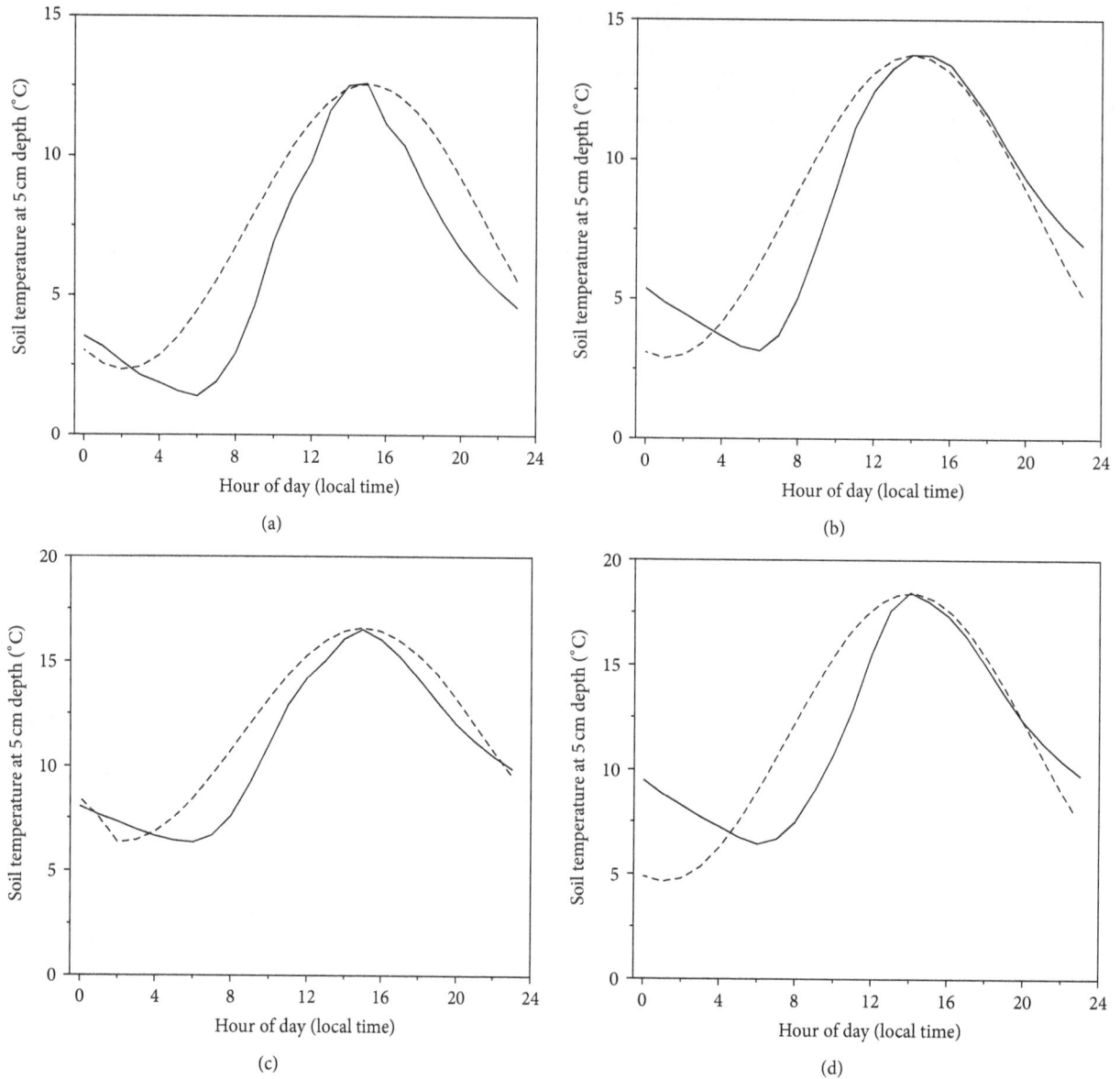

FIGURE 3: Comparison between 3-day daily mean diurnal soil temperature (solid line, 5 cm) and the corresponding sinusoidal temperature (dashed line) with the same daily range for May (a), June (b), July (c), and August (d).

Development Programme of China (973 Programme) (no. 2010CB951402), the Strategic Priority Research Program (XDB030303) and One Hundred People Plan (O927581001) of the Chinese Academy of Sciences, and the Chinese National Natural Science Foundation Commission (41271089), as well as independent Grants from State Key Laboratory of Cryospheric Sciences (SKLCS-ZZ-2012-2-2).

References

[1] D. S. Schimel, "Terrestrial ecosystems and the carbon cycle," *Global Change Biology*, vol. 1, no. 1, pp. 77–91, 1995.

[2] R. A. Houghton, "Balancing the global carbon budget," *Annual Review of Earth and Planetary Sciences*, vol. 35, pp. 313–347, 2007.

[3] M. Heimann and M. Reichstein, "Terrestrial ecosystem carbon dynamics and climate feedbacks," *Nature*, vol. 451, no. 7176, pp. 289–292, 2008.

[4] J. W. Raich, C. S. Potter, and D. Bhagawati, "Interannual variability in global soil respiration, 1980–94," *Global Change Biology*, vol. 8, no. 8, pp. 800–812, 2002.

[5] A. Heinemeyer and N. P. McNamara, "Comparing the closed static versus the closed dynamic chamber flux methodology: implications for soil respiration studies," *Plant and Soil*, vol. 346, no. 1, pp. 145–151, 2011.

[6] S. Olajuyigbe, B. Tobin, M. Saunders, and M. Nieuwenhuis, "Forest thinning and soil respiration in a Sitka spruce forest in Ireland," *Agricultural and Forest Meteorology*, vol. 157, pp. 86–95, 2012.

[7] C. E. Hicks Pries, E. A. G. Schuur, and K. G. Crummer, "Thawing permafrost increases old soil and autotrophic respiration

Diurnal Characteristics of Ecosystem Respiration of Alpine Meadow on the Qinghai-Tibetan Plateau: Implications
for Carbon Budget Estimation

123

in tundra: partitioning ecosystem respiration using $\delta^{13}C$ and $\Delta^{14}C$," *Global Change Biology*, vol. 19, no. 2, pp. 649–661, 2013.

[8] J. Priess and H. Fölster, "Carbon cycle dynamics and soil respiration of forest under natural degradation in the Gran Sabana," *Interciencia*, vol. 19, pp. 317–322, 1994.

[9] B. Meda, C. R. Flechard, K. Germain, P. Robin, C. Walter, and M. Hassouna, "Greenhouse gas emissions from the grassy outdoor run of organic broilers," *Biogeosciences*, vol. 9, no. 4, pp. 1493–1508, 2012.

[10] J. Wang, G. Wang, Y. Wang, and Y. Li, "Influences of the degradation of swamp and alpine meadows on CO_2 emission during growing season on the Qinghai-Tibet Plateau," *Chinese Science Bulletin*, vol. 52, no. 18, pp. 2565–2574, 2007 (Chinese).

[11] Q.-W. Hu, Q. Wu, G.-M. Cao, D. Li, R.-J. Long, and Y.-S. Wang, "Growing season ecosystem respirations and associated component fluxes in two alpine meadows on the Tibetan Plateau," *Journal of Integrative Plant Biology*, vol. 50, no. 3, pp. 271–279, 2008.

[12] C. Jiang, G. Yu, H. Fang, G. Cao, and Y. Li, "Short-term effect of increasing nitrogen deposition on CO_2, CH_4 and N_2O fluxes in an alpine meadow on the Qinghai-Tibetan Plateau, China," *Atmospheric Environment*, vol. 44, no. 24, pp. 2920–2926, 2010.

[13] X. Lin, S. Wang, X. Ma et al., "Fluxes of CO_2, CH_4, and N_2O in an alpine meadow affected by yak excreta on the Qinghai-Tibetan plateau during summer grazing periods," *Soil Biology and Biochemistry*, vol. 41, no. 4, pp. 718–725, 2009.

[14] X. Lin, Z. Zhang, S. Wang et al., "Response of ecosystem respiration to warming and grazing during the growing seasons in the alpine meadow on the Tibetan plateau," *Agricultural and Forest Meteorology*, vol. 151, no. 7, pp. 792–802, 2011.

[15] Z. H. Zhang, J. C. Duan, S. P. Wang et al., "Effects of seeding ratios and nitrogen fertilizer on ecosystem respiration of common vetch and oat on the Tibetan plateau," *Plant Soil*, vol. 362, pp. 287–299, 2013.

[16] T. Zhu, S. Cheng, H. Fang, G. Yu, J. Zheng, and Y. Li, "Early responses of soil CO_2 emission to simulating atmospheric nitrogen deposition in an alpine meadow on the Qinghai Tibetan Plateau," *Acta Ecologica Sinica*, vol. 31, no. 10, pp. 2687–2696, 2011 (Chinese).

[17] M. A. Liebig, S. L. Kronberg, J. R. Hendrickson, X. Dong, and J. R. Gross, "Carbon dioxide efflux from long-term grazing management systems in a semiarid region," *Agriculture, Ecosystems and Environment*, vol. 164, pp. 137–144, 2013.

[18] O. M. Raspopov, V. A. Dergachev, J. Esper et al., "The influence of the de Vries (~ 200-year) solar cycle on climate variations: results from the Central Asian Mountains and their global link," *Palaeogeography, Palaeoclimatology, Palaeoecology*, vol. 259, no. 1, pp. 6–16, 2008.

[19] S. Yi, Z. Zhou, S. Ren et al., "Effects of permafrost degradation on alpine grassland in a semi-arid basin on the Qinghai-Tibetan Plateau," *Environmental Research Letters*, vol. 6, no. 4, Article ID 045403, 2011.

[20] W. J. Liu, S. Y. Chen, X. Qin et al., "Storage, patterns, and control of soil organic carbon and nitrogen in the northeastern margin of the Qinghai-Tibetan Plateau," *Environmental Research Letters*, vol. 7, no. 3, Article ID 035401, 2012.

[21] G. D. Cheng and S. L. Wang, "On the zonation of high altitude permafrost in China," *Journal of Glaciology and Geocryology*, vol. 4, pp. 1–17, 1982 (Chinese).

[22] F. Conen and K. A. Smith, "A re-examination of closed flux chamber methods for the measurement of trace gas emissions from soils to the atmosphere," *European Journal of Soil Science*, vol. 49, no. 4, pp. 701–707, 1998.

[23] L. Xu, M. D. Furtaw, R. A. Madsen, R. L. Garcia, D. J. Anderson, and D. K. McDermitt, "On maintaining pressure equilibrium between a soil CO_2 flux chamber and the ambient air," *Journal of Geophysical Research D*, vol. 111, no. 8, Article ID D08S10, 2006.

[24] G. M. Cao, Y. N. Li, J. X. Zhang, and X. Q. Zhao, "Effect of soil circumstances biogeochemical factors on carbon dioxide emission from Mollic-Gryic Cambisols," *Acta Agrestia Sinica*, vol. 9, pp. 307–312, 2001 (Chinese).

[25] J. X. Zhang, G. M. Cao, D. W. Zhou, Q. W. Hu, and X. Q. Zhao, "The carbon storage and carbon cycle among the atmosphere, soil, vegetation and animal in the Kobresia humilis alpine meadow ecosystem," *Acta Ecologica Sinica*, vol. 23, pp. 627–633, 2003 (Chinese).

[26] X. Tang, S. Liu, G. Zhou, D. Zhang, and C. Zhou, "Soil-atmospheric exchange of CO_2, CH_4, and N_2O in three sub-tropical forest ecosystems in southern China," *Global Change Biology*, vol. 12, no. 3, pp. 546–560, 2006.

[27] X. Li, H. Fu, D. Guo, X. Li, and C. Wan, "Partitioning soil respiration and assessing the carbon balance in a Setaria italica (L.) Beauv. Cropland on the Loess Plateau, Northern China," *Soil Biology and Biochemistry*, vol. 42, no. 2, pp. 337–346, 2010.

[28] J. Yan, D. Zhang, G. Zhou, and J. Liu, "Soil respiration associated with forest succession in subtropical forests in Dinghushan Biosphere Reserve," *Soil Biology and Biochemistry*, vol. 41, no. 5, pp. 991–999, 2009.

[29] K. L. Thomas, J. Benstead, S. H. Lloyd, and D. Lloyd, "Diurnal oscillations of gas production and effluxes (CO_2 and CH_4) in cores from a peat bog," *Biological Rhythm Research*, vol. 29, no. 3, pp. 247–259, 1998.

[30] S. K. Sheppard and D. Lloyd, "Diurnal oscillations in gas production (O_2, CO_2, CH_4, and N_2) in soil monoliths," *Biological Rhythm Research*, vol. 33, no. 5, pp. 577–591, 2002.

[31] R. S. Jassal, T. A. Black, Z. Nesic, and D. Gaumont-Guay, "Using automated non-steady-state chamber systems for making continuous long-term measurements of soil CO_2 efflux in forest ecosystems," *Agricultural and Forest Meteorology*, vol. 161, pp. 57–65, 2012.

[32] Y. Geng, Y. Wang, K. Yang et al., "Soil respiration in tibetan alpine grasslands: belowground biomass and soil moisture, but not soil temperature, best explain the large-scale patterns," *PLoS One*, vol. 7, no. 4, Article ID e34968, 2012.

11

Shesher and Welala Floodplain Wetlands (Lake Tana, Ethiopia): Are They Important Breeding Habitats for *Clarias gariepinus* and the Migratory *Labeobarbus* Fish Species?

Wassie Anteneh,[1] Eshete Dejen,[2] and Abebe Getahun[3]

[1] *Department of Biology, College of Science, Bahir Dar University, P.O. Box 79, Bahir Dar, Ethiopia*
[2] *FAO-Sub Regional Office for Eastern Africa, P.O. Box 5536, Addis Ababa, Ethiopia*
[3] *Fisheries and Aquatic Science Stream, Faculty of Life Sciences, Addis Ababa University, P.O. Box 1176, Addis Ababa, Ethiopia*

Correspondence should be addressed to Wassie Anteneh, wassie74@gmail.com

Academic Editors: S. Brucet and K. Halačka

This study aims at investigating the spawning migration of the endemic *Labeobarbus* species and *C. gariepinus* from Lake Tana, through Ribb River, to Welala and Shesher wetlands. The study was conducted during peak spawning months (July to October, 2010). Fish were collected through overnight gillnet settings. A total of 1725 specimens of the genus *Labeobarbus* (13 species) and 506 specimens of *C. gariepinus* were collected. Six species of *Labeobarbus* formed prespawning aggregation at Ribb River mouth. However, no *Labeobarbus* species was found to spawn in the two wetlands. More than 90% of the catch in Welala and Shesher wetlands was contributed by *C. gariepinus*. This implies that these wetlands are ideal spawning and nursery habitats for *C. gariepinus* but not for the endemic *Labeobarbus* species. Except *L. intermedius*, all the six *Labeobarbus* species (aggregated at Ribb River mouth) and *C. gariepinus* (spawning at Shesher and Welala wetlands) were temporally segregated.

1. Introduction

The contemporary *Labeobarbus* species of Lake Tana (Ethiopia) form the only known remaining intact species flock of large cyprinid fishes, since the one in Lake Lanao in the Philippines has almost disappeared due to destructive fishing [1]. The vast majority of cyprinids occur in rivers, but some *Labeobarbus* and *Labeo* species are adapted to a lacustrine environment [2]. However, these lake-dwelling cyprinids spawn in rivers, by undertaking a single annual breeding migration up rivers [3]. This spawning strategy makes the large African cyprinids vulnerable for modern fisheries, since the fishermen target spawning aggregations at river mouths by effectively blocking them off from the lake, preventing mature individuals from reaching the upstream spawning areas [4, 5]. Although the causes for the decline are not properly identified, the migratory riverine spawning species of *Labeobarbus* in Lake Tana have undergone drastic decline (>75% in biomass and 80% in number) from 1991 to 2001 [6, 7].

The other commercially important species in Lake Tana, *C. gariepinus* [8], at the beginning of the rainy season (June-July), moves through the littoral areas towards the inundated floodplains and upstream inflowing rivers for spawning [7, 8]. *Clarias gariepinus* is the most dominant species during the rainy season upstream of the turbid Ribb River probably due to the availability of extended floodplain [9]. When the water level starts to decrease (October–December), *C. gariepinus* migrates back through the littoral zone towards the pelagic zone (Lake Tana). *Clarias gariepinus* is targeted by the commercial gillnet fishery when migrating between the floodplains (spawning areas) and the lake [7]. Although the large, older individuals proved to be vulnerable for increased mortality by the commercial gillnet fishery, it is known that, compared with *Labeobarbus* spp., *C. gariepinus* is only moderately susceptible to fishing pressure in Lake Tana. This is because *C. gariepinus* is found to be more resilient [10]. In the last decades, as a result of the low monetary value and poor preference to this species by the Ethiopians, it was not selectively

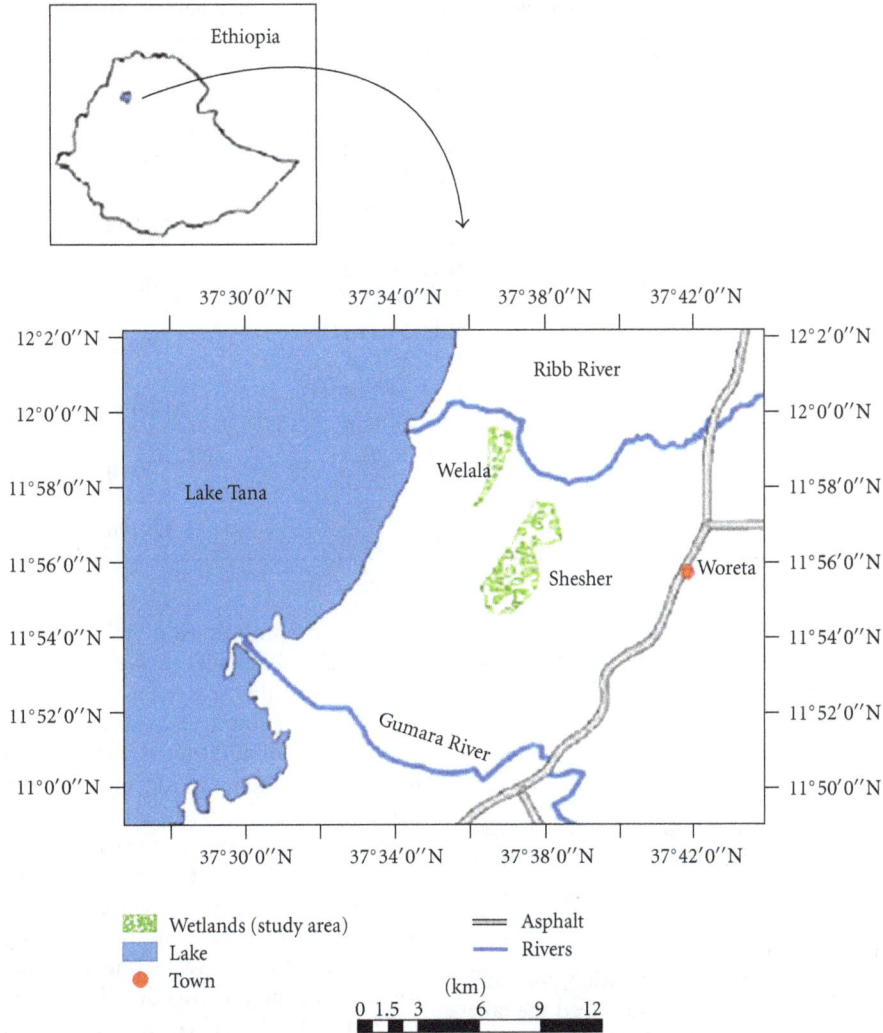

FIGURE 1: Map of Lake Tana and Ribb River and associated Shesher and Welala floodplain wetlands (after Atnafu et al. [18]).

targeted by the commercial gillnet fishery and is mainly landed as bycatch [7]. However, according to Atnafu [11], *C. gariepinus* recently has become highly preferred fish by the commercial fishermen in Lake Tana area for dry fish export, especially to Sudan.

Various studies [9, 12–17] showed that seven (*L. macrophthalmus, L. truttiformis, L. megastoma, L. brevicephalus, L. tsanensis, L. platydorsus,* and *L. acutirostris*) of the 15 endemic *Labeobarbus* species migrate more than 50 km up rivers during the rainy season to spawn in fast flowing, clear and gravel bed streams. However, mass spawning migrations for the remaining eight *Labeobarbus* species (*L. nedgia, L. dainellii, L. gorguari, L. longissimus, L. intermedius, L. gorgorensis, L. surkis,* and *L. crassibarbis*) were missing from all tributaries studied so far. According to de Graaf et al. [14], these missing species may spawn in the lake or adjacent floodplain wetlands.

The Shesher and Welala wetlands are located 3–5 km away from Lake Tana (Figure 1) and are valuable for the local community. They provide fishes, water, and grazing for

livestock. They also harbor large diversity of bird species including internationally endangered and threatened ones [18]. They are the buffering zones of Lake Tana [17]. However, due to unsustainable farming activities by local farmers, the existence of these floodplain wetlands and associated ecological services as well as socioeconomic importance is under threat [18, 19]. It was observed that the local farmers were draining and pumping the water to expand farming land. Another potential threat is the large dam under construction on Ribb River that could minimize the water overflowing to these wetlands [20]. To have management plans for the two wetlands and also to conduct environmental impact assessment studies for all future development projects around the Lake Tana are strongly recommended.

Probably due to remoteness and inaccessibility, data about the ecological importance of the two prominent wetlands, Welala and Shesher, for the migratory fishes of Lake Tana are totally absent. Ribb River charges these wetlands as it overflows during the rainy months (July to October) and form direct connections with the lake during the rainy season

TABLE 1: Sampling sites, estimated distance from the lake, coordinates, elevation, and bottom types of the sampling sites.

Sampling site	Estimated distance (km) from the lake	Coordinates	Elevation	Bottom type
Ribb River mouth	—	N 11°59′54.2″ E 37°33′06.1″	1788 m	Mud and Silt (mixed)
Welala I	3.5	N 11°58′59.1″ E 37°36′12.3″	1789 m	Mud
Welala II	3.5	N 11°58′38.2″ E 37°36′34.9″	1789 m	Mud
Welala III	3.5	N 11°58′14.1″ E 37°36′18.6″	1789 m	Mud
Shesher I	4	N 11°58′25.3″ E 37°37′16.4″	1791 m	Mud
Shesher II	4	N 11°57′01.3″ E 37°37′27.5″	1791 m	Mud
Shesher III	4	N 11°56′58.2″ E 37°37′35.3″	1791 m	Mud

through fringes. Therefore, the aim of this study was to investigate whether *Labeobarbus* species and *C. gariepinus* use these wetlands as spawning and/or nursery habitats.

2. Methods

2.1. Description of Study Area. Lake Tana, Ethiopia's largest lake and the source of Blue Nile River, has a surface area of *ca.* 3200 km². It is situated in the northwestern highlands at an altitude of approximately 1800 m. It is a shallow (maximum depth 14 m, mean 8 m) lake. More than 60 small seasonal tributaries and seven perennial rivers (Gumara, Ribb, Megech, Gelgel Abbay, Gelda, Arno-Garno, and Dirma) feed the lake [17]. The only outflowing river is the Blue Nile; however, the ichthyofauna is isolated from the lower Blue Nile by a 40 m waterfall located 30 km from Lake Tana. Fogera and Dembia floodplains are the largest wetlands of the country and border Lake Tana in the eastern and northern parts, respectively. Welala and Shesher wetlands (Figure 1) are located in the Fogera floodplain.

Fogera *Woreda* (district) is one of the ten Woredas bordering Lake Tana and is found in South Gondar Administrative Zone. It is situated at 11°58′00″ N latitude and 37°41′00″ E longitude [18]. Woreta, capital of the Fogera Woreda is found 620 km from Ethiopia's capital city, Addis Ababa and 55 km from Bahir Dar, the regional capital, (Figure 1). Ribb River originates from Gunna Mountains, at an altitude of above 3000 m and has the length of *ca.* 130 km and drainage area of about 1790 km² [21]. In its lower and middle reaches, the river flows over the extensive alluvial Fogera Floodplains. The river meanders and flows slowly over this floodplain, and this resulted in river channel deposition and overflowing of riverbanks and charging water to Welala and Shesher during the rainy season.

According to the information from the local people, Shesher dries usually in February or March; whereas, Welala dries in April or May. In some years, when there is high overflow from Ribb River, Welala never dries throughout the year

(personal communication with the local people). This is because Welala is smaller in size and deeper (maximum depth 2.5 m in the rainy season) as compared to Shesher which is wider and shallower (maximum depth 1.75 m in the rainy season). Location, distance from the lake, elevation and bottom types of the sampling sites are summarized in Table 1. Coordinates and elevations were assessed with a GPS. The bottoms of these two wetlands are muddy (Table 1) and nowadays, during the post rainy season, the local farmers drain the water by digging canals from these two wetlands to get fertile land (the muddy bottom) for crop production [18].

The drastic changes in the areas of Shesher and Welala wetlands in the last two decades are shown in Figure 2. In 1987, the total surface area of Shesher and Welala was *ca.* 1557 and 298 hectares, respectively (Figure 2(a)) [19]. Whereas, in 2008 the surface area of Shesher shrunk to 136 hectares (91% shrinkage) and Welala shrunk to 159 hectares (47% shrinkage) (Figure 2(b)) [19]. These wetlands are shrinking at an alarming rate, mainly because of unsustainable farming practices by the local inhabitants [18]. The local farmers drain the water of these wetlands to expand their farmland and pump water for irrigation. The large irrigation dam under construction on Ribb River is another potential threat. This dam prevents overflow from Ribb River into the wetlands and disrupts the connection with Lake Tana which is vital for the survival of these wetlands [18–20].

2.2. Sampling. Sampling took place from July 2010 to October 2010. Three sampling sites were selected in each wetland, two shore sites and one site at the middle. In the shore sites, gillnets were set at the mouth of the inflow from Ribb River overflow and at the outlets to Ribb River main channel. The outflow from Ribb first enters to Shesher, from the north, and when Shesher is filled, it overflows to Welala, and finally to Lake Tana (Figure 1). Besides Welala and Shesher wetlands, samples were collected at Ribb River mouth. Fish and physicochemical parameters were collected nine times at the seven selected sites, once in July (third week), three

Shesher and Welala Floodplain Wetlands (Lake Tana, Ethiopia): Are They Important Breeding Habitats for Clarias
gariepinus and the Migratory Labeobarbus Fish Species?

127

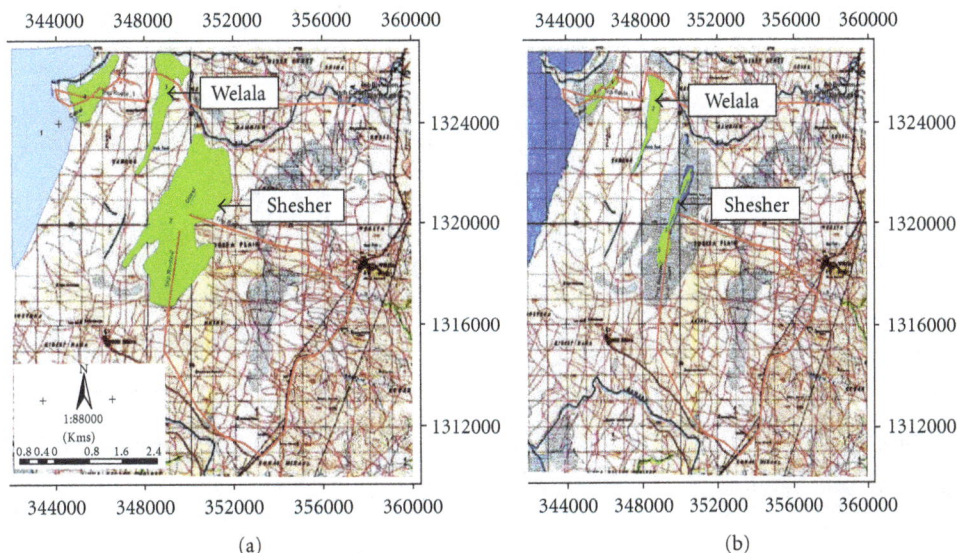

FIGURE 2: Extent of Shesher and Welala wetlands in 1987 (a) and their current (2008) extent (b) (after Nemomissa [19]).

times in August (first, second, and fourth weeks), three times in September (first, third, and fourth weeks) and twice in October (first and fourth weeks).

2.3. Physicochemical Parameter Measurements. At all the sampling sites, observation of bottom type and measurements of depth and Secchi depth (using 30 cm diameter Secchi-disk) were taken. Similarly, dissolved oxygen, temperature, and pH were measured (using probes) at the surfaces of all sampling sites and at all times, in the morning immediately after overnight gillnet catch collection.

2.4. Fish Collection. Multifilament gillnets (6, 8, 10, 12, and 14 cm stretched mesh size) with a panel length of 100 m and depth of 1.5 m were used. Gillnets were set usually at 6:00 PM, and catches were collected in the following morning at about 6:00 AM. All of the fishes caught were identified to species level with immediate inspection (for *C. gariepinus*) and with the help of identification key developed by Nagelkerke and Sibbing [22] for *Labeobarbus* species. After identification to the species level, each fish was dissected; the gonads were examined visually and sexed. The gonad maturity stage of each specimen of *Labeobarbus* species was determined visually, using the key developed by Nagelkerke [17], but for *C. gariepinus* the gonad maturity stage was determined according to Wudneh [8].

2.5. Data Processing. All the statistical computations were done using Minitab version 14 and SPSS version 11 software. Pairwise comparison of dissolved oxygen content (mgL^{-1}), temperature (°C) and vertical transparency or Secchi depth (cm) of the seven sampling sites were compared through one-way analysis of variance (one-way ANOVA) followed by Bonferroni's post hoc tests for multiple comparisons if significant variance was evident. One-way ANOVA was also used to investigate temporal segregation of *Labeobarbus* species

aggregating at Ribb River mouth and *C. gariepinus* spawning in Shesher and Welala wetlands. Only fish with ripe and spent gonads were considered for temporal segregation analysis as reproductively immature fishes would not be expected to show temporal variation in aggregation and migration. Catch per unit of effort (CpUE) was defined number of fish per overnight gillnet setting.

3. Result

3.1. Physicochemical Parameters. There was a highly significant overall variation on dissolved oxygen ($F_{(6,56)} = 33.85$; $P < 0.001$) but not in temperature ($F_{(6,56)} = 2.15$; $P > 0.05$), pH ($F_{(6,56)} = 0.83$; $P > 0.05$), and vertical transparency ($F_{(6,56)} = 0.47$; $P > 0.05$) among the sampling sites (Tables 2 and 3). All the sampling sites in Shesher and Ribb River mouth have significantly higher dissolved oxygen concentration ($P < 0.001$; Table 3) than the sites in Welala. The pairwise comparison also showed that the sampling sites of Shesher, except site II, have higher dissolved oxygen concentration ($P < 0.001$) than Ribb River mouth (Table 3). The average dissolved oxygen concentration of Shesher (when the three sites pooled), 6.13 ± 0.22, is higher than Welala, 5.14 ± 0.06 (Table 2).

3.2. Fish Species Composition and Abundance. A total of 2403 fish specimens were collected from Ribb River mouth and Shesher and Welala floodplain wetlands. From this catch, 1725 (72%) specimens were contributed by *Labeobarbus* species, followed by *C. gariepinus* (21%, 506 specimens), *Oreochromis niloticus* (7.1%, 170 specimens) and *Varicorhinus beso* (with two specimens only). *Clarias gariepinus* and *Labeobarbus* spp. were dominant at the two wetlands and Ribb River mouth, respectively (Figure 3).

A total of 469 (250 and 219 from Shesher and Welala, resp.) fish specimens were collected from all the sampling

TABLE 2: Mean ± SE (Standard Error) values of oxygen concentration, temperature, pH, and Sechi-disk depth at the sampling sites. N refers to number of samplings.

Site name	N	Oxygen (mgL^{-1})	Temp. (°C)	pH	Secchi-disk depth (cm)
Ribb River mouth	9	5.66 ± 0.09	21.2 ± 0.14	6.97 ± 0.03	4.56 ± 1.04
Welala I	9	5.12 ± 0.05	21.5 ± 0.22	7.02 ± 0.03	5.33 ± 0.76
Welala II	9	5.15 ± 0.05	20.9 ± 0.21	7.01 ± 0.01	5.32 ± 0.51
Welala III	9	5.14 ± 0.08	21.1 ± 0.23	6.99 ± 0.10	6.11 ± 0.92
Shesher I	9	6.17 ± 0.37	21.6 ± 0.22	6.96 ± 0.02	6.22 ± 1.06
Shesher II	9	6.02 ± 0.09	21.7 ± 0.24	7.02 ± 0.01	6.11 ± 1.01
Shesher III	9	6.19 ± 0.19	21.4 ± 0.21	7.03 ± 0.05	6.00 ± 0.83

TABLE 3: Pairwise comparison of dissolved oxygen concentrations (mgL^{-1}) among sampling sites. Abbreviation used: RM = river mouth.

	Ribb RM	Welala I	Welala II	Welala III	Shesher I	Shesher II	Shesher III
Ribb RM	×						
Welala I	**	×					
Welala II	**	NS	×				
Welala III	**	NS	NS	×			
Shesher I	**	***	***	***	×		
Shesher II	NS	***	***	***	NS	×	
Shesher III	**	***	***	***	NS	NS	×

$^{*}P < 0.05$; $^{**}P < 0.01$; $^{***}P < 0.001$; NS: not significant; $P > 0.05$.

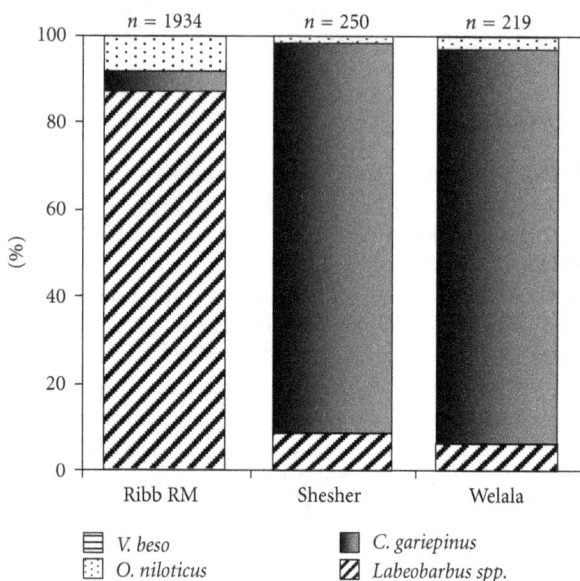

FIGURE 3: Percentage composition of fish collected at Ribb River mouth and its associated wetlands: Shesher and Welala. Data from all sites (for Shesher and Welala) and months pooled. "n" refers to the number of fish.

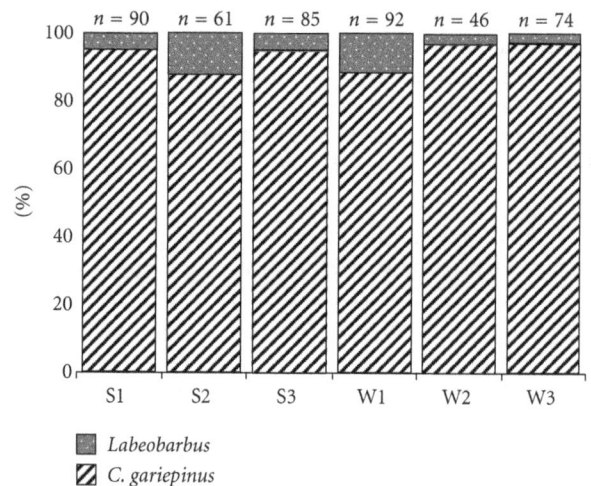

FIGURE 4: Percentage contribution of Labeobarbus spp. and C. gariepinus collected from the six different sampling sites of Shesher and Welala wetlands. Pooled data collected from July to October. "S" and "W" stand for Shesher and Welala, respectively. "n", number of specimens.

sites of Shesher and Welala wetlands. Clarias gariepinus was the most numerous (Figure 4) fish species and contributed about 93% (421 specimens) of the total catch from all sampling sites of the two wetlands. However, Labeobarbus species were incidentally caught at Shesher and Welala wetlands (Figure 4). Only 27 (6.2% of the total wetland catch)

specimens of Labeobarbus were collected from the wetlands. The Labeobarbus spp. collected from the two wetlands include L. acutirostris (1 specimen), L. brevicephalus (8 specimens), L. megastoma (3 specimens), L. intermedius (14 specimens), and L. tsanensis (1 specimen). Together with C. gariepinus and Labeobarbus spp., only ten specimens of O. niloticus were collected from the two wetlands.

Shesher and Welala Floodplain Wetlands (Lake Tana, Ethiopia): Are They Important Breeding Habitats for Clarias
gariepinus and the Migratory Labeobarbus Fish Species?

129

TABLE 4: The abundance of the six most dominant *Labeobarbus* species collected from Ribb River mouth, and their catch from Shesher and Welala wetlands.

Species	Number of specimens		
	Ribb River mouth	Shesher	Wolala
L. brevicephalus	206	10	3
L. intermedius	675	7	10
L. megastoma	273	2	2
L. platydorsus	81	0	0
L. truttiformis	196	0	0
L. tsanensis	164	1	0

TABLE 5: One-way ANOVA result on the catch data of the six *Labeobarbus* species from Ribb River mouth and *C. gariepinus* from Shesher and Welala wetlands, for differences in the four spawning months (July–October).

Species	Source of variation	MS	$F_{(3,5)}$ value	P
L. brevicephalus	Month	695.19	17.04	**
L. intermedius	Month	2216.72	5.24	NS
L. megastoma	Month	873.41	65.51	***
L. platydorsus	Month	103.69	6.23	*
L. truttiformis	Month	589.46	36.02	**
L. tsanensis	Month	176.13	8.32	*
C. gariepinus	Month	391.54	33.57	**

$^*P < 0.05$; $^{**}P < 0.01$; $^{***}P < 0.001$; NS: not significant, $P > 0.05$.

From the 15 endemic species of *Labeobarbus* described in Lake Tana, 13 species were collected from Ribb River mouth. However, from these 13 *Labeobarbus* species, only six species (Table 4) were the most dominant contributing nearly 95% (1595 specimens) of the total *Labeobarbus* catch from Ribb River mouth. Seven species (*L. acutirostris, L. crassibarbis, L. gorgorensis, L. longissimus, L. macrophthalmus, L. nedgia,* and *L. surkis*) were incidentally captured and contributed less than 8% of the *Labeobarbus* catch from Ribb River mouth. Two species, *L. dainellii* and *L. gorguari*, were totally missing from the catches of the river mouth.

3.3. Gonad Maturity Stages.
From the total of 1668 specimens of the six *Labeobarbus* species collected from Ribb River mouth, only 101 (6%) were immature (gonad stages II and III); whereas 1563 (93%) specimens were ripe (gonad stages IV and V), and 14 (1%) were spent (gonad stage VII) (Figure 5(a)). However, no running (gonad stage VI) *Labeobarbus* specimen was collected at the river mouth (Figure 5(a)). However, from the total of 88 specimens of *C. gariepinus* collected from Ribb River mouth only 11 (12.5%) were ripe or spawning (gonad stage IV), 77 (87.5%) were immature (gonad stage I–III) and only one specimen was spent (gonad stage V). Of the gonads of *C. gariepinus* collected from Shesher and Welala wetlands, 133 (32%) were ripe, 197 (46.5%) were immature, and 91 (22.5%) were spent (Figure 5(b)).

3.4. CpUE and Temporal Segregation.
Collectively, the peak CpUE value for the six *Labeobarbus* species aggregating at Ribb River mouth was observed in September (Figure 6(a)). However, the peak CpUE for *C. gariepinus* collected from Shesher and Welala wetlands was in July (Figure 6(b)), and the slope of the graph remained negative for the whole study period.

The monthly catches of the six *Labeobarbus* species aggregating at Ribb River mouth, except *L. intermedius*, showed significant temporal segregation ($P < 0.05$; Table 5). Significant temporal segregation ($P < 0.05$; Table 5) was also evident for *C. gariepinus* spawning at Shesher and Welala wetlands. The details of temporal segregation trends and apparent overlaps among the six *Labeobarbus* species aggregating at Ribb River mouth is shown in Figure 7. *Labeobarbus megastoma* and *L. intermedius* were apparently the first to appear

at Ribb River mouth for prespawning aggregation (during July). The peak CpUE for these two species was in September (Figure 7), but it declined from August to October. Whereas, a reverse trend was observed for *L. brevicephalus*, CpUE increases from August to October. The other three species: *L. platydorsus, L. truttiformis,* and *L. tsanensis* aggregate during August and September, but their CpUE declined sharply during October.

4. Discussion

The water temperature and pH values obtained in the present study (Table 2) from Shesher and Welala wetlands lie within the same range as Lake Tana's [23]. This is due to the fact that these wetlands have hydrological connections with Lake Tana. Dissolved oxygen was significantly higher ($P < 0.05$) in Shesher than in Welala and Ribb River mouth sampling sites. This is probably due to high mixing by wind, since Shesher is shallow and has no shore vegetation cover. Similarly, high dissolved oxygen concentration was obtained in Shesher by Atnafu et al. [18]. However, as compared to the *Labeobarbus* spawning streams in Gumara [9] and Megech [15], tributaries of Lake Tana, these two wetlands have lower dissolved oxygen concentration, high turbidity, and lack gravel substrate.

Prespawning aggregations and upstream migrations of *Labeobarbus* species to the tributary rivers of Lake Tana was intensively studied in the last two decades [9, 12–16, 24–26]. These studies showed that seven species of *Labeobarbus*, after making brief prespawning aggregations at the river mouths, migrate more than 50 km up rivers and spawn in clear, fast flowing, well-oxygenated, and gravel bed small streams. Almost all African *Labeobarbus*, whether lake dwelling or riverine, require these conditions in their spawning grounds [3]. However, unlike the other African *Labeobarbus*, eight of the 15 species in Lake Tana, are absent in all the seven perennial tributaries. The most acceptable assumption is that like many other cyprinid genera, the eight missing (species not found to spawn in river tributaries) *Labeobarbus* species most probably breed in the lake and adjacent floodplain wetlands [14]. The use of marginal vegetation of the lake's shore and

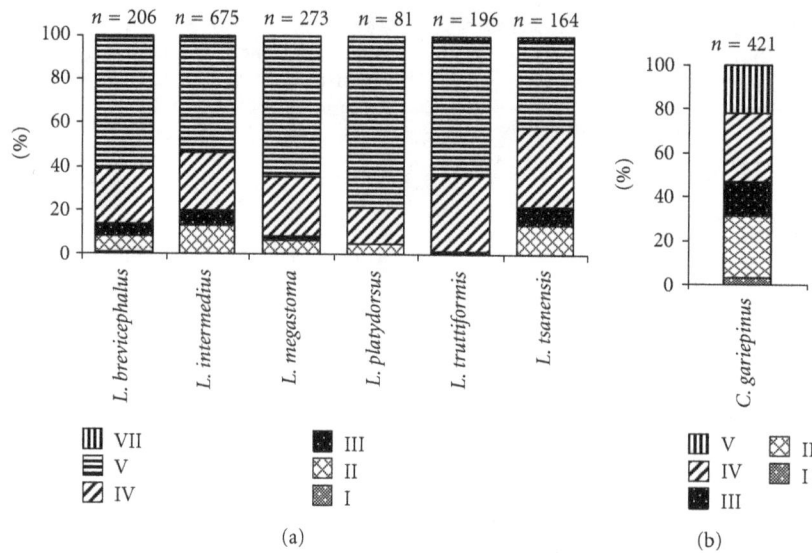

FIGURE 5: Gonad stages of immature (gonad stages I–III), ripe (gonad stages IV and V) and spent (gonad stage VII) of the six most abundant *Labeobarbus* species aggregating at Ribb River mouth (a), and *C. gariepinus* collected from Shesher and Welala wetlands (b). Note that gonad maturity stage V is ripe for *Labeobarbus* spp., but spent for *C. gariepinus*.

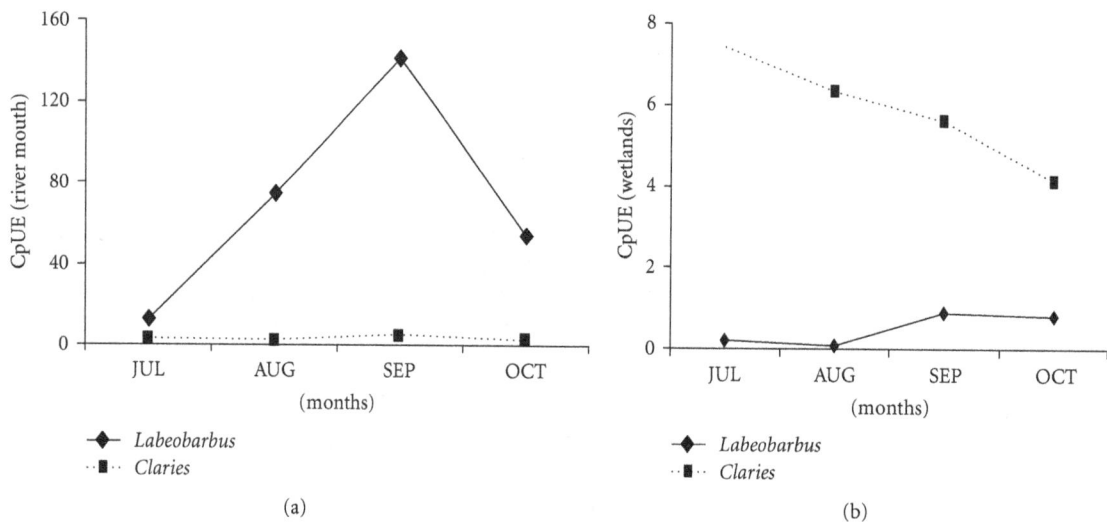

FIGURE 6: CpUE of *Labeobarbus* and *Clarias* at Ribb River mouth (a) and in Shesher and Welala Wetlands (b).

adjacent floodplain wetlands shelters from predators and provides high densities of prey for larvae and juveniles [27].

Contrary to the above assumption, in the present study, these missing *Labeobarbus* species (*L. nedgia, L. dainellii, L. gorguari, L. longissimus, L. intermedius, L. gorgorensis, L. surkis,* and *L. crassibarbis*) were not found to spawn in the most prominent adjacent floodplain wetlands, Shesher and Welala. This is probably because none of the requirements for *Labeobarbus* spawning were satisfied in the adjacent floodplain wetlands. During the spawning months, these wetlands were so turbid, poorly oxygenated, and the bottom is muddy, instead of gravel (Table 1). Unlike *C. gariepinus* and Nile tilapia, *Labeobarbus* species in Lake Tana are ecologically specialized and highly vulnerable [7]; hence, it is unlikely that the larvae of *Labeobarbus* will survive under these poor spawning

ground conditions. Another explanation could be the existence of *C. gariepinus* in mass in these two wetlands (Figures 3 and 4) that predate on fish. The larvae and juveniles of *Labeobarbus* could be preyed upon by *C. gariepinus* [8]; hence, coexistence of *C. gariepinus* and early life stages of *Labeobarbus* in this relatively small wetlands would result in high mortality rates of juvenile *Labeobarbus*.

In the present study, a large number of *C. gariepinus* were collected from Shesher and Welala Wetlands. *Clarias gariepinus* is well adapted to the environmental conditions of the wetlands, it is highly tolerant to low oxygen levels and high turbidity. Although the CpUE of *C. gariepinus* during our sampling declines from July to October, the presence of a relatively high percentage of spent implies that a substantial proportion of the fish may not drift back to the lake, rather

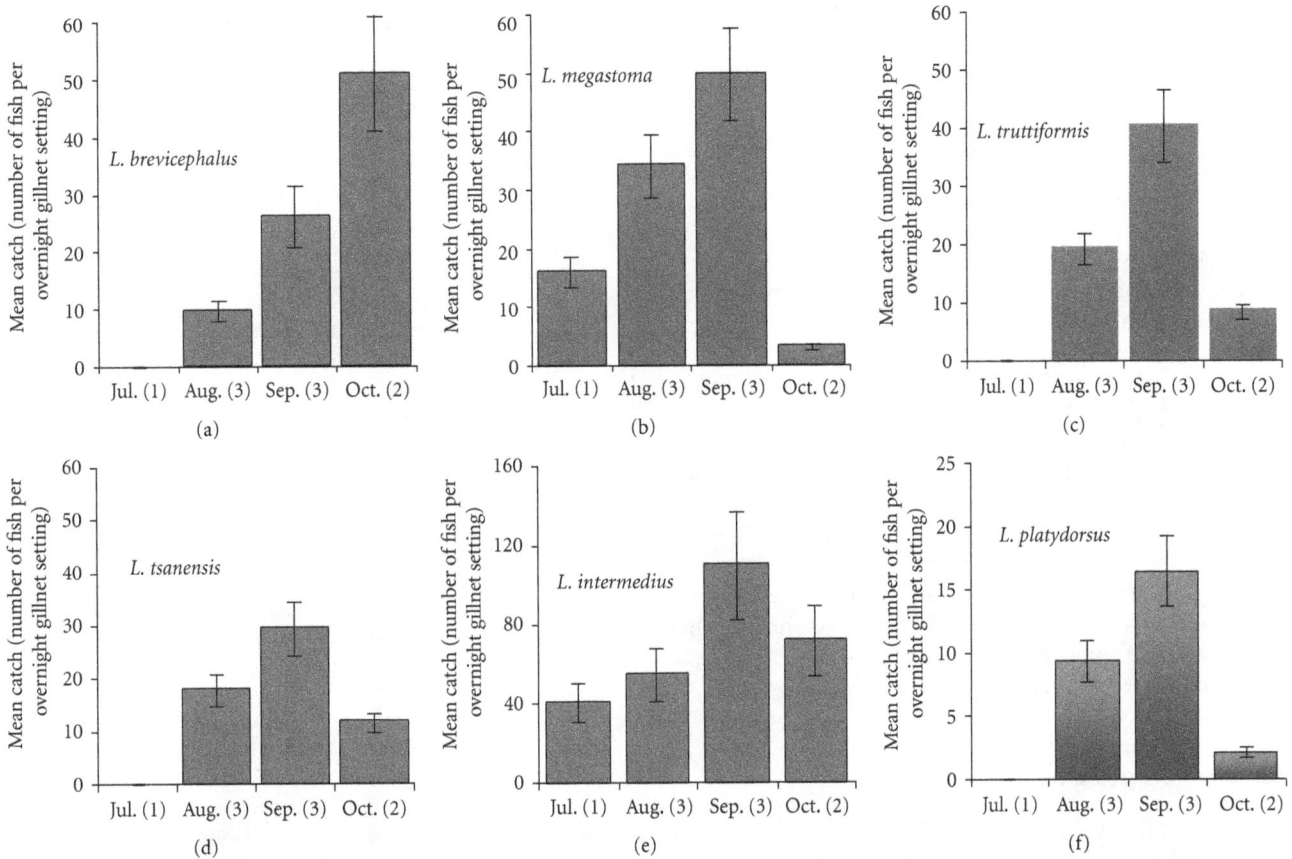

FIGURE 7: Mean abundance (number of fish per overnight gillnet setting) with 95% confidence limits (CL) of the six *Labeobarbus* species aggregating at Ribb River mouth during the spawning months (July to October).

they spend the rest of their life, after spawning, in these wetlands. The presence of a relatively high proportion of immature individuals (46.5%) in our catch indicates the presence of feeding migration of *C. gariepinus* to these wetlands as well. It was indicated by the local people that they observe mass of juveniles of *C. gariepinus* in these wetlands during the postrainy season (end of October and November). Most of the dry *C. gariepinus* exported to neighbouring countries such as Sudan is obtained from Shesher and Welala wetlands [11], and intensive fishing activities by local people using seine nets takes place in February and March. Probably, most of the juveniles of *C. gariepinus* in the nearby shallow inundated floodplains move to these wetlands for growth, since these wetlands get dry usually in April or May. What the local people intensively catch are those immature feeding migrants that stayed in these wetlands, young of the year (juveniles), and those spent *C. gariepinus* that remained in the wetlands.

Unlike *C. gariepinus*, only few specimens of *Labeobarbus* species aggregating at Ribb River mouth have reproductively immature gonads, more than 80% were ripe. However, no running (gonad stage VII; shedding eggs and sperm) *Labeobarbus* was collected from Ribb River mouth. This supports the suggestion made by Palstra et al. [9] which states that, if spawning maturity is only reached when the fish arrives at its spawning ground (more than 30 km up rivers), there could

be a fine-tuning between homing and gonad development. Figure 7 shows the details of temporal segregation and overlapping during the spawning months (July to October) among the six *Labeobarbus* species aggregating at Ribb river mouth. All these six *Labeobarbus*, except *L. intermedius*, were temporally segregated in their spawning aggregation at Ribb River mouth in the four spawning months. *Labeobarbus intermedius* is the only *Labeobarbus* species in Lake Tana that spawns throughout the year, and ripe individuals are always common in the river mouths [14]. *Labeobarbus megastoma* migrates from the lake to the river mouth, starting in July, and the CpUE declined after September. Three species, *L. truttiformis*, *L. platydorsus* and *L. tsanensis*, start to aggregate in August and CpUE reached peak in September and then started to decline. The last species to aggregate was *L. brevicephalus*, its peak is in October. Strong temporal segregation during prespawning aggregation at the river mouths among the migratory riverine *Labeobarbus* species was reported by de Graaf et al. [14]. In addition to the six species found to aggregate at Ribb River mouth in the present study, de Graaf et al. [14] observed two more species, *L. acutirostris* and *L. macrophthalmus* in four tributaries of Tana (Gelgel Abbay, Gelda, Gumara and Ribb) from July to October [14]. Moreover, these two species were also found to migrate more than 30 km in Gumara River upstream [9]. But, similar to the present study, both species did not form prespawning ag-gregations in other tributaries, Dirma and Megech Rivers [25] and Arno-Garno River [16]. These irregularities proba-bly originate from the traditional fish migration study meth-ods used (e.g., CpUE data). Other modern fish migration study methods such as radio-tracking may supplement the existing information and clarify the secrecy of the spawning grounds of those eight *Labeobarbus* species not found in the tributaries of rivers.

The hypothesis that the eight Labeobarbus species that do not migrate to rivers for spawning may spawn in adjacent floodplain wetlands [14] was not supported in the present study. The other possibility is that they may spawn in the rocky shores of Lake Tana [14]. Although the absence of these species in the tributary rivers and adjacent wetlands does not automatically mean they spawn in the lake, lake spawning now seems the best option. However, we again strongly re-commend the application of radio-tracking or other depend-able methods to investigate the actual spawning place(s) of these eight missing *Labeobarbus* species. Since Lake Tana and its shore wetlands are under heavy human pressure, mapping the spawning habitat is essential to conserve this unique *Labeobarbus* species flock.

Acknowledgment

The authors thank the Bahir Dar Fish and Aquaculture Re-search Center for the logistics support. The authors also need to thank Asratu Wondie (fishing assistance), Endalamaw Asres (boat driver), Ayenew Gedif (boat driver), and Getnet Temesgen (fishing assistance). They also thank Dereje Tew-abe for his support in the field works. The research was fund-ed by World Bank-financed Ethiopian-Nile Irrigation and Drainage Project Coordination Office, Ministry of Water Resources, Addis Ababa, Ethiopia, and International Found-ation for Science (IFS) Grt A/4922-1 (W. Anteneh).

References

[1] I. Kornfield and K. E. Carpenter, "Cyprinids of Lake Lanao, Philippines: taxonomic validity, evolutionary rates and specia-tion scenarios," in *Evolution of Fish Species Flocks*, A. A. Echelle and I. Kornfield, Eds., pp. 69–83, Orono Press, 1984.

[2] P. H. Skelton, D. Tweddle, and P. Jackson, "Cyprinids of Africa," in *Cyprinid Fishes, Systematics, Biology and Exploita-tion*, I. J. Winfield and J. S. Nelson, Eds., pp. 211–233, Chapman & Hall, London, UK, 1991.

[3] T. Tómasson, J. A. Cambray, and P. B. N. Jackson, "Reproduc-tive biology of four large riverine fishes (Cyprinidae) in a man-made lake, Orange River, South Africa," *Hydrobiologia*, vol. 112, no. 3, pp. 179–195, 1984.

[4] R. Ogutu-Ohwayo, "The decline of the native fishes of lakes Victoria and Kyoga (East Africa) and the impact of introduced species, especially the Nile perch, *Lates niloticus*, and the Nile tilapia, *Oreochromis niloticus*," *Environmental Biology of Fishes*, vol. 27, no. 2, pp. 81–96, 1990.

[5] P. B. O. Ochumba and J. O. Manyala, "Distribution of fishes along the Sondu-Miriu River of Lake Victoria, Kenya with special reference to upstream migration, biology and yield,"

[6] M. de Graaf, M. A. M. Machiels, T. Wudneh, and F. A. Sibbing, "Declining stocks of Lake Tana's endemic *Barbus* species flock (Pisces, Cyprinidae): natural variation or human impact?" *Biological Conservation*, vol. 116, no. 2, pp. 277–287, 2004.

[7] M. de Graaf, P. A. M. van Zwieten, M. A. M. Machiels et al., "Vulnerability to a small-scale commercial fishery of Lake Ta-na's (Ethiopia) endemic *Labeobarbus* compared with African catfish and Nile tilapia: an example of recruitment-overfish-ing?" *Fisheries Research*, vol. 82, no. 1–3, pp. 304–318, 2006.

[8] T. Wudneh, *Biology and management of fish stocks in Bahir Dar Gulf, Ethiopia*, Ph.D. thesis, Wageningen University, The Netherlands, 1998.

[9] A. P. Palstra, M. de Graaf, and F. A. Sibbing, "Riverine spawn-ing and reproductive segregation in a lacustrine cyprinid spec-ies flock, facilitated by homing?" *Animal Biology*, vol. 54, no. 4, pp. 393–415, 2004.

[10] K. P. C. Goudswaard and F. Witte, "The catfish fauna of Lake Victoria after the Nile perch upsurge," *Environmental Biology of Fishes*, vol. 49, no. 1, pp. 21–43, 1997.

[11] N. Atnafu, *Assessment of Ecological and Socio-Economic Impor-tance of Fogera Floodplains: the case of Wolala and Shesher Wetlands*, M.S. thesis, Bahir Dar University, Bahir Dar, Ethiopia, 2010.

[12] L. A. J. Nagelkerke and F. A. Sibbing, "Reproductive segrega-tion among the *Barbus intermedius* complex of Lake Tana, Ethiopia. An example of intralacustrine speciation?" *Journal of Fish Biology*, vol. 49, no. 6, pp. 1244–1266, 1996.

[13] Y. Y. Dgebuadze, M. V. Mina, S. S. Alekseyev, and A. S. Golub-tsov, "Observations on reproduction of the Lake Tana barbs," *Journal of Fish Biology*, vol. 54, no. 2, pp. 417–423, 1999.

[14] M. de Graaf, E. D. Nentwich, J. W. M. Osse, and F. A. Sibbing, "Lacustrine spawning: is this a new reproductive strategy among 'large' African cyprinid fishes?" *Journal of Fish Biology*, vol. 66, no. 5, pp. 1214–1236, 2005.

[15] W. Anteneh, A. Getahun, and E. Dejen, "The lacustrine species of *Labeobarbus* of Lake Tana (Ethiopia) spawning at Megech and Dirma Tributary Rivers," *SINET*, vol. 31, pp. 21–28, 2008.

[16] S. Gebremedhin, *Spawning migration of Labeobarbus species of Lake Tana to Arno-Garno River, Ethiopia*, M.S. thesis, Bahir Dar University, Bahir Dar, Ethiopia, 2011.

[17] L. A. J. Nagelkerke, *The barbs of Lake Tana, Ethiopia: morpho-logical diversity and its implications for taxonomy, trophic resource partitioning and fisheries*, Ph.D. thesis, Wageningen University, The Netherlands, 1997.

[18] N. Atnafu, E. Dejen, and J. Vijverberg, "Assessment of the Ecological Status and Threats of Welala and Shesher Wetlands, Lake Tana Sub-Basin (Ethiopia)," *Journal of Water Resource and Protection*, vol. 3, pp. 540–547, 2011.

[19] S. Nemomissa, *Delineation of wetlands of Fogera floodplain and scaling their biological and physical connections to Rib River and Lake Tana*, Ministry of Water Resources, Addis Ababa, Ethiopia, 2008.

[20] E. Dejen, A. Getahun, and W. Anteneh, *Detailed Implementa-tion Manual for Environmental and Social Management Plan of Ribb River Fisheries Resource*, Ministry of Water and Energy, Addis Ababa, Ethiopia, 2011.

[21] RIDP_ESIA, "Environmental and social impact assessment of the Ribb irrigation and drainage project," 2010, http://www.mowr.gov.et/attachmentfiles/Downloads/RIDP_ESIA_vol2.pdf.

[22] L. A. J. Nagelkerke and F. A. Sibbing, "The large barbs (*Barbus* spp., Cyprinidae, Teleostei) of Lake Tana (Ethiopia), with

Aquaculture & Fisheries Management, vol. 23, no. 6, pp. 701–719, 1992.

Shesher and Welala Floodplain Wetlands (Lake Tana, Ethiopia): Are They Important Breeding Habitats for Clarias
gariepinus and the Migratory Labeobarbus Fish Species?

133

a description of a new species, *Barbus osseensis*," *Netherlands Journal of Zoology*, vol. 50, no. 2, pp. 179–214, 2000.

[23] E. Dejen, J. Vijverberg, L. A. J. Nagelkerke, and F. A. Sibbing, "Temporal and spatial distribution of microcrustacean zoo-plankton in relation to turbidity and other environmental factors in a large tropical lake (L. Tana, Ethiopia)," *Hydrobiologia*, vol. 513, pp. 39–49, 2004.

[24] S. S. Alekseyev, Y. Y. Dgebuadze, M. V. Mina, and A. N. Mironovsky, "Small "large barbs" spawning in tributaries of Lake Tana: what are they?" *Folia Zoologica*, vol. 45, no. 1, pp. 85–96, 1996.

[25] A. Getahun, E. Dejen, and W. Anteneh, *Fishery studies of Ribb River, Lake Tana basin, Ethiopia. Final Report E1573*, vol. 2, Ethiopian Nile Irrigation and Drainage Project Coordination Office, Ministry of Water Resources, Ethiopia, 2008.

[26] W. Anteneh, *The spawning migration and reproductive biology of Labeobarbus (Cyprinidae: Teleostei) of Lake Tana to Dirma and Megech Rivers*, M.S. thesis, Addis Ababa University, Addis Ababa, Ethiopia, 2005.

[27] C. A. Mills, "Reproduction and life history," in *Cyprinid Fishes, Systematics, Biology and Exploitation*, I. J. Winfield and J. S. Nelson, Eds., pp. 483–529, Chapman & Hall, London, UK, 1991.

Effluents of Shrimp Farms and Its Influence on the Coastal Ecosystems of Bahía de Kino, Mexico

Ramón H. Barraza-Guardado,[1,2] **José A. Arreola-Lizárraga,**[3] **Marco A. López-Torres,**[2] **Ramón Casillas-Hernández,**[1] **Anselmo Miranda-Baeza,**[4] **Francisco Magallón-Barrajas,**[3] **and Cuauhtemoc Ibarra-Gámez**[1]

[1] *Instituto Tecnológico de Sonora (ITSON), 85000 Ciudad Obregón, SON, Mexico*
[2] *Departamento de Investigaciones Científicas y Tecnológicas de la Universidad de Sonora (DICTUS),*
  83000 Hermosillo, SON, Mexico
[3] *Centro de Investigaciones Biológicas del Noroeste, S.C. (CIBNOR, S.C.), 85454 Guaymas, SON, Mexico*
[4] *Universidad Estatal de Sonora (US), 85800 Navojoa, SON, Mexico*

Correspondence should be addressed to José A. Arreola-Lizárraga; aarreola04@cibnor.mx

Academic Editors: F. Amezcua-Martínez, M. G. Frias-Espericueta, J. R. Ruelas-Inzunza, M. F. Soto-Jimenez, and M. Teichberg

The impact on coastal ecosystems of suspended solids, organic matter, and bacteria in shrimp farm effluents is presented. Sites around Bahía de Kino were selected for comparative evaluation. Effluent entering Bahia Kino (1) enters Laguna La Cruz (2). A control site (3) was outside the influence of effluents. Water quality samples were collected every two weeks during the shrimp culture period. Our data show that the material load in shrimp farm effluents changes biogeochemical processes and aquatic health of the coastal ecosystem. Specifically, the suspended solids, particulate organic matter, chlorophyll *a*, viable heterotrophic bacteria, and *Vibrio*-like bacteria in the bay and lagoon were two- to three-fold higher than the control site. This can be mitigated by improvements in the management of aquaculture systems.

1. Introduction

Worldwide, brackish-water aquaculture production (4.7 million tons) consisted of crustaceans (57%), freshwater fishes (19%), diadromous fishes (15%), marine fishes (7%), and marine mollusks (2%) in 2010; more than 99 percent of the crustaceans were marine shrimps [1]. It shows the importance of research about the effect of shrimp farming on the environment [2], with water pollution from shrimp pond effluents as the most common complaint [3–5]. This activity depends directly or indirectly on a range of coastal and marine ecosystem services some of which may be used at rates that are not sustainable [6, 7].

Most of shrimp production is carried out in ponds. The most common shrimp aquaculture systems use inland ponds that are near or on the coast. Water is discharged from these shrimp ponds to coastal ecosystem as part of the water exchange when ponds are drained. The main components in the shrimp farm effluents are organic matter mainly in particulate form from different sources, as well as nitrogen and phosphorus in both organic and inorganic forms, and suspended solids [8, 9].

Production systems in the culture of marine shrimp, semi-intensive or intensive, lead to significant increases in the levels of nutrients, phytoplankton biomass, organic matter, and suspended solids in the environment receiving the farm's effluents [10–13]. In addition, it has been reported that water quality shows short term increases in parameters of water bodies receiving shrimp discharge waters, but other studies indicate that there are no significant differences over background levels on an annual basis [14, 15]. The impact of pond effluents on adjacent ecosystems is variable and depends on various factors, including the magnitude of the discharge, the chemical composition of the pond effluents, and the specific characteristics of the environment that receives the discharge, such as circulation and dilution rates [16].

The characterization of the shrimp farm's effluents in terms of water quality is very important to gauge the environmental health of an ecosystem in order to achieve a better regulation of the industry [17, 18]. Further, the continuous monitoring of the physical, chemical, and biological parameters of pond, effluent, and inlet waters helps not only to predict and control negative conditions for shrimp farming but also avoids environmental damages and collapse of the production process [19].

In Mexico, about 97% of shrimp aquaculture ponds are located around the Gulf of California in the states of Baja California, Baja California Sur, Sonora, Sinaloa, and Nayarit. Beginning in the mid-1980s, the Gulf of California ecoregion experienced an increase in shrimp aquaculture and became the second largest producer in the western hemisphere [20]. Recently, Mexico was the sixth largest producer of shrimp culture worldwide ~110,000 t [1]. The State of Sonora contributed with 37% of this production in 2011, and specifically Bahía de Kino region which is the main shrimp producer [21].

The goals of this study were to examine the effluent loads from shrimp farms, and its influence on the coast in Bahía de Kino. We measured concentrations of total suspended solids, particulate organic matter, chlorophyll *a*, viable heterotrophic bacteria, and *Vibrio*-like bacteria in the effluent and in the coastal ecosystems. We then assessed the influence of these effluents on the coastal ecosystems.

2. Materials and Methods

2.1. Study Area. Bahía de Kino is located on the Gulf of California in the State of Sonora, Mexico ($28°47'$N, $111°54'$W). Into this bay, effluents from 1,350 hectares (effluents rate ~ 160,000 $m^3 ha^{-1} yr^{-1}$) from four shrimp farms that operate from April through October are delivered into the bay, about 2 km south of the mouth of the Laguna de la Cruz. Seawater is taken directly from the open sea for shrimp farming (Figure 1). Bahía de Kino has a seasonal pattern of water temperature, a maximum of 32.2°C in August and minimum of 15.6°C in January. Salinity varies from 35–38.3 [22]. This region has a dry desert climate with evaporation of ~3000 $mm yr^{-1}$ which exceeds rainfall of <300 $mm yr^{-1}$ [23].

2.2. Sampling and Measurements. Water quality was sampled every two weeks at 12 sampling stations in June, September, and October (2009), the period when the most effluents and organic matter from partial and final harvest of shrimp takes place.

A control station outside the influence of the shrimp effluents (~6 km) was located at Isla Alcatraz; four stations were established in the southern part of the bay near the outlet of the effluents; three stations were established in the effluent channel; and four stations were established inside Laguna la Cruz (Figure 1).

Water samples were taken between 07:00 and 13:00 h. At each station, water was collected near the surface in 1 L plastic bottles; these samples were used to measure suspended solids, chlorophyll *a*, and pH. Samples for study of bacteria were

collected in sterile 120 mL Whirl pack bags. All samples were transported to the laboratory in ice.

At each station, temperature, dissolved oxygen, and salinity were measured (YSI field oxygen meter 85, YSI, Yellow Springs, OH); water transparency was measured with a Secchi disk, and pH was measured in the laboratory with pH meter (model 220A, Hanna Instruments, Woonsocket, RI).

2.3. Water Quality Analysis. The water samples collected for the analysis of suspended solids and organic matter were filtered through precombusted and preweighed Whatman GF/C (Whatman International Ltd.) glass fibers. Then, the filters were dried in an oven at 60°C for 24 h and weighed to determine total suspended solids (TSS), for particulate organic matter (POM) estimation (ash-free dry weight), filters and suspended material were then ignited to constant weight at 550°C for 8 h and weighed again. Inorganic suspended solids (ISSs) were estimated by subtracting the value of POM on the TSS [24]. Samples for analysis of chlorophyll *a* were collected by filtration through Whatman GF/C glass fibre filters, extracted with 90% v/v acetone, and measured by spectrophotometry, according to [25].

For the quantification of culturable bacteria, including viable heterotrophic bacteria (VHB), and *Vibrio*-like bacteria (VLB), the spread plate method [26, 27] in Marine Agar 2216 and Thiosulfate-citrate-bile-sucrose agar (TCBS, DIFCO, U.S.A.) was used. A 10-fold dilution series were prepared with 3.0% of sterile seawater diluted with distilled water. The plates were incubated for 48 ± 2 h to 30 ± 2°C, and the bacterial colonies were counted and reported as colony-forming units for mL^{-1} (CFU mL^{-1}).

2.4. Statistical Data Analysis. We used multivariate, multidimensional, and nonparametric scaling of data which was transformed ($\log x + 1$) and standardized to determine whether the sites (island, effluent, bay, and lagoon) were similar in terms of water quality. Statistical software (Primer-E 6.0, Primer-E, Ivybridge, UK) was used to perform the analyses.

The data grouped by area (island, effluent, bay, and lagoon) were subjected to a normality and equal variance test [28]. After this, parametric tests were made through a one way ANOVA with its respective Tukey test of multiple comparisons and also a nonparametric test using one way ANOVA and a Kruskal-Wallis *a posteriori* test with a 95% level of significance [29]. The NCSS program (2007) was used for statistical analysis [30].

3. Results

The result from the similarity analysis for the multivariate method of nonparametric multidimensional scaling (nMDS) indicated a difference of similarity among the study areas (Figure 2). Effluents and water quality parameters near the island had lower similarity, while the bay and lagoon had similar water quality conditions, but differed from the water quality conditions of the effluent.

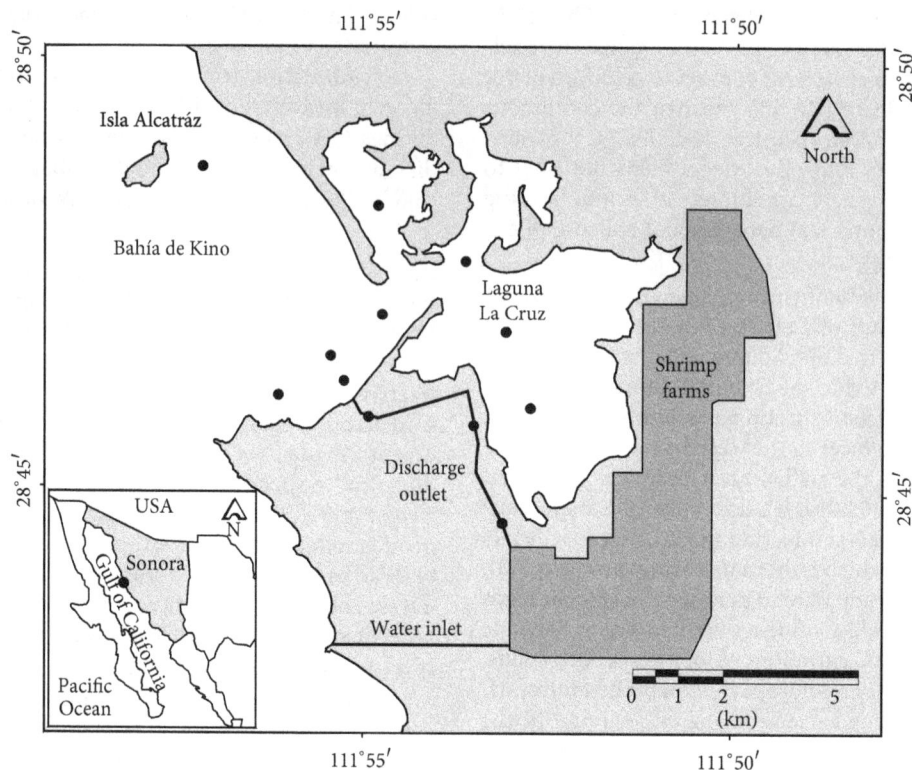

FIGURE 1: Bahía Kino area, including Laguna La Cruz and Isla de Alcatraz, showing seawater inlet to shrimp farms and discharge channel to the bay.

Transform: log($X + 1$)
Normalise
Resemblance: D1 euclidean distance

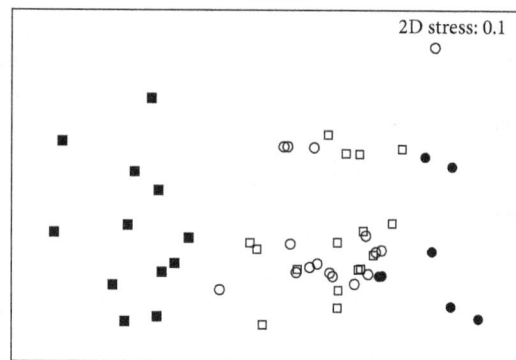

- ■ Effluent   □ Bay
- ● Island   ○ Lagoon

FIGURE 2: nMDS ordination of water quality parameters for the study areas. Temperature, salinity, dissolved oxygen, suspended solids, particulate organic matter, chlorophyll *a*, *Vibrio*-like bacteria, and viable heterotrophic bacteria were included in the multivariate analysis.

The average water temperature during all the time of the study on the island was 27.3 ± 3.3°C, effluent of 24.8 ± 5.3°C, at the bay of 25.5 ± 4.6°C, and the lagoon of 26.6 ± 4.4°C, without showing any statistical differences among areas ($H = 5.27$; $P = 0.15$) (Table 1). However, a seasonal behavior was

TABLE 1: Average values (±SD) of physicochemical parameters for different parts of the study area during discharge of effluent.

Areas	Temperature (°C)	Salinity	Dissolved oxygen (mg L^{-1})	pH
Island	27.3 ± 3.3[a]	36.7 ± 0.4[a]	6.6 ± 1.2[b]	8.3 ± 0.1[a]
Effluent	24.8 ± 5.3[a]	39.3 ± 1.3[b]	4.5 ± 1.4[a]	8.2 ± 0.2[a]
Bay	25.5 ± 4.6[a]	37.2 ± 0.6[a]	5.7 ± 0.9[b]	8.2 ± 0.2[a]
Lagoon	26.6 ± 4.4[a]	37.2 ± 0.4[a]	6.0 ± 0.6[b]	8.2 ± 0.1[a]

Different letters among areas for each parameter indicate significant differences ($P < 0.05$; $n = 96$).

observed when high temperatures were recorded in June and September, and the lowest in October.

Salinity values were found significantly higher ($F = 74.5$; $P < 0.001$) in the effluent (39.3 ± 1.3) over the island (36.7 ± 0.4), bay (37.2 ± 0.6), and lagoon (37.2 ± 0.4) (Table 1). It was noted that the bay, lagoon, and the control area (island) remained similar in salinities among them.

In the effluent, significantly lower values of DO (4.5 ± 1.3 mg L^{-1}) ($F = 20.2$; $P < 0.001$) were found in comparison to the rest of the other study areas (Table 1). No significant differences were found ($P > 0.05$) of DO among the island, bay, and lagoon areas. The pH did not show statistical differences among the areas studied ($H = 4.41$; $P = 0.22$) (Table 1).

Water transparency had average values on the island, effluent, bay, and lagoon of 2.5 ± 0.8, 0.2 ± 0.1, 0.9 ± 0.4,

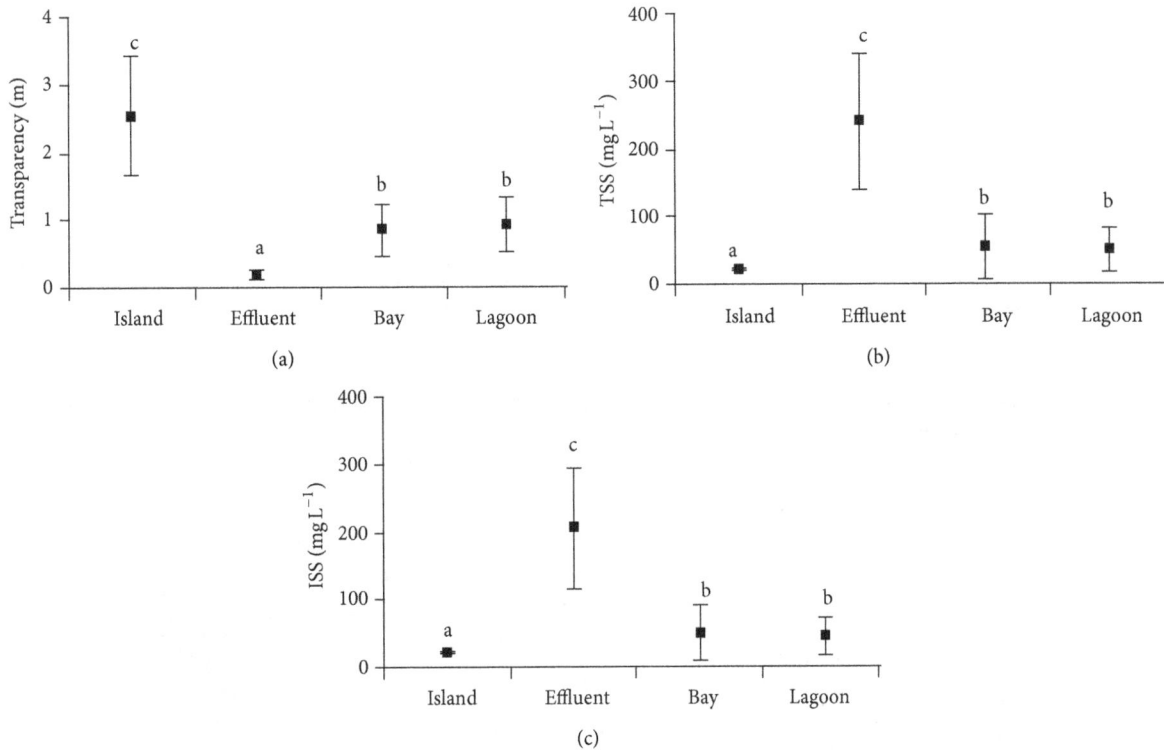

FIGURE 3: Average values (±SD) of selected variables. (a) Water transparency, (b) total suspended solids, and (c) inorganic suspended solids. Different letters among zones for each variable indicate significant differences ($P < 0.05$).

and 0.9 ± 0.4 m, respectively (Figure 3). Significantly lower values were recorded in the effluent, and higher on the island ($F = 34.05$; $P < 0.001$). Water transparency both in the bay and lagoon had intermediate values, not statistically different, but significantly lower than the island (Figure 3).

The average concentration of total suspended solids (TSS) was found to be 26.7 ± 1.2 mg L^{-1} in waters near the island (control area), 233.2 ± 95.7 mg L^{-1} in the effluent area, 56.2 ± 45.1 mg L^{-1} in the bay, and 52.7 ± 30.6 mg L^{-1} in the lagoon. The inorganic suspended solids (ISSs) showed the same pattern as TSS. Both TSS and ISS presented averages significantly (TSS; $H = 33.15$ and $P < 0.001$; ISS: $H = 33.14$ and $P = 0.001$) lower in the island and higher in the effluent, with intermediate values in both bay and lagoon, but significantly higher than near the island (Figure 3).

Particulate organic matter (POM) had average concentrations near the island of 4.62 ± 0.51 mg L^{-1}, 26.1 ± 9.2 mg L^{-1} in effluent zone, 7.1 ± 4.2 mg L^{-1} in the bay, and 7.0 ± 2.7 mg L^{-1} in the lagoon. Significantly higher values ($H = 31.07$; $P < 0.001$) of POM were observed in the effluent zone and lower ones near the island; while POM values in both bay and lagoon remained similar, it was observed that the lagoon presented higher levels than the island (Figure 4).

Chlorophyll a concentration in the control area (island) had average values of 2.3 ± 0.7 mg m^{-3}, a value which is significantly ($H = 38.72$; $P < 0.001$) lower when it is compared with the other areas (Figure 4). By contrast, significantly higher values ($P < 0.001$) were found in the effluent zone during the complete culture period with

21.9 ± 6.7 mg m^{-3}. Chl a of bay (4.8 ± 1.4 mg m^{-3}) and lagoon (7.5 ± 3.7 mg m^{-3}) was maintaining similar values.

The average concentration of viable heterotrophic bacteria (VHB) in the control area (island) was of $5.64 \times 10^2 \pm 0.35 \times 10^1$ CFU mL–1, while effluent was of $9.07 \times 10^3 \pm 0.24 \times 10^1$ CFU mL–1, bay $1.76 \times 10^3 \pm 0.20 \times 10^1$ CFU mL–1, and lagoon $3.11 \times 10^3 \pm 0.18 \times 10^1$ CFU mL–1. Significantly higher concentrations ($H = 45.76$; $P < 0.001$) were quantified in the effluent zone and lower ones near the island. There was no difference in VHB between island and bay, whereas concentrations in the lagoon were higher than both island and bay. The *Vibrio*-like bacteria (VLB) presented the same as than the viable heterotrophic bacteria ($H = 50.6$; $P < 0.001$) (Figure 5).

4. Discussion

The multivariate analysis (nMDS) showed that water conditions varied considerably among the four areas, especially between the sites in the effluent channel and the other areas. The effluent had no effect on water quality in the control area. The bay and lagoon are similar because there is a large exchange of materials between the two areas, mainly from tidal exchange. Our results suggested that pollutants from the shrimp farm influence environmental conditions in the bay and lagoon.

4.1. Water Quality of Shrimp Farm's Effluents. Low dissolved oxygen (DO) in the effluent results from a heavy load of

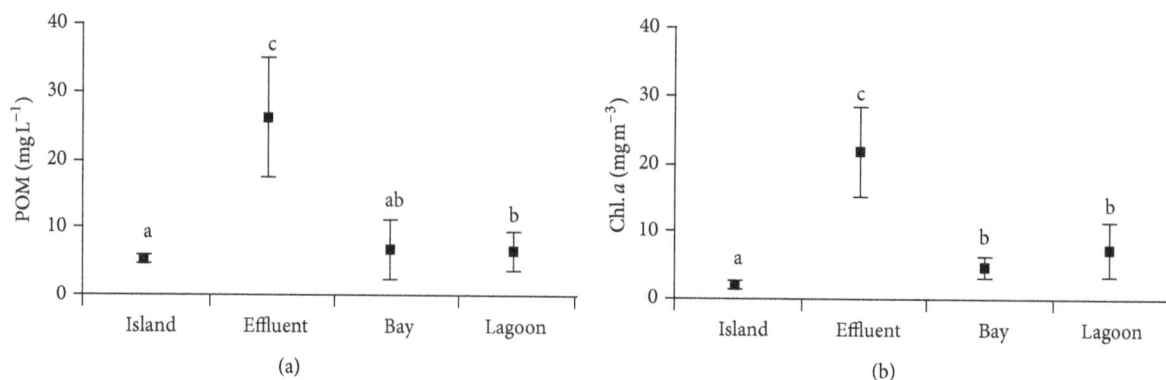

FIGURE 4: Average values (±SD) of selected variables. (a) Particulate organic matter and (b) chlorophyll a. Different letters among zones for each variable indicate significant differences ($P < 0.05$).

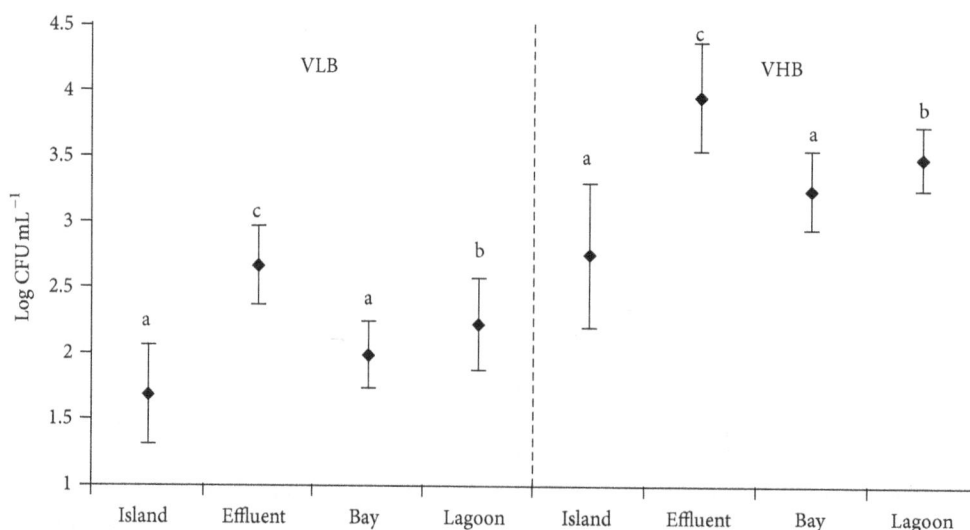

FIGURE 5: Average concentration (±SD) of *Vibrio*-like bacteria (VLB) and viable heterotrophic bacteria (VHB) in the different areas of study. Different letters among zones for each group of bacteria indicate significant differences ($P < 0.05$).

organic matter generated in the shrimp farms [8, 9, 31, 32], which includes unconsumed shrimp food, detritus, phytoplankton, zooplankton, and bacteria. The level of oxygen at the discharge outlet at the bay was $4.5 \pm 1.4\,\mathrm{mg\,L^{-1}}$, within the range of water quality standards for shrimp farm effluents recommended by the Global Aquaculture Alliance [33].

Effluent was highly saline because of the high evaporation rates in the shrimp ponds in this subtropical desert region, which is consistent with [34] who report salinity 40–42 in effluents in this region. The suspended solids (TSS and ISS) increased during the course of effluent water flow because the bottom sediments were resuspended and the channel walls were eroded. The effluent was less transparent, caused by the concentrations of suspended solids, POM, and phytoplankton biomass.

Water quality standards for shrimp farm effluents recommended by the Global Aquaculture Alliance for TSS has a standard of $<100\,\mathrm{mg\,L^{-1}}$ and a target standard of $<50\,\mathrm{mg\,L^{-1}}$ [33]. Effluents discharging into Bahía de Kino had TSS concentrations of $233.2 \pm 95.7\,\mathrm{mg\,L^{-1}}$.

High phytoplankton biomass and organic matter observed in the wastewater is promoted by inorganic fertilizers in the cultivation system. With current management practices, the ponds are rich in phytoplankton and organic matter, and this water is later discharged [11, 13, 15, 35].

Microorganisms in general and bacteria, in particular, are key elements in the marine ecosystems operation and react quickly to changes in environmental parameters [36, 37]. The quantity of bacteria is an important variable in monitoring of shrimp farm effluent. *Vibrio* bacteria are opportunistic pathogens in shrimp farms at the larval and grow-out stages; it is the most serious pathogen, causing up to 100% mortality [38–40].

In summary, our results showed that shrimp farm effluents reaching the bay have higher concentrations of TSS, POM, phytoplankton biomass, and bacteria than the bay; this is consistent with observations in other studies [10–13, 15]. The control site at the island had low concentrations of TSS, POM, phytoplankton biomass, and bacteria, thus, the effluent loadings are still a good indicator of likely impact.

The knowledge of those loads is useful for understanding the responses from water bodies receiving shrimp farm effluents.

4.2. Influences on Coastal Ecosystems. Sustainability of shrimp culture requires maintenance of good water quality in the adjacent coastal region. Our results showed that the suspended solids, POM, Chl *a*, VHB, and VLB in the bay and lagoon were two- to three-fold higher than the control site. Environmental problems from shrimp farm effluents are associated with water pollution and diseases.

An excess of organic matter discharged into the bay and lagoon induces a higher demand of dissolved oxygen which negatively affects ecosystems by hypoxia. Variations in effluents with low concentrations of dissolved oxygen in the bay and lagoon could be explained by winds pattern [41], tidal mixing, coastal circulation along the coast of the Gulf of California [42], and water exchange time for the lagoon (21 days) [43]. These results suggest that the system was assimilating the organic matter discharged. However, the marginal rate of assimilation by the system does not indicate an absence of ecological impact. High impact may occur during the night in the area surrounding the discharge in the bay, creating hypoxic events, mainly during the summer when winds are less intense, water temperature is higher, and dissolved oxygen is, on average, lower. This represents a potential negative effect on biogeochemical processes and aquatic life.

One of the key environmental concerns about shrimp farming is the discharge of waters with high levels of nutrients into adjacent body waters. Pond water is continuously exchanged (5–30% d–1), with drainage through a ditch that brings back diluted waste water to the bay and inlet channel. With 1,350 ha of shrimp ponds in operation, yielding about $2.5 \, t \, ha^{-1}$ and disposing of $72 \, kg \, N$ and $13 \, kg \, P$ waste for each ton of harvested shrimp [44], we estimate a load of $\sim 243 \, t \, N \, yr^{-1}$ and $\sim 44 \, t \, P \, yr^{-1}$. This contributes to significantly high levels of Chl *a* in the bay, compared to the control site. Nutrient loads have been linked to lower diversity of phytoplankton species and nuisance algae blooms, with impacts the ecological health of coastal ecosystems [45]. In a nearby area in April 2003, a harmful algal bloom of *Chattonella marina*, *C. cf. ovata*, *Gymnodinium catenatum*, and *G. sanguineum* caused a massive die-off of fish and mollusks [46]; hence, there is a risk potential of harmful algal blooms.

Most shrimp diseases are bacterial and viral. Most bacterial diseases are caused by *Vibrio* spp. Vibriosis outbreaks is a serious problem in intensive shrimp ponds in this region [47]. These pathogenic bacteria can affect other cultivated marine populations, such as oysters, in this region, as well as species native to the area. When the aquatic environment is enriched by accumulating organic matter, several species of *Vibrio* spp. grow rapidly, not because it has a high growth rate, but because it is adapted to oxygen-deficient conditions [48]. *Vibrio* spp. affects fish, crustaceans, and cultivated shellfish. The important decade-old oyster farming activity (*Crassostrea gigas* and *C. corteziensis*) surrounding the bay and lagoon [49, 50] could be impacted by the shrimp farm effluent. These results indicate the need for increased research to determine the consequences of bacterial loads and waste nutrients into the coastal ecosystems.

5. Conclusions

Shrimp farm effluent provides significantly high salinity, suspended solids, organic particulate matter, chlorophyll *a*, and bacteria to coastal ecosystems, as well as reduced dissolved oxygen and transparency. Effluent produced changes in water quality in the bay and lagoon. Accumulation of solids, organic matter, and bacterial biomass affects environmental conditions and processes of these ecosystems. There is still insufficient knowledge of how effluents are affecting coastal ecosystems and how it affects aquaculture activities. Current water quality of Bahía de Kino appears to result from inadequate management of shrimp ponds. Our results suggest that both in Mexico and worldwide efforts at waste prevention and minimization at the source and onsite treatment and reuse of effluent would reduce losses of coastal ecosystem services.

Conflict of Interests

The authors declare they have not conflict of interests.

Acknowledgments

Leopoldo Encinas Bracamontes, Ulises Becerra Lamadrid, Fulgencio García Ochoa, and Oscar Acosta González of the UEK of DICTUS provided technical support during sampling. María del Refugio López Tapia provided water analysis, and David Urías Laborín of CIBNOR Guaymas prepared the map. Instituto Tecnológico de Sonora, Universidad de Sonora, Consejo Nacional de Ciencia y Tecnología (CONACYT), and Comité de Sanidad Acuícola del Estado de Sonora provided funding of this project. Miguel Cordoba-Matson provided initial English review. Ira Fogel provided comprehensive editorial services. The first author would also like to thank CONACYT for the grant that has allowed him to pursue his doctoral studies.

References

[1] FAO, *The State of World Fisheries and Aquaculture*, Food and Agriculture Organization, Rome, Italy, 2012.

[2] G. Sarà, "Ecological effects of aquaculture on living and non-living suspended fractions of the water column: a meta-analysis," *Water Research*, vol. 41, no. 15, pp. 3187–3200, 2007.

[3] C. E. Boyd, "Guidelines for aquaculture effluent management at the farm-level," *Aquaculture*, vol. 226, no. 1–4, pp. 101–112, 2003.

[4] P. T. Anh, C. Kroeze, S. R. Bush, and A. P. J. Mol, "Water pollution by intensive brackish shrimp farming in south-east Vietnam: causes and options for control," *Agricultural Water Management*, vol. 97, no. 6, pp. 872–882, 2010.

[5] T. D. Bui, J. Luong-Van, and C. M. Austin, "Impact of shrimp farm effluent on water quality in coastal areas of the world heritage-listed Ha Long Bay," *American Journal of Environmental Sciences*, vol. 8, no. 2, pp. 104–116, 2012.

[6] R. L. Naylor, R. J. Goldburg, J. H. Primavera et al., "Effect of aquaculture on world fish supplies," *Nature*, vol. 405, no. 6790, pp. 1017–1024, 2000.

[7] L. Deutsch, S. Gräslund, C. Folke et al., "Feeding aquaculture growth through globalization: exploitation of marine ecosystems for fishmeal," *Global Environmental Change*, vol. 17, no. 2, pp. 238–249, 2007.

[8] J. S. Hopkins, R. D. Hamilton II, P. A. Sandifer, C. L. Browdy, and A. D. Stokes, "Effect of water exchange rate on production, water quality, effluent characteristics and nitrogen budgets of intensive shrimp ponds," *Journal of the World Aquaculture Society*, vol. 24, no. 3, pp. 304–320, 1993.

[9] C. E. Boyd, "Farm effluent during draining for harvest," *Global Aquaculture*, vol. 3, pp. 26–27, 2000.

[10] A. D. McKinnon, L. A. Trott, D. M. Alongi, and A. Davidson, "Water column production and nutrient characteristics in mangrove creeks receiving shrimp farm effluent," *Aquaculture Research*, vol. 33, no. 1, pp. 55–73, 2002.

[11] X. Biao, D. Zhuhong, and W. Xiaorong, "Impact of the intensive shrimp farming on the water quality of the adjacent coastal creeks from Eastern China," *Marine Pollution Bulletin*, vol. 48, no. 5-6, pp. 543–553, 2004.

[12] L. D. de Lacerda, A. G. Vaisman, L. P. Maia, C. A. R. Silva, and E. M. S. Cunha, "Relative importance of nitrogen and phosphorus emissions from shrimp farming and other anthropogenic sources for six estuaries along the NE Brazilian coast," *Aquaculture*, vol. 253, no. 1–4, pp. 433–446, 2006.

[13] A. P. Cardozo, V. O. Britto, and C. Odebrecht, "Temporal variability of plankton and nutrients in shrimp culture ponds vs. adjacent estuarine water," *Pan-American Journal of Aquatic Sciences*, vol. 6, no. 1, pp. 28–43, 2011.

[14] L. A. Trott and D. M. Alongi, "Variability in surface water chemistry and phytoplankton biomass in two tropical, tidally dominated mangrove creeks," *Marine and Freshwater Research*, vol. 50, no. 5, pp. 451–457, 1999.

[15] L. A. Trott and D. M. Alongi, "The impact of shrimp pond effluent on water quality and phytoplankton biomass in a tropical mangrove estuary," *Marine Pollution Bulletin*, vol. 40, no. 11, pp. 947–951, 2000.

[16] F. PÁez-Osuna, "The environmental impact of shrimp aquaculture: causes, effects, and mitigating alternatives," *Environmental Management*, vol. 28, no. 1, pp. 131–140, 2001.

[17] GESAMP, "The ecological effects of coastal aquaculture wastes," Report Studies GESAMP 57, IMO/FAO/UNESCO-IOC/WMO/WHO/IAEA/UN/-UNEP Joint Group of Experts on the Scientific Aspects of Marine Environmental Protection, Rome, Italy, 1996.

[18] J. Fuchs, J. L. M. Martin, and N. T. An, "Impact of tropical shrimp aquaculture on the environment in Asia and the Pacific," *European Commission Fisheries Cooperative Bulletin*, vol. 12, no. 4, pp. 9–13, 1999.

[19] N. C. Ferreira, C. Bonetti, and W. Q. Seiffert, "Hydrological and Water Quality Indices as management tools in marine shrimp culture," *Aquaculture*, vol. 318, no. 3-4, pp. 425–433, 2011.

[20] F. Páez-Osuna, A. Gracia, F. Flores-Verdugo et al., "Shrimp aquaculture development and the environment in the Gulf of California ecoregion," *Marine Pollution Bulletin*, vol. 46, no. 7, pp. 806–815, 2003.

[21] CONAPESCA, *Estadístico de Pesca y Acuacultura*, Comisión Nacional de Acuacultura y Pesca, Mazatlán, Sinaloa, México, 2011.

[22] J. M. Grijalva-Chon and R. H. Barraza-Guardado, "Distribution and abundance of postlarvae and juveniles of shrimps of the genus Penaeus in Kino Bay and La Cruz Lagoon, Sonora, Mexico," *Ciencias Marinas*, vol. 18, no. 3, pp. 153–169, 1992.

[23] E. García, *Modificaciones al Sistema de Clasificación Climática de Koppen (Para Adaptarlo a las Condiciones Climáticas de México*, Instituto de Geografía, UNAM, México, 5th edition, 2004.

[24] J. D. H. Strickland and T. R. Parsons, *Practical Handbook of Seawater Analysis*, Fisheries Research Board of Canada Bulletin, 2nd edition, 1972.

[25] T. R. Parsons, Y. Maita, and C. M. Lalli, *Manual Chemical and Biological Methods For Seawater Analysis*, Pergamon Press, New York, NY, USA, 1984.

[26] N. Neufeld, "Procedures for the bacteriological examination of seawater and shellfish," in *Laboratory Procedures For the Examination of Seawater and Shellfish*, A. E. Greenberg and D. A. Hunt, Eds., pp. 37–63, American Public Health Association (A.P.H.A.), Washington, DC, USA, 1985.

[27] L. S. Clesceri, A. E. Greenberg, and A. D. Eaton, *Methods For the Examination of Water and Wastewater*, American Public Health Association (A.P.H.A.), American Water Works Association (A.W.W.A.), and Water Environment Federation (W.E.F.), Washington, DC, USA, 1998.

[28] W. J. Conover, *Nonparametric Statistics*, John Wiley & Sons, New York, NY, USA, 2nd edition, 1980.

[29] J. H. Zar, *Biostatistical Analysis*, Prentice Hall, Englewood, NJ, USA, 2nd edition, 1984.

[30] NCSS, *Cruncher Statistical System User Guide*, Statistical & Power Analysis Software, North Caroline, NC, USA, 2007.

[31] A. Miranda-Baeza, D. Voltolina, M. A. Brambila-Gámez, M. G. Frías-Espiricueta, and J. Simental, "Effluent characteristics and nutrient loading of a semi-intensive shrimp farm in NW México," *Life and Environment*, vol. 57, no. 1-2, pp. 21–27, 2007.

[32] R. Casillas-Hernández, H. Nolasco-Soria, T. García-Galano, O. Carrillo-Farnes, and F. Páez-Osuna, "Water quality, chemical fluxes and production in semi-intensive Pacific white shrimp (Litopenaeus vannamei) culture ponds utilizing two different feeding strategies," *Aquacultural Engineering*, vol. 36, no. 2, pp. 105–114, 2007.

[33] C. E. Boyd and D. Gautier, "Effluent composition and water quality standards," *Global Aquaculture*, vol. 3, no. 5, pp. 61–66, 2000.

[34] H. A. González-Ocampo, L. F. Beltrán-Morales, C. Cáceres-Martínez et al., "Shrimp aquaculture environmental diagnosis in the semiarid coastal zone in Mexico," *Fresenius Environmental Bulletin*, vol. 15, no. 7, pp. 1–11, 2006.

[35] A. C. Ruiz-Fernández and F. Páez-Osuna, "Comparative survey of the influent and effluent water quality of shrimp ponds on Mexican farms," *Water Environment Research*, vol. 76, no. 1, pp. 5–14, 2004.

[36] T. E. Ford, "Response of marine microbial communities to anthropogenic stress," *Journal of Aquatic Ecosystem Stress and Recovery*, vol. 7, no. 1, pp. 75–89, 2000.

[37] H. W. Paerl, J. Dyble, L. Twomey, J. L. Pinckney, J. Nelson, and L. Kerkhof, "Characterizing man-made and natural modifications of microbial diversity and activity in coastal ecosystems," *International Journal of General and Molecular Microbiology*, vol. 81, no. 1–4, pp. 487–507, 2002.

[38] C. Kwei-Lin, "Prawn cultured in Taiwan. What went wrong?" *World Aquaculture*, vol. 20, pp. 19–20, 1989.

[39] N. Bhaskar and T. M. R. Setty, "Incidence of vibrios of public health significance in the farming phase of tiger shrimp (Penaeus monodon)," *Journal of the Science of Food and Agriculture*, vol. 66, no. 2, pp. 225–231, 1994.

[40] K.-K. Lee, S.-R. Yu, F.-R. Chen, T.-I. Yang, and P.-C. Liu, "Virulence of *Vibrio alginolyticus* isolated from diseased tiger prawn, *Penaeus monodon*," *Current Microbiology*, vol. 32, no. 4, pp. 229–231, 1996.

[41] A. Parés-Sierra, A. Mascarenhas, S. G. Marinone, and R. Castro, "Temporal and spatial variation of the surface winds in the Gulf of California," *Geophysical Research Letters*, vol. 30, no. 6, pp. 1–4, 2003.

[42] M. F. Lavin and S. G. Marinone, "An overview of the physical oceanography of the Gulf of California," in *Processes in Geophysical Fluid Dynamics*, O. U. Velasco-Fuentes, J. Sheinbaum, and J. Ochoa, Eds., pp. 173–204, Kluwer Academic Publishers, Dodrecht, The Netherlands, 2003.

[43] M. Botello-Ruvalcaba, E. Valdez-Holguín, Estero La Cruz, and Sonora, "Comparison of Carbon, Nitrogen and Phosphorus fluxes in Mexican coastal lagoons," in *LOICZ Reports & Studies No. 10*, S. V. Smith, S. Ibarra-Obando, P. R. Boudreau, and V. F. Camacho-Ibar, Eds., pp. 21–24, LOICZ Core Project Office, Netherlands Institute for Sea Research (NIOZ), Texe, The Netherlands, 1997.

[44] R. Casillas-Hernández, F. Magallón-Barajas, G. Portillo-Clarck, and F. Páez-Osuna, "Nutrient mass balances in semi-intensive shrimp ponds from Sonora, Mexico using two feeding strategies: trays and mechanical dispersal," *Aquaculture*, vol. 258, no. 1–4, pp. 289–298, 2006.

[45] R. Alonso-Rodríguez and F. Páez-Osuna, "Nutrients, phytoplankton and harmful algal blooms in shrimp ponds: a review with special reference to the situation in the Gulf of California," *Aquaculture*, vol. 219, no. 1–4, pp. 317–336, 2003.

[46] J. García-Hernández, L. García-Rico, M. E. Jara-Marini, R. Barraza-Guardado, and A. H. Weaver, "Concentrations of heavy metals in sediment and organisms during a harmful algal bloom (HAB) at Kun Kaak Bay, Sonora, Mexico," *Marine Pollution Bulletin*, vol. 50, no. 7, pp. 733–739, 2005.

[47] P. Menasveta, "Improved shrimp growout systems for disease prevention and environmental sustainability in Asia," *Reviews in Fisheries Science*, vol. 10, no. 3-4, pp. 391–402, 2002.

[48] G. Aguirre-Guzmán, Y. Labreuche, D. Ansquer et al., "Proteinaceous exotoxins of shrimp-pathogenic isolates of Vibrio penaeicida and Vibrio nigripulchritudo," *Ciencias Marinas*, vol. 29, no. 1, pp. 77–88, 2003.

[49] R. H. Barraza-Guardado, J. Chávez-Villalba, H. Atilano-Silva, and F. Hoyos-Chairez, "Seasonal variation in the condition index of Pacific oyster postlarvae (*Crassostrea gigas*) in a land-based nursery in Sonora, Mexico," *Aquaculture Research*, vol. 40, no. 1, pp. 118–128, 2008.

[50] J. Chávez-Villalba, A. Arreola-Lizárraga, S. Burrola-Sánchez, and F. Hoyos-Chairez, "Growth, condition, and survival of the Pacific oyster *Crassostrea gigas* cultivated within and outside a subtropical lagoon," *Aquaculture*, vol. 300, no. 1–4, pp. 128–136, 2010.

13

Evaluation of the Impacts of Land Use on Water Quality: A Case Study in The Chaohu Lake Basin

Juan Huang,[1] Jinyan Zhan,[1] Haiming Yan,[1] Feng Wu,[1] and Xiangzheng Deng[2,3]

[1] State Key Laboratory of Water Environment Simulation, School of Environment, Beijing Normal University, Beijing 100875, China
[2] Institute of Geographic Science and Natural Resource Research, CAS, Beijing 100101, China
[3] Center for Chinese Agricultural Policy, CAS, Beijing 100101, China

Correspondence should be addressed to Jinyan Zhan; zhanjy@bnu.edu.cn

Academic Editors: J. Bai and B. Cui

It has been widely accepted that there is a close relationship between the land use type and water quality. There have been some researches on this relationship from the perspective of the spatial configuration of land use in recent years. This study aims to analyze the influence of various land use types on the water quality within the Chaohu Lake Basin based on the water quality monitoring data and RS data from 2000 to 2008, with the small watershed as the basic unit of analysis. The results indicated that there was significant negative correlation between forest land and grassland and the water pollution, and the built-up area had negative impacts on the water quality, while the influence of the cultivated land on the water quality was very complex. Besides, the impacts of the landscape diversity on the indicators of water quality within the watershed were also analyzed, the result of which indicated there was a significant negative relationship between them. The results can provide important scientific reference for the local land use optimization and water pollution control and guidance for the formulation of policies to coordinate the exploitation and protection of the water resource.

1. Introduction

The land use within the watershed has great impacts on the water quality of rivers. The water quality of rivers may degrade due to the changes in the land cover patterns within the watershed as human activities increase [1, 2]. Changes in the land cover and land management practices have been regarded as the key influencing factors behind the alteration of the hydrological system, which lead to the change in runoff as well as the water quality [3, 4].

There have been three waves of the research that tried to reveal the effects of the land use and land cover change on the quality of surface water [5, 6]. The researchers have started to study the linkage between land cover and the river water quality in order to investigate the effects of morphological features of watersheds on the turbidity, dissolved oxygen and temperature of the river water since the early 1960s [7]. The second wave of researches on this topic emerged in the 1970s, focusing on the analysis at the watershed scale [8]. The third wave of these studies have started to take advantage of the remote sensing, GIS, and multivariate analysis to explore the influence of the land cover on the suspended sediment, nutrients and ecological integrity of the stream [9–14].

Related research in China started from the 1980s and mainly focused on the role of the macroscopic characteristics of nonpoint source pollution and urban runoff pollution and the quantitative calculation of pollution loading [15–17]. For example, the research carried out by Guo indicated that it is necessary to take into account both the land use and the land cover pattern simultaneously in the study of the impacts of land use on the water quality [18]. Besides, the export coefficient model and RS and GIS techniques have been applied in the study of the effects of the land use change on the nonpoint source pollution load in the upper reach of Yangtze River [19], and the result indicated that the grassland played a dominant role in influencing the TN and TP in Jinsha River, while cultivated land played a key role in other parts of the study area. There was also research about the relationship between land use types and water quality in Xin'anjiang River based on ArcGIS [20]. The results showed that cultivated

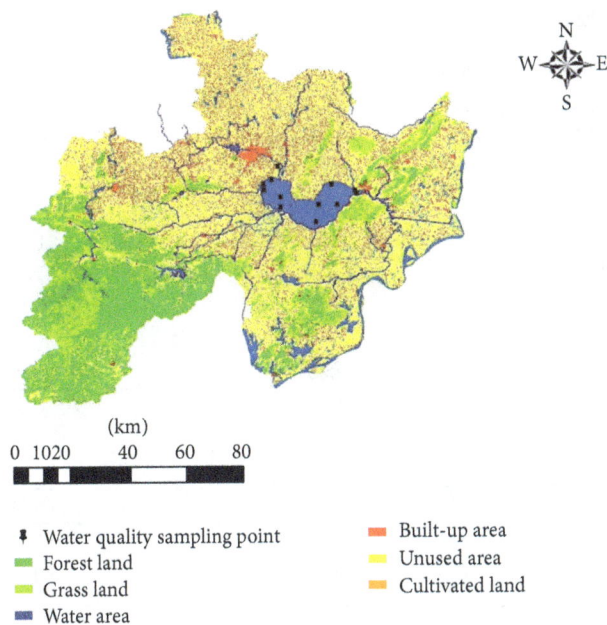

FIGURE 1: The Chaohu Lake Basin. Water quality points are shown in the figure. Upstream catchment of each water quality sampling point and land use types were delineated.

land, grassland, and forest land had the most significantly important impacts on TN, TP, and fecal coliform bacteria. On the whole, the previous studies in China have focused on only several lakes such as Tai Lake [21] and Dianchi Lake [22]. Besides, these researches have only taken into account the impacts of the composition of land use types within these basins on the water quality. Only few studies have considered the effects of spatial patterns of land use on the water quality, which could provide evidence on landscape planning and land use management. Therefore, The primary objectives of this paper were (1) to describe the water quality change in the Chaohu Lake Basin from 2000 to 2008; (2) to investigate the land use change in the study area; and (3) to identify the relationship between land use change and water quality.

2. Materials and Methods

2.1. *Study Area.* The Chaohu Lake Basin (Figure 1) is located in the central part of Anhui Province and between $117°16'46'16''$–$117°51'54'51''$E, $30°43'28'43''$–$31°25'28'25''$N. The Chaohu Lake belongs to the drainage system in the lower reaches of the Yangtze River, and it is the fifth largest freshwater lake in China, with a total watershed area of $13350\,km^2$. The total annual inflow from 33 rivers is $4.8 \times 10^9\,m^3\,year^{-1}$ and the total outflow is $3.4 \times 10^9\,m^3\,year^{-1}$. A large portion of the inflow is from Nanfei River, Hangbu River, and Yuxi River. The average annual temperature in the Chaohu Lake Basin ranges between 15°C and 16°C, with a mean annual rainfall of 1100 mm. The Chaohu Lake Basin is one of the most densely populated regions in Anhui Province [23], with a population of more than 9.65 million. The Chaohu Lake Basin also plays an important role in the

local economic development, accounting for 24.65% of GDP in Anhui Province.

2.2. *Water Quality Situation and Data Resources.* The Chaohu Lake Basin was known as the land of fish and rice in the early stage when the local ecological environment was very good and the water quality of Chaohu Lake used to be very fine. However, the hydrological conditions and downstream ecosystems of this lake have been altered since the establishment of Chaohu Dam on Yuxi River [24, 25], with the amount of annual water exchange volume decreasing from $13.6 \times 10^8\,m^3$ to $1.6 \times 10^8\,m^3$ [26] and the average annual water level falling from 4.3 m to 2.9 m. Besides, with the rapid development of local economy and social activity, the wetlands in the Chaohu Lake Basin were reclaimed or occupied [27]. And water quality of Chaohu Lake has continuously deteriorated due to the large amount of pollution discharge from the local industry, agriculture, and daily life since the late 1970s [28]. Along with that is the outbreak of water bloom, which has aroused great concern of the government. The Chaohu Lake has reached the eutrophic state, with the high concentration of nutrient salts and rapid growth of algae.

The water quality of Chaohu Lake has consciously improved to some degree. In 1997, the Ministry of Environmental Protection of China promulgated the water quality of the five biggest fresh water lakes, among which the situation of Chaohu Lake was the worst. During 1996–1999, the data from water quality monitoring points around Chaohu Lake showed that the percentage of water exceeding the level V had decreased from 80% to 60%. Since 2000, water quality of the main tributaries of Chaohu Lake, including Nanfei River, Shiwuli River, Pai River, and Shuangqiao River, was always exceeding the level V, with ammonia nitrogen as the key pollutant. But the water quality of other tributaries was better, generally fluctuating between level III and level IV (Table 1).

In 2003, Chaohu Lake reached the meso-eutrophication the whole, with the total phosphorus (TP) and total nitrogen (TN) as the main pollutants. The water quality of Chaohu Lake improved slightly in 2005, reaching the meso-eutrophic state on the whole. The mean value of CODmn and TN succeeded in achieving the goal of the tenth Five-year Plan in Chaohu Lake, but the mean value of TP failed. For the moment, Chaohu Lake is known as one of the three most polluted freshwater lakes in China [29]. Since Chaohu Lake serves as the primary drinking water source of Hefei City, it has been ranked as the key lake to be managed.

In this study, the Chaohu Lake Basin was divided into nine watersheds according to the local river systems and the monitoring points were set up over there (Figure 1). There are many variables of water quality available from these monitoring points, including pellucidity, TP, TN, DO, BOD5, and CODcr. But we only selected TP, TN, DO, NH3-N, and CODmn measured in every month from 2000 to 2008. Other water quality variables were also important; we chose these five not because they were more important than others in Chaohu Lake, but because many researches had chosen them, and these water quality variables had the complete data. The average annual values of these variables were also used in view

TABLE 1: Water quality of Chaohu Lake between 2000 and 2007.

River	2000	2001	2002	2003	2004	2005	2006	2007
Nanfei River	Bad V	Bad V	Bad V	Bad V	Bad V	V	V	Bad V
Shiwuli River	Bad V	Bad V	Bad V	Bad V	Bad V	Bad V	Bad V	Bad V
Pai River	Bad V	Bad V	Bad V	Bad V	Bad V	Bad V	Bad V	Bad V
Hangbu River	IV	II	III	III	IV	IV	III	IV
Baishitian River	IV	III	IV	III	IV	III	IV	IV
Zhao River	III	III	III	III	IV	IV	IV	IV
Tuogao River	III	III	IV	III	III	IV	III	III
Yuxi River	IV	III	III	IV	II	IV	III	IV
Shuangqiao River	Bad V	Bad V	Bad V	Bad V	Bad V	Bad V	Bad V	Bad V

TABLE 2: Descriptive statistics for the study area, including water quality and land use characteristics.

Categories	Variable	Obs	Mean	Min	Max	S.D.
	CODmn (mg/L)	81	5.18	2.7	7.8	1.09
	NH3–N (mg/L)	81	0.54	0.00	1.61	0.39
Water quality	TP (mg/L)	81	0.19	0.07	0.57	0.19
	TN (mg/L)	81	2.23	1.04	6.48	1.11
	DO (mg/L)	81	8.22	6.69	9.65	0.58
	Cultivated land (%)	81	0.47	0.32	0.80	0.12
	Forest land (%)	81	0.02	0.00	0.06	0.02
Land use	Grassland (%)	81	0.02	0.00	0.67	0.02
	Water area (%)	81	0.36	0.00	0.58	0.16
	Built-up area (%)	81	0.11	0.03	0.33	0.08
Landscape metrics	Shannon	81	2.83	1.76	3.85	0.50

S.D.: standard deviation.

of the seasonal variations of algal species and water quality in Chaohu Lake [30].

2.3. Land Use Data.
The land use data, which was extracted from the Lantsat TM images (from 2000 to 2008), was provided by the Data Center of the Chinese Academy of Sciences. There are six kinds of land use types, that is, the cultivated land, forest land, grassland, water area, built-up area and unused land. The Landsat ETM images in 2000 and 2005 were interpreted at a scale of 1 : 100,000 and the overall interpretation accuracy of the land use categories reached 92.7% according to the field survey and random sampling check conducted by Data Center of the Chinese Acadamy of Sciences (CAS) [31, 32]. The watershed boundaries were delimitated based on the DEM data with the "automatic delineation utility" in BASINS. The Chaohu Lake Basin was divided into nine small watersheds; then we used GIS tools to calculate the area of each land use type within each subwatershed. Based on that we got the proportion of each land use area within each sub-watershed.

2.4. Spatial Patterns of Land Use.
Landscape pattern change is mainly caused by the change in land cover and land use change [33]. The landscape ecologists and other researchers have developed numerous metrics to investigate the effects of the landscape pattern on the ecological processes [34]. In view of the multicollinearity among metrics and the erratic behaviors of some metrics across scales, we selected Shannon's diversity index (SHDI) as the indicator of landscape metric use in this study. SHDI indicates the patch diversity in a landscape based on the information theory, and it is calculated with the following form:

$$\text{SHDI} = -\sum_{i=1}^{m} \left(\text{pi} \ln \text{pi} \right), \tag{1}$$

where pi is the proportion of the landscape occupied by land use type i and m is the number of land use type present in the landscape.

The SHDI is a sensitive indicator to analyze the diversity and heterogeneity of the same landscape in different times. The big value of SHDI means that the land use pattern is various and the degree of fragmentation is high. We calculated the percentages of the five land use types and the SHDI in these nine sub-watersheds and then analyzed the relationship between the SHDI and the indicators of water quality.

3. Results and Discussions

3.1. Descriptive Statistics of Measures.
As showed in Table 2, the average CODmn concentration in 2000 and 2008 was 5.18 mg/L, with the concentration of NH3-N, TP, TN, and

TABLE 3: LUCC of the Chaohu Lake Basin between 1995 and 2005.

Year	Statistics variable	Cultivated land	Forest land	Grassland	Water area	Built-up land	Nonuse land
1995	Area (km^2)	17153.11	5781.29	1817.64	2009.12	2165.91	1.32
	Proportion (%)	59.30%	19.98%	6.28%	6.95%	7.49%	0.00%
2000	Area (km^2)	16960.22	5770.15	1817.34	2015.09	2364.22	1.32
	Proportion (%)	58.63%	19.95%	6.28%	6.97%	8.17%	0.00%
2005	Area (km^2)	16850.80	5764.96	1816.65	2019.72	2474.93	1.32
	Proportion (%)	58.25%	19.93%	6.28%	6.98%	8.56%	0.00%

DO being 0.54 mg/L, 0.19 mg/L, 2.23 mg/L, and 8.22 mg/L, respectively. The standard deviation of indicators of water quality was generally very small.

The most important indicators of the water quality are the CODmn and TN, the average concentration of which reached 5.18 mg/L and 2.23 mg/L, respectively. The two indicators also vary greatly among the sub-watersheds, with their standard deviations being 1.09 and 1.11, respectively. There are main cultivated land and water areas in the Chaohu Lake Basin, accounting for 47% and 36% of the total area, respectively. Besides, the cultivated land and water area also vary most greatly among small watersheds, with their standard deviations reaching 0.12 and 0.16, respectively.

3.2. Water Quality Change in the Study Area. According to the data from water quality monitoring points, water quality change in the study area was shown in Figure 2. In general, the water quality has improved from 2000 to 2008. Change trend of NH3-N and TP was small. And the change trend of DO was upward. The rest has changed a lot among this period; however, the beginning value was close to the finishing value.

3.3. Land Use Change. The overall state of land use in Chaohu Lake Basin was extracted based on the land use images (Table 3). The Chaohu Lake Basin is dominated by agriculture, and the cultivated land accounted for almost 60 percentage of the total area from 1995 to 2005. The total area of cultivated area has changed from 17153 km^2 in 1995 to 16850 km^2 in 2005, decreasing by about 1.7 percentage, while the area of built-up land increased by about 14 percentage. The urban expansion is the main driving factor of the decrease of cultivated land, and this kind of conversion would change the wetland soil which served as natural sinks and filtration system [35, 36]. The forest land accounted for approximately 20% of the total land area in the Chaohu Lake Basin, which was very stable in these years. The proportion of the grasslandand and water area in the Chaohu lake Basin was about 6.28% and less than 7%, respectively, both of which were very stable during 1995 and 2005. In this study, The unused land was not included in this study since it accounted for almost 0% of the total area.

3.4. Relationship between Land Use and Water Quality. The land use data and water quality monitoring data were analyzed by Stata. The model we used is an econometric model

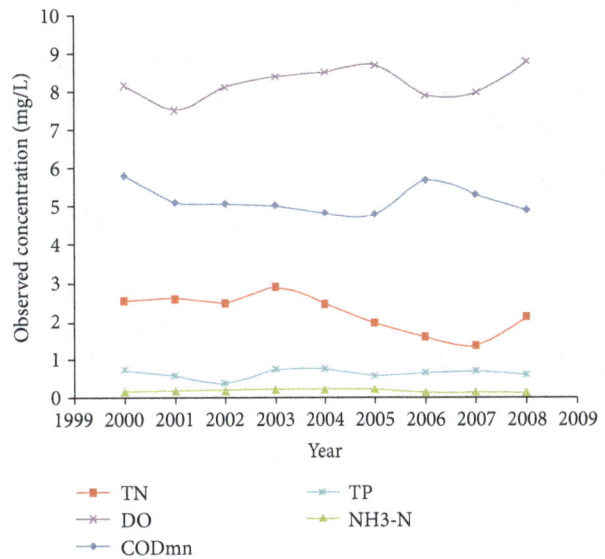

FIGURE 2: Water quality change from 2000 to 2008 in the study area.

based on the research of four lakes watersheds in Hanyang district [18]:

$$\text{NPS} = \exp\left(\beta 1 \times \text{land1} + \beta 2 \times \text{land2} + \cdots \beta i \times \text{land}i\right), \quad (2)$$

where NPS means the water quality variables in the study area, α is a constant, and β means the correlation between land use area (%) and water quality variables. When $\beta i > 0$, it means that land use type i has a positive effect on the indicators of water quality. If $\beta i < 0$, it means that land use type i has a negative effect on the indicators of water quality.

Since we got the data, the panel data has been used so as to comprehensively and completely reflect the relationship between the water quality and land use types in the Chaohu Lake Basin. Our panel data analysis was about 9 small watersheds in 2000–2008. Given the robustness and accuracy of the econometric analysis, we used both the fixed effect model and the random effect model so as to make a comparison, and finally the fixed effect model was selected according to the result of the Hausman test. Then we got the forms to

TABLE 4: The correlation coefficients between different land uses and water quality variables at the scale of the whole watershed.

Variables (%)	ln (TN) coef.	ln (TP) coef.	ln (CODmn) coef.	ln (NH3–N) coef.	ln (DO) coef.
Cultivated land	−0.46	−0.42	−0.08	0.31	0.06
Forest land	−4.11**	−9.59***	−5.47***	−9.26***	0.31
Grassland	−6.52***	−7.87***	−2.93***	−8.83***	0.54
Water area	−0.30	0.26	0.29*	1.22**	0.0029
Built-up area	0.86	1.54**	0.70**	2.60***	−0.27*
Constant	1.20***	−1.44***	1.69***	−1.04***	2.08***
R-squared	0.49	0.75	0.80	0.69	0.36

***$P < 0.01$, **$P < 0.05$, *$P < 0.1$.

TABLE 5: The correlation coefficients between landscape matrix and water quality variables at the scale of the whole watershed.

Variables (%)	ln (TN) coef.	ln (TP) coef.	ln (CODmn) coef.	ln (NH3–N) coef.	ln (DO) coef.
Shannon	−0.410***	−0.633***	−0.296***	−0.650***	0.0435***
Constant	1.866***	−0.00560	2.463***	1.232***	1.981***

***$P < 0.01$, **$P < 0.05$, *$P < 0.1$.

describe the relationship between land use types and water quality according to the analysis results of Stata (Table 4):

$$\ln (TN) = -0.47C - 4.11F - 6.52G - 0.3W + 0.86B + 1.2,$$

$$\ln (TP) = -0.42C - 9.59F - 7.87G - 0.26W + 1.54B - 1.44,$$

$$\ln (CODmn) = -0.08C - 5.47F - 2.93G + 0.29W + 0.7B + 1.69,$$

$$\ln (NH3 - N) = 0.31C - 9.26F - 8.83G + 1.22W + 2.6B - 1.04,$$

$$\ln (DO) = 0.06C + 0.31F + 0.54G + 0.0029W - 0.27B + 2.08,$$

(3)

where C is cultivated land area (%), F is forest area (%), G is grassland area (%), W is water area (%), B is built-up area (%).

There was a positive relationship between the cultivated land area (%) and the concentration of NH3-N and DO. This is mainly due to the developed agriculture in the Chaohu Lake Basin and the emission of NH3-N from the exposure of soil surface resulting from the agricultural practices and the application of chemical fertilizers [37]. Besides, the concentrations of TP and TN are negatively related with the cultivated land area (%). On the one hand, the fertilizers used in the cultivated land will get into the runoff and flow into the river and ultimately pollute the river water. On the other hand, the vegetation in the surface soil of the cultivated land can absorb, retain the pollutants. As a result, the cultivated land plays a complicated role in influencing the water quality in the Chaohu Lake Basin.

The forest land and grassland both have significant positive influence on the water quality. The areas of the forest land and grassland were negatively related to TP, TN, NH3-N, and the CODmn and was positively related to DO. Many researches had shown similar results [15, 37].The significant negative relationship between the forest land and grassland area and TP, TN, CODmn, and NH3-N indicates that the the forest land and grassland played a key role in reducing the nitrogen pollutants and phosphorus pollutants and played a controlling role in regrating the water quality. The vegetation and soil in the forest land and grassland can effectively reduce the nutrient salts brought into the river by the surface runoff since they play an important role in reducing the surface runoff, conserving the water and soil, and absorbing the pollutants. Therefore, the increase of the forest land and grassland area will reduce the concentration of TP, TN, and oxygen-consuming substances, increase the concentration of dissolved oxygen, and consequently improve the water quality.

The built-up area played a negative role in influencing the water quality on the whole. The built-up area was positively related to TP, TN, NH3-N, and CODmn and was negatively related to DO, indicating that the increase of the built-up area tends to degrade the water quality. Mouri et al. found similar relationship between concentration of TN and the area of built-up area [38], which is in agreement with that of Amiri and Nakane, who analyzed the relationship between TN, DO, NH3-N, and built-up area [37]. This could be the result of the increase of the nutrient concentration. The dense population density and economic activities both concentrated in the built-up area, which leads to very serious pollution. Besides, there is a lot of impermeable surface in the built-up area, which will contribute to the increase of surface runoff and may increase the concentration of nutrient salts in the river and consequently degrade the water quality within the watershed. There was a positive relationship between the built-up area (%) and TP and CODmn. The increase of built-up area was the result of transformation from the land with natural vegetation that could prevent the soil erosion. Since the vegetation can protect the soil from raindrops and tends to slow down the movement of runoff and allows the excessive surface water to infiltrate into soil, the conversion of the land with vegetation into built-up area will aggravate the soil erosion and consequently increase the amount of TP into the runoff.

3.5. Relationship between Spatial Patterns of Land Use and Water Quality. The relationship between SHDI and water quality was revealed in Table 5. SHDI had a negative relationship with NH3-N, TP, CODmn, and TN. According to the results of Stata analysis, we can know that the relationship between SHDI to CODmn, TP, TN, and NH3-N was very significant. The significant negative relationship mean that the landscape diversity in the Chaohu Lake Basin is closely related to the water quality. The higher the SHDI is, the greater the diversity landscape is and the slighter the deterioration of water quality is. As the landscape diversity increases, the landscape heterogeneity increases and consequently makes the patches of each landscape type more evenly distributed.

4. Conclusion

Studying the relationship between the proportion of land use types and water quality in the Chaohu Lake Basin in this study indicated that built-up land was generally positively related to the indicators of water quality, and the forest land and grass land and water area were negatively related with the water quality variables, while the influence of the cultivated land on the water quality was very complex. Additionally, the built-up land, grassland, and forest land had significant influence on some indicators of water quality. The regression result of the landscape indicators and the indicators of water quality suggested that SHDI was negatively related to most of the water quality variables, indicating that the increase of landscape diversity can contribute to the improvement of water quality.

According the result mentioned previously and the current conditions of the local water quality in the Chaohu Lake Basin, it is necessary to increase the area of forest land, grassland, and water area in the local land use planning. Since the forest land is more closely related to the local water quality, it is specially important to increase the area of forest land. Besides, the growth rate of the urban land should be slowed down under the condition of guaranteeing the minimum land area needed by the city development within the watershed. In addition, it is necessary to increase the landscape diversity because the greater the landscape diversity is, the more evenly patches of each kind are distributed and the more the water pollution will be alleviated.

The results of these studies can provide scientific reference for the local land use optimization and water pollution control and assist the formulation of policies for coordinating the water resource exploitation and protection. In addition,

the previous researches have indicated that their landscape diversity has impacts on the water quality within the watershed, but it is still necessary to add some other ecological indicators and analyze their influence on the water quality. In particular, we should expand the method of analysing the relationship between land use and water quality but not the simple regression. Our study focuses on the effect of land use types and landscape patterns on water quality in the study area. However, there are many factors related to water quality, such as the climate, precipitation, and density of population. In the future work, we will refine the method and indicators to deeply reveal the reasons causing water quality change within a watershed.

Acknowledgments

This research was supported by the Natural Science Foundation of China (no. 71225005) and the State Major Project for Water Pollution Control and Management of China (2009ZX07106-001). Data support from the projects funded by the Chinese Academy of Sciences (KZZD-EW-08; GJHZ1312) and Exploratory Forefront Project for the Strategic Science Plan in IGSNRR, CAS is also greatly appreciated.

References

[1] E. Ngoye and J. F. Machiwa, "The influence of land-use patterns in the Ruvu river watershed on water quality in the river system," *Physics and Chemistry of the Earth A, B, C*, vol. 29, no. 15–18, pp. 1161–1166, 2004.

[2] L. Sliva and D. D. Williams, "Buffer zone versus whole catchment approaches to studying land use impact on river water quality," *Water Research*, vol. 35, no. 14, pp. 3462–3472, 2001.

[3] S. T. Y. Yong and W. Chen, "Modeling the relationship between land use and surface water quality," *Journal of Environmental Management*, vol. 66, no. 4, pp. 377–393, 2002.

[4] J. Bai, H. Ouyang, R. Xiao et al., "Spatial variability of soil carbon, nitrogen, and phosphorus content and storage in an alpine wetland in the Qinghai-Tibet Plateau, China," *Australian Journal of Soil Research*, vol. 48, no. 8, pp. 730–736, 2010.

[5] G. Yang and J. Wang, *Economic Development of Taihu Watershed*, Science Press, Beijing, China, 2003.

[6] U. S. Tim and R. Jolly, "Evaluating agricultural nonpoint-source pollution using integrated geographic information systems and hydrologic/water quality model," *Journal of Environmental Quality*, vol. 23, no. 1, pp. 25–35, 1994.

[7] R. C. Harrel and T. C. Dorris, "Stream order, morphometry, physico-chemical conditions, and community structure of benthic macroinvertebrates in an intermittent stream system," *American Midland Naturalist*, vol. 80, no. 1, pp. 220–251, 1968.

[8] F. H. Bormann, G. E. Likens, and J. S. Eaton, "Biotic regulation of particulate and solution losses from a forest ecosystem," *Bioscience*, vol. 19, no. 7, pp. 600–610, 1969.

[9] N. E. Roth, J. D. Allan, and D. L. Erickson, "Landscape influences on stream biotic integrity assessed at multiple spatial scales," *Landscape Ecology*, vol. 11, no. 3, pp. 141–156, 1996.

[10] K. B. Jones, A. C. Neale, M. S. Nash et al., "Predicting nutrient and sediment loadings to streams from landscape metrics: a multiple watershed study from the United States Mid-Atlantic Region," *Landscape Ecology*, vol. 16, no. 4, pp. 301–312, 2001.

[11] D. S. Ahearn, R. W. Sheibley, R. A. Dahlgren, M. Anderson, J. Johnson, and K. W. Tate, "Land use and land cover influence on water quality in the last free-flowing river draining the western Sierra Nevada, California," *Journal of Hydrology*, vol. 313, no. 3-4, pp. 234–247, 2005.

[12] G. Sylaios, N. Stamatis, A. Kallianiotis, and P. Vidoris, "Monitoring water quality and assessment of land-based nutrient loadings and cycling in Kavala Gulf," *Water Resources Management*, vol. 19, no. 6, pp. 713–735, 2005.

[13] C. Tafangenyasha and L. T. Dube, "An investigation of the impacts of agricultural runoff on the water quality and aquatic organisms in a lowveld sand river system in Southeast Zimbabwe," *Water Resources Management*, vol. 22, no. 1, pp. 119–130, 2008.

[14] A. Haidaryy, B. J. Amiri, J. Adamowski, N. Fohrer, and K. Nakane, "Assessing the impacts of four land use types on water quality of Wetlands in Japan," *Water Resources Management*, vol. 27, no. 7, pp. 2217–2229, 2013.

[15] B. Fu, K. Ma, H. Zhou, and L. Chen, "The impact of land use change on soil nutrition distribution of Losses Pleatau," *Science Bulletin*, vol. 43, no. 22, pp. 2444–2448, 1998.

[16] B. Fu, L. Chen, and K. Ma, "The effect of land use chang on the regional environment in the Yangjuangou catchment in the loess plateau of China," *Acta Geographic Sinica*, vol. 54, no. 3, pp. 241–246, 1999.

[17] H. Li and J. Shen, "The establishment and case study of the model for nonpoint source pollution for watershed," *Acta Scientiae Circumstantiae*, vol. 17, no. 2, pp. 140–147, 2009.

[18] Q. H. Guo, K. M. Ma, and Y. Zhang, "Impact of land use pattern on lake water quality in urban region," *Acta Eclologica Sinica*, vol. 29, no. 2, pp. 776–787, 2009 (Chinese).

[19] R. Liu, Z. Yang, Z. Shen, and X. Wu, "Relationship and sumulation information system of land use and cover change and nonpoint source pollution in Yangtze River Basin," *Resources and Environment in the Yantze Basin*, vol. 15, no. 3, pp. 372–377, 2006 (Chinese).

[20] F. Cao, X. Li, D. Wang, Y. Zhao, and Y. Wang, "Effects of land use structure on water quality in Xin' anjiang River," *Environmental Science*, vol. 34, no. 7, pp. 2582–2587, 2012 (Chinese).

[21] X. Yu and G. Yang, "Land use/cover change of catchment and its water quality effects—a case of Xitiaoxi catchment in Zhejiang province," *Resources and Environment in the Yangtze Basin*, vol. 12, no. 3, pp. 211–217, 2003 (Chinese).

[22] J. Sun, X. Cao, and Y. Huang, "Effect of land use on inflow rivers water quality in lake Dianchi watershed," *China Environmental Science*, vol. 31, no. 12, pp. 2052–2057, 2011 (Chinese).

[23] S. Tsujimura, H. Tsukada, H. Nakahara, T. Nakajima, and M. Nishino, "Seasonal variations of Microcystis populations in sediments of Lake Biwa, Japan," *Hydrobiologia*, vol. 434, no. 1–3, pp. 183–192, 2000.

[24] S. A. Brandt, "Classification of geomorphological effects downstream of dams," *Catena*, vol. 40, no. 4, pp. 375–401, 2000.

[25] J. Bai, R. Xiao, K. Zhao, and H. Gao, "Arsenic and heavy metal pollution in wetland soils from tidal freshwater and salt marshes before and after the flow-sediment regulation regime in the Yellow River Delta, China," *Journal of Hydrology*, vol. 450-451, no. 11, pp. 244–253, 2012.

[26] Q. Tu, *The Researches on the Eutroph-Ication in Chaohu Lake*, University of Science and Technology of China Press, Heifei, China, 1990 (Chinese).

[27] J. Bai, Z. Yang, B. Cui, H. Gao, and Q. Ding, "Some heavy metals distribution in wetland soils under different land use types along a typical plateau lake, China," *Soil and Tillage Research*, vol. 106, no. 2, pp. 344–348, 2010.

[28] G. Shang and J. Shang, "Causes and control countermeasures of eutrophication in Chaohu Lake, China," *Chinese Geographical Science*, vol. 15, no. 4, pp. 348–354, 2005 (Chinese).

[29] W. Zwisler, N. Selje, and M. Simon, "Seasonal patterns of the bacterioplankton community composition in a large mesotrophic lake," *Aquatic Microbial Ecology*, vol. 31, no. 3, pp. 211–225, 2003.

[30] L. Yang, X. Han, P. Sun, W. Yan, and Y. Li, "Canonical correspondence analysis of algae community and its environmental factors in the. Lake Chaohu, China," *Journal of Agro-Environment Science*, vol. 30, no. 5, pp. 952–958, 2011.

[31] X. Deng, *Analysis of Land Use Conversions*, China Land Press, Beijing, China, 2008 (Chinese).

[32] J. Liu, Z. Zhang, D. Zhuang et al., "A study on the spatial-temporal dynamic changes of land-useand driving forces analyses of China in the 1990s," *Geographical Research*, vol. 22, no. 2, pp. 1–12, 2003 (Chinese).

[33] J. Bai, Q. Lu, J. Wang et al., "Landscape pattern evolution processes of alpine wetlands and their driving factors in the Zoige plateau of China," *Journal of Mountain Science*, vol. 10, no. 1, pp. 54–67, 2013.

[34] P. O'Neil, "Selection on flowering time: an adaptive fitness surface for nonexistent character combinations," *Ecology*, vol. 80, no. 3, pp. 806–820, 1999.

[35] H. Gao, J. Bai, R. Xiao, P. Liu, J. Wei, and J. Wang, "Levels, sources and risk assessment of trace elements in wetland soils of a typical shallow freshwater lake, China," *Stochastic Environmental Research and Risk Assessment*, vol. 27, no. 1, pp. 275–284, 2013.

[36] J. Bai, R. Xiao, K. Zhang, H. Gao, B. Cui, and X. Liu, "Soil organic carbon affected by land use in young and old reclaimed regions of a coastal estuary wetland, China," *Soil Use and Management*, vol. 29, no. 1, pp. 57–64, 2013.

[37] B. J. Amiri and K. Nakane, "Modeling the linkage between river water quality and landscape metrics in the Chugoku district of Japan," *Water Resources Management*, vol. 23, no. 1, pp. 931–956, 2009.

[38] G. Mouri, S. Takizawa, and T. Oki, "Spatial and temporal variation in nutrient parameters in stream water in a rural-urban catchment, Shikoku, Japan: effects of land cover and human impact," *Journal of Environmental Management*, vol. 92, no. 7, pp. 1837–1848, 2011.

14

An Investigation into Occasional White Spot Syndrome Virus Outbreak in Traditional Paddy Cum Prawn Fields in India

Deborah Gnana Selvam, K. M. Mujeeb Rahiman, and A. A. Mohamed Hatha

Department of Marine Biology, Microbiology, and Biochemistry, Cochin University of Science and Technology, Lakeside Campus, Cochin 682 016, India

Correspondence should be addressed to A. A. Mohamed Hatha, mohamedhatha@gmail.com

Academic Editors: R. Bastida and U. S. Gaipl

A yearlong (September 2009–August 2010) study was undertaken to find out possible reasons for occasional occurrence of White Spot Syndrome Virus (WSSV) outbreak in the traditional prawn farms adjoining Cochin backwaters. Physicochemical and bacteriological parameters of water and sediment from feeder canal and four shrimp farms were monitored on a fortnightly basis. The physicochemical parameters showed variation during the two production cycles and between the farms studied. Dissolved oxygen (DO) content of water from feeder canal showed low oxygen levels (as low as 0.8 mg/L) throughout the study period. There was no disease outbreak in the perennial ponds. Poor water exchange coupled with nutrient loading from adjacent houses resulted in phytoplankton bloom in shallow seasonal ponds which led to hypoxic conditions in early morning and supersaturation of DO in the afternoon besides considerably high alkaline pH. Ammonia levels were found to be very high in these ponds. WSSV outbreak was encountered twice during the study leading to mass mortalities in the seasonal ponds. The hypoxia and high ammonia content in water and abrupt fluctuations in temperature, salinity and pH might lead to considerable stress in the shrimps triggering WSSV infection in these traditional ponds.

1. Introduction

Across the globe, aquaculture industry is one of the rapidly growing industries in food sector. Aquaculture production in India alone has increased by 72% (2.1 million metric ton) within a decade from 1990 to 2000, compared to the 150% increase in production worldwide. Shrimp production trend in India has shown a steady rise from 1990 to 2003 [1]. Shrimp production from coastal aquaculture during 2004 stood at approximately 120,000 tonnes. Farmed shrimp accounts for about 60% of shrimp exported from the country [2]. The penaeid shrimps have tend to dominate brackish-water culture due to high-value, short production cycles and accessible technologies. Production has increased almost exponentially since the mid-1970s and now accounts for about 58 per cent of aquaculture production from brackish water (72% by value) [3].

The sustainability and development of shrimp aquaculture are largely at stake, as significant ecological and pathological problems are increasing in vast majority of the shrimp producing countries. The production is regularly and seriously affected by problems linked to environmental degradation and to infectious and noninfectious diseases [4]. Of the infectious diseases, bacterial and viral infections, either as single or multiple pathogen conditions, caused most of the production losses. White spot syndrome virus (WSSV) first appeared as an epidemic in penaeid shrimp farms in 1993 in China and then quickly spread in Asian countries and subsequently to all over the world [5]. WSSV has resulted in high mortality in many cultured penaeid shrimp species and huge shrimp production losses. White spot syndrome virus (WSSV) has been reported to cause severe mortalities of cultured penaeid shrimp in several parts of Asia including India [6–8]. There is still no efficient measure to control the disease [9] and the loss caused by this virus has been estimated to be several million dollars in different parts of India. The most surprising feature of this virus is its wide range of potential hosts such as several species of penaeid shrimp, a wide range of other decapods including crabs and more distantly related crustaceans such as the freshwater prawn, *Macrobrachium rosenbergii*. Crabs and planktonic shrimp species which are common inhabitants of shrimp ponds can serve as asymptomatic carriers of WSSV, thus posing a serious threat to the shrimp farming industry. It has been suggested that copepods and perhaps even aquatic insect larvae may also be infected [10]. In India, WSSV has been reported in noncultured crustaceans of shrimp farms like Pest shrimp, *Acetes* sp., *Macrobrachium rosenbergii*, Mud crab *Scylla serrata*, Pest crab *Sesarma oceanica*, and *Pseudograpsus intermedius* [11].

Tiger prawn, *Penaeus monodon* and Indian white prawn *Fenneropenaeus indicus* are the major cultivated species in the paddy cum prawn culture system along Cochin backwaters. Paddy cum prawn culture is a modified farm with occasional inputs of seeds and without any control of water quality parameters. Disease outbreaks are intermittent which are mostly looked at from viral outbreak point, especially WSSV. Though the Cochin backwaters and the feeder canal linked to them have undergone considerable change over the years, no attempt has been made to study the key water quality parameters that are important to the survival of shrimps. Stressors, which are usually related to the phyisiochemical properties of both water and pond bottom which compromise the shrimp immune system thus increasing the rate of WSSV infection [12].

The objective of the study was to identify possible environmental trigger for WSSV outbreak in the paddy cum prawn fields along Cochin backwaters by regular monitoring of the water and sediment quality parameters of the feeder canal and couple of seasonal and perennial ponds.

2. Materials and Methods

2.1. Study Area and Shrimp Farming Practices.
Sampling site is located along Cochin backwaters at Edavanakkad, Cochin, Kerala. Shrimp farming in this region is generally a kind of extensive aquaculture, which is carried out in paddy fields adjoining the Cochin backwaters. This type of shrimp farming is known as paddy-cum-prawn culture and usually carried out soon after the harvesting of the paddy. The decaying paddy straw used to provide enough detritus food for the prawn larvae which used to enter the paddy fields during high tide. The larvae are retained in these ponds for 4-5 months, during which no controls in environmental parameters are exerted. Tidal amplitude between high and low tide is used to partially drain the pond water through a sluice gate, where net is placed and prawns are caught.

For the present study two types of farms were selected: seasonal and perennial ponds of different sizes and depths. The farms in this area are well connected to the Feeder canal system that brings water to this area from Cochin backwater. Exchange of water is carried out in the farms through the sluice gate according to the tidal fluctuation. For the present study, water, sediment, and shrimp samples were collected from Feeder canal, 2 seasonal ponds, and 2 perennial ponds.

2.2. Collection of Samples.
Water and sediment samples were collected on a fortnightly basis for water quality parameters, while for bacteriological analysis, samples were collected on a monthly basis. For analysis, water samples were collected in sterile bottles from a depth of 0.1 to 0.5 meter from the surface and sediment samples were collected with mud sampler and transferred to sterile jars. Water and sediment samples were collected from different locations of each sampling site and the samples were pooled before analysis.

2.3. Analysis of the Physicochemical Parameters.
Temperature, salinity, and pH of the water were measured *in situ* using centigrade thermometer, salinity refractometer (Atago, Japan), and hand-held digital pH meter (Eutech, Singapore), respectively. Temperature and pH of sediment were also measured *in situ* using centigrade thermometer and hand-held digital pH meter (Eutech, Singapore). The dissolved oxygen, alkalinity, total hardness, total ammonia nitrogen, nitrite, and inorganic phosphate of water samples were estimated as per APHA (1998) method. Nitrate was estimated by Resorcinol method [13].

2.4. Bacteriological Analysis.
Samples for bacteriological analysis were kept in icebox soon after collection and brought to the laboratory. Analysis was completed within 4 hr of collection. Water samples were serially diluted up to 10^{-3} and sediment samples up to 10^{-5} using sterile distilled water. For the analysis of shrimp samples, they were weighed, homogenized using sterile mortar and pestle, and serially diluted to 10^{-6}. Aseptic procedures were strictly followed during processing of samples.

2.5. Estimation of Total Viable Count.
Aliquots of 0.1 mL samples from each dilution were spread plated in triplicate on tryptone soya agar (TSA) or 1/2 strength Zobell's marine agar (1/2 ZMA) or ZMA depending on the salinity of the water samples for the enumeration of total aerobic heterotrophic bacteria, which is expressed as total viable count (TVC). The plates were then incubated at 30°C for 48 h. After incubation, plates with 30 to 300 colonies were selected for counting and isolation of bacteria.

2.6. Estimation of Total Coliform.
Aliquots of 0.1 mL sample from each dilution were spread plated in triplicate on Mac Conkey agar for the enumeration of total coliform (TC). The plates were incubated at 37°C for 48 h. After incubation, plates were selected for counting, typical coliform like colonies were counted and expressed as TC.

2.7. Estimation of Total Vibrio Count.
Aliquots of 0.1 mL samples from each dilution were spread plated in triplicate on thiosulphate citrate bile sucrose (TCBS) agar for the enumeration of *Vibrio* like organism, which is expressed as total *Vibrio* like organism (TVLO). The plates were incubated at 30°C for 48 h. After incubation, plates were selected for counting, typical *Vibrio* like colonies were counted and expressed as TVLO.

2.8. Estimation of Total Aeromonas Count. Aliquots of 0.1 mL samples from each dilution were spread plated in triplicate on starch ampicillin (SA) agar for the enumeration of *Aeromonas* like organism, which is expressed as total *Aeromonas* like organism (TALO). The plates were incubated at 30°C for 48 h. After incubation, plates were selected for counting. Typical *Aeromonas* like colonies were counted and expressed as TALO.

2.9. Isolation and Identification of Total Heterotrophic Bacterial Isolates. After recording the morphological characters and pigmentation, representative types constituting at least 20–40 numbers of colonies were selected from each plate and restreaked onto TSA, 1/2 ZMA or ZMA plates to ensure purity. All the purified isolates were maintained on TSA or 1/2 ZMA or ZMA slants for further characterisation and identified to generic level using the taxonomic key for identification by [14–17]. The various tests carried out included Gram staining, spore staining, motility test, Kovac's oxidase test, oxidation fermentation test, catalase test, acid production in glucose, and Voges-Proskauer test.

2.10. Isolation and Identification of Vibrio Isolates. *Vibrio* colonies were isolated according to the colouration of the colonies in TCBS plates. The isolated colonies were restreaked onto TSA plates to ensure purity. All the purified isolates were maintained on TSA slants for further characterisation and identified up to species level using the following biochemical test as per U.S. Food and Drug Administration's Bacteriological Analytical manual (USFDA BAM). The following tests were used to characterise the vibrios: oxidase test, arginine dihydrolase test, ornithine decarboxylase, lysine decarboxylase test, growth in 0%, 3%, 6%, 8%, and 10% NaCl, growth at 42°C, acid from Sucrose, D-cellobiose, lactose, arabinose, D-mannose, and D-mannitol, ONPG (Ortho Nitrophenyl-β-galactoside), test and Voges-Proskauer test.

2.11. Isolation and Identification of Aeromonas Isolates. After recording the starch utilisation capacity on SA agar, colonies were selected from plate and restreaked onto TSA plates to ensure purity. All the purified isolates were maintained on TSA slants for further characterisation and identified to species level using the following biochemical test as per analytical profile index 20E (API 20E). The tests used were ONPG (orthonitrophenyl-β-galactoside) test, arginine dihydrolase test, lysine decarboxylase test, ornithine decarboxylase test, citrate utilization test, H_2S production test, urease test, tryptophan deaminase (TDA) test, indole production test, Voges Proskauer test, gelatinase test, and acid from glucose, mannitol, inositol, sorbitol, rhamnose, sucrose, amygdalin, melibiose, arabinose, and oxidase production.

2.12. Detection of White Spot Syndrome Virus (WSSV) from Shrimp Samples. The deproteinised genomic DNA of the shrimp was prepared according to the method for preparation of genomic DNA from mammalian tissue [18]. Briefly, 200 mg muscle tissue excised from the abdomen of the shrimp was rapidly frozen and crushed to a fine powder.

The processed tissue was placed in 2.4 mL digestion buffer (100 mM NaCl, 10 mM Tris-HCl, pH 8, 25 mM EDTA, pH 8, 0.5% sodium dodecyl sulfate, 0.1 mg/mL proteinase K) and incubated at 65°C for 12 to 18 h. The digest was deproteinised by successive phenol/chloroform/isoamyl alcohol extractions, recovered by ethanol precipitation, and dried and resuspended in 0.1 X TE buffer at 65°C for 30 min, and then stored at 4°C until use for PCR.

Oligonucleotide primers (146F and 146R) were used for the amplification of WSBV DNA fragments. Primers 146F and 146R were designed on the basis of the DNA sequence of a cloned WSBV 1461-bp *Sal* I DNA fragment in recombinant plasmid (pms146). The following are the sequences of the primers: 146F1, 5′-ACT ACT AAC TTC AGC CTA TCT AG-3′; 146R1, 5′-TAA TGC GGG TGT AAT GTT CTT ACG A-3′. With this primer set, a 1447-bp fragment is expected to be amplified from WSBV genomic DNA. The internal primers (146F2, 5′-GTA ACT GCC CCT TCC ATC TCC A-3′; and 146R2, 5′-TAC GGC AGC TGC TGC ACC TTG T-3′) were used to confirm that the amplified fragment was indeed from the WSBV 941-bp *Sal* I DNA fragment. The deproteinised DNA samples used for amplification totaled 0.1 to 0.3 pg in a 100 μL reaction mixture containing 10 mM Tris-HCl, pH 9 at 25°C, 50 mM KCl, 1.5 mM $MgCl_2$, 0.1% Triton X-100, 200 μM each of dNTP, 100 pmol each of primer, 2.5 units of Taq DNA polymerase. The amplification was performed in a thermal station for 1 cycle of 94°C for 4 min, 55°C for 1 min, 72°C for 3 min; and then 39 cycles of 94°C for 1 min, 55°C for 1min, 72°C for 3 min; plus a final 5 min extension at 72°C after 40 cycles. The PCR products were analyzed in 1% agarose gel containing ethidium bromide at a concentration of 0.5 μg/mL and visualized under ultraviolet transillumination [19].

3. Results

3.1. Water Quality Parameters. The temperature of water remained within the acceptable limit required for prawn culturing for most of the culture period. The minimum temperature encountered was 27°C (July, 2010) and it reached up to 37°C (April, 2010) without any further increase throughout the study period. The sediment temperature was always found to be equal or one or two degrees lesser than that of water (Tables 1 and 2). pH of the sediment ranged between 6.38 to 8.36 which falls within the limit that is necessary for the development of shrimps. The mean values of physicochemical parameters of water recorded during the study period are given in Table 1. Though the pH of water was normal in all the ponds, an alkaline pH of 9.05 (which coincided with heavy algal growth) was encountered in seasonal pond-2 at the start of the culture period (October, 2009). But as the culturing progressed, it gradually decreased and reached a normal level. There was a sharp increase in pH (9.7) again during the monsoon season (July, 2010) and remained high for the rest of the study period. The salinity of water was very low in September (0 ppt) which gradually increased and reached a maximum of 30.8 ppt in February. It is attributed to the variations in the salinity of the water from monsoon to the postmonsoon season. Though the salinity

TABLE 1: Water quality parameters of the feeder canal and the shrimp farms (mean value ± SD).

Parameter	Feeder canal	Perennial pond-1	Perennial pond-2	Seasonal pond-1	Seasonal pond-2
Temperature (°C)	29.95 ± 2.01 (27–36)*	30.75 ± 2.19 (28–36)	30.65 ± 2.23 (28–37)	30.45 ± 1.95 (28–36)	31.10 ± 2.26 (28–37)
Salinity (ppt)	11.45 ± 9.48 (0–30.00)	10.51 ± 7.92 (0–27.50)	10.41 ± 9.51 (0–30.80)	10.82 ± 9.61 (0–30.30)	10.63 ± 8.07 (2.97–29.00)
Dissolved oxygen (mg/L)	3.16 ± 1.59 (0.8–7.42)	5.23 ± 1.62 (2.0–7.01)	6.70 ± 1.79 (3.71–10.30)	6.62 ± 2.82 (2.88–13.00)	8.31 ± 2.80 (4.50–13.80)
pH	7.30 ± 0.2 (6.90–7.90)	7.60 ± 0.3 (7.27–8.40)	8.00 ± 0.5 (6.61–8.90)	7.68 ± 0.5 (6.86–8.60)	8.50 ± 0.8 (7.30–9.70)
Alkalinity (mgCaCO$_3$/L)	62.40 ± 15.10 (24–84)	67.00 ± 15.23 (36–92)	62.60 ± 22.3 (16–92)	60.00 ± 18.17 (20–88)	68.20 ± 17.53 (20–88)
Total hardness (mgCaCO$_3$/mL)	1684.2 ± 1230.192 (300–3820)	1695.8 ± 1066.57 (300–3600)	1647.0 ± 1276.354 (368–4200)	1695.0 ± 1328.06 (400–4400)	1698.0 ± 1172.346 (580–3900)
TAN (mg/L)	0.1958 ± 0.15 (0.00394–0.64600)	0.1758 ± 0.15 (0.05168–0.66348)	0.2498 ± 0.4 (0.00608–1.87112)	0.1647 ± 0.15 (0.01596–0.70680)	0.1287 ± 0.16 (0.00608–0.65892)
Nitrite (mg/L)	0.0149 ± 0.008 (0.00286–0.032186)	0.0102 ± 0.008 (0.00078–0.03059)	0.0214 ± 0.010 (0.00390–0.046284)	0.01504 ± 0.010 (0.000782–0.029792)	0.0185 ± 0.02 (0.00442–0.044688)
Nitrate (mg/L)	0.1748 ± 0.09 (0.057860–0.428164)	0.1205 ± 0.062 (0.02893–0.219868)	0.1930 ± 0.08 (0.046288–0.352946)	0.1397 ± 0.07 (0.052074–0.266156)	0.1379 ± 0.08 (0.005786–0.266156)
Phosphate (mg/L)	0.1840 ± 0.109 (0.05410–0.40254)	0.2728 ± 0.34 (0.06924–0.39168)	0.2928 ± 0.18 (0.12334–0.87425)	0.1998 ± 0.12 (0.03462–0.42631)	0.1658 ± 0.10 (0.01948–0.37437)

*Value in parentheses indicates range.

TABLE 2: Temperature and pH (Mean ± SD) of the sediment of feeder canal and the ponds.

Parameter	Feeder canal	Perennial pond-1	Perennial pond-2	Seasonal pond-1	Seasonal pond-2
Temperature (°C)	29.5 ± 1.88 (28–33)*	30.5 ± 2.02 (28–34)	30 ± 1.8 (28–33)	30.41 ± 1.5 (28–33)	30.5 ± 1.8 (28–33)
pH	7.6 ± 0.52 (6.11–7.8)	7.3 ± 0.31 (7–8)	7.3 ± 0.35 (7-8)	7.18 ± 0.52 (6.25–7.9)	7.36 ± 0.38 (6.9–8.1)

*Value in parentheses indicates range.

was low initially, good growth of the shrimps could be observed, and there was no mortality during that period.

Dissolved oxygen levels were relatively low in the feeder canal (as low as 0.8 mg/L), and in the case of the other ponds, they were within the levels conducive for shrimp aquaculture. In seasonal pond-2, the dissolved oxygen levels showed much fluctuations (4.5 to 13.8 mg/L) and alkaline pH (9.7) which coincided with algal blooms. Thus, seasonal hypoxic conditions were observed in this pond during our study. Total hardness level was found to be between 300–4400 mg CaCO$_3$ mg/L. It was found to increase with increasing salinity and reached a maximum during the summer season and declining as the monsoon started.

The levels of ammonia were higher than the acceptable limit (0.01 mg/L) for penaeid shrimps. The levels are very high in the month of November (perennial pond-2 recording 1.87 mg/L) and towards the end, the levels are slightly higher. In the shrimp ponds and the feeder canal, TAN values were between 0.0039–1.87 mg/L. The nitrite levels recorded during the study are very low and were well below the safe level (1.28 mg/L) of nitrite recommended for *Penaeus monodon*

by Law [20]. Phosphate and nitrate values recorded were between 0.01–0.87 mg/L and 0.005–0.428 mg/L, respectively.

3.2. Bacteriology of the Ponds and Feeder Canal. Figures 1 and 2 show the mean count of different groups of bacteria enumerated from water and sediment during the study period. The bacteriological analysis showed the THB load to be between 0.2×10^3 to 9×10^4 cfu/mL in water, and in the sediment, it was between 1.3×10^4 to 1.79×10^7 cfu/g. In water, TVLO count was found to be between $0–3.1 \times 10^4$ cfu/mL, and TALO count was between 0.07×10^3–1.43×10^4 cfu/mL. TVLO and TALO of the sediment were $0–2.61 \times 10^4$ and 0.8×10^3–10.3×10^4 cfu/g, respectively. In all the sampling stations, sediment was found to have higher microbial load than the water samples. TVC count was found to be generally higher in the feeder canal in contrast to both the perennial and seasonal ponds. In seasonal pond-2, TVLO were sometimes absent though the inoculum size (0.1 mL) was the same as in other ponds. The TC count of water and sediment were $0–2 \times 10^4$ cfu/mL and $0–1.4 \times 10^5$ cfu/g, respectively. Coliform count and Total *Vibrio*-like organisms

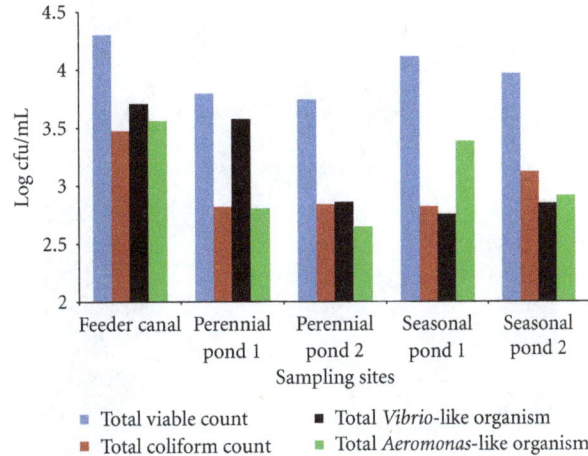

FIGURE 1: Mean value of bacteriological count of water from different sampling sites.

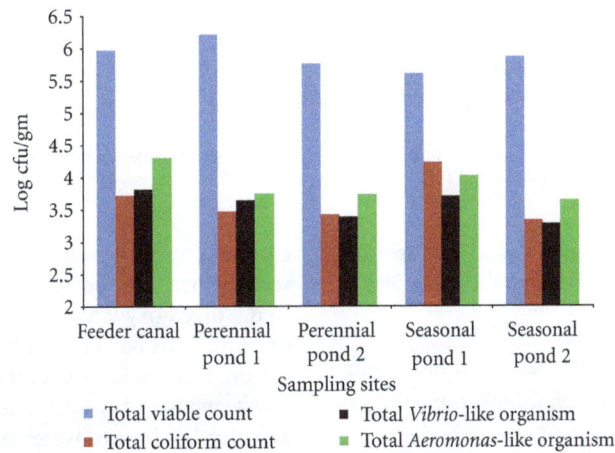

FIGURE 2: Mean value of bacteriological count of sediment from different sampling sites.

count were also found to be higher in sediments than water. The same pattern was observed in the case of total *Aeromonas* count too.

Table 3 shows the different genera of bacteria isolated from water, sediment, and shrimp in the months of September, 2009 and December, 2009. Genera of heterotrophic bacteria encountered were *Acinetobacter, Bacillus, Corynebacteria, Klebsiella, Kurthia, Listeria, Micrococcus, Moraxella, Staphylococcus, Alcaligenes, Cytophaga, Vibrio. Vibrio* sp, and *Aeromonas* sp. isolated from the feeder canal, perennial pond-1, perennial pond-2 are given in Table 4. Further species were characterization of Aeromonas and Vibrio isolates revealed the presence of *Aeromonas hydrophila* and various species of Vibrio such as *V. fluvialis, V. hollisae, V. vulnificus, V. furnissii, V. metschnikovii,* and *V. fischeri.*

3.3. WSSV Outbreak and Detection of WSSV by PCR. During the study period, WSSV outbreak occurred twice, once in the postmonsoon season (December, 2009) and once in the monsoon season (June–July, 2010), which was confirmed by PCR (Figure 3). Complete mortality was observed in

TABLE 3: Genera of heterotrophic bacteria identified from water, sediment, and shrimps in the feeder canal and shrimp farms.

Sl. no.	Bacterial genera identified from		
	Water	Sediment	Shrimp
1	*Acinetobacter*	*Alcaligenes*	*Alcaligenes*
2	*Bacillus*	*Bacillus*	*Bacillus*
3	*Corynebacteria*	*Corynebacteria*	*Corynebacteria*
4	*Klebsiella*	*Cytophaga*	*Listeria*
5	*Kurthia*	*Klebsiella*	*Micrococcus*
6	*Listeria*	*Listeria*	*Staphylococcus*
7	*Micrococcus*	*Micrococcus*	*Vibrio*
8	*Moraxella*	*Staphylococcus*	
9	*Staphylococcus*		

seasonal pond-2 during the outbreak in June, 2010. There was no significant increase in the number of presumptive Vibrionaceae during the WSSV outbreak in December or in June, 2010. The result showed that the shrimps from both

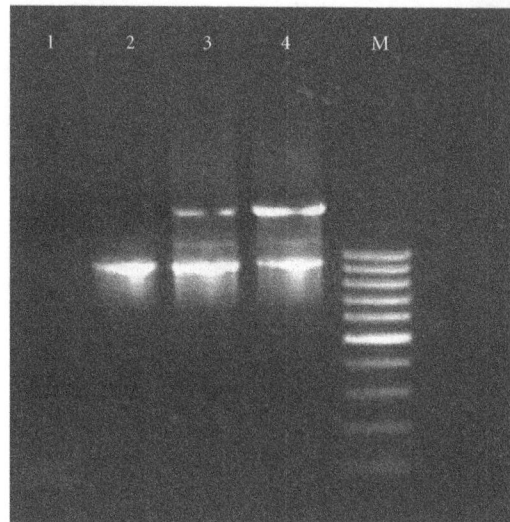

Lane 1: Negative control
Lane 2: Sample
Lane 3 and 4: Positive controls
M: Molecular weight marker (100 bp)

FIGURE 3: Results of PCR for WSSV from diseased shrimps from seasonal pond-1.

TABLE 4: Occurrence of *Aeromonas hydrophila* and *Vibrio* sp. from water, sediment, and shrimps in feeder canal and shrimp farms.

Aeromonas/Vibrio sp. identified	Sample
Aeromonas hydrophila	Water, sediment, and shrimp
V. hollisae	Water, sediment, and shrimp
V. fluvialis	Shrimp
V. metschnikovii, *V. fischeri*	Sediment
V. furnissii, *V. vulnificus*	Water

the seasonal ponds were under WSSV attack. The shrimp seasonal pond-1 also had parasitic *Zoothamnium* attack.

4. Discussion

Water and soil conditions influence shrimp production greatly that it is necessary to manage the pond properly for better results. Fluctuation of pH around the neutral value during culture period has been reported earlier, and it does not affect the growth of the shrimps, as the estuary is well buffered against serious changes in pH [21]. Results of the study revealed considerable dissolved oxygen stress in the feeder canal which is closely integrated to the seasonal ponds and the perennial ponds in the study region. Since the major seed supply to the farm is through natural water getting to the pond during high tide. These are normally retained in the pond and fattened to a marketable size. Over the years, the depth of the feeder canal has reduced considerably due to siltation and effluent from the nearby seafood preprocessing units and markets are drained into the feeder canal, which can considerably increase the nutrient load in the feeder

canal and resulting in the reduction of dissolved oxygen. The seeds which gain entry into the pond might have subjected to dissolved oxygen stress and become susceptible to WSSV/opportunistic bacterial pathogens. Another important observation was periodical cyanobacterial bloom (identified as *Anabaenopsis* sp.—unpublished data) in the seasonal pond 2, which has limited water exchange. The dissolved oxygen in the pond was found to vary considerably during the day with hypoxic condition in the early morning (1.23–1.64 mg/L) and supersaturated dissolved oxygen conditions in the afternoon. This pond also shows alkaline pH to the tune of 9.7. WSSV affected shrimps were frequently encountered in these ponds. pH of the water samples were found to be within the acceptable range in the feeder canal, seasonal pond 1 and the perennial ponds, while that of the sediment from feeder canal occasionally showed acidic nature. Bottom soil quality is an important factor as the shrimps spend most of their time burrowing in the bottom and also ingest some of it [22]. Decreasing sediment pH or acidification of sediment (7.0–6.0) increases the hemolymph osmotic pressure of shrimps which is an indicator of variation in osmotic regulation. A significant decrease in hemolymph OP was seen as the water pH decreased from 7.0 to 6.5 [23].

Many studies have reported growth of *Penaeus monodon* in fresh water and low salinity as low as 2 ppt [22, 24–26]. Our observations have shown that the DO levels of the inlet canal is almost always very low and the canal is highly polluted by the wastes that are discharged into it from homes along the banks as well as seafood processing units in that area. Shrimp farmers have often expressed concerns over the polluted water surrounding their farms, as pollution by industrial, commercial, and urban contaminations [27] can slow growth, increase disease outbreaks, and accelerates the mortality rate of shrimps. Low levels of DO (<3.7 mg/L) when combined with high NH_3-N (>0.5 mg/L) has shown to be harmful for P. semisulcatus [28]. L. E. Burnett and K. G. Burnett [29] suggested hypoxia results in a depression of the generalized innate immune response in Paleomonetes pugio and Penaeus vannamei on the basis of measurements of circulating hemocytes and survival of shrimp exposed to Vibrio.

Guerrero-Galván et al. [30] have reported that during the rainy season, dissolved oxygen level tend to decrease, as the feeding rates and shrimp and phytoplankton biomass were increasing until harvest. A similar decrease in dissolved oxygen levels were seen during the monsoon season (diurnal changes in dissolved oxygen levels 1.24 mg/L at 7.00 a.m and around saturation in the afternoon) along with high phytoplankton biomass.

Presence of total ammonia nitrogen (TAN) can be attributed to the sludge that is deposited at the bottom of the ponds. In the ponds studied, there is no periodic removal of the settled sludge during the production cycle. During the past, the seasonal ponds were regularly used for paddy cultivation after a crop of shrimp, during which plowing and tilling operation were practiced before paddy cultivation. This was helpful in removing the sludge and utilization of it for growing paddy. Deepening of the perennial ponds was also prevented by farm owners, and the removed sludge and sediment were used for growing plants along the bunds of the farms. However, paddy cultivation and drying of perennial ponds are abandoned due to the unavailability of farm labour. Several studies have proved the contribution of settled sludge to ammonia production in shrimp growout ponds [31]. Yearsley et al. [32] have found that settled sludge that is not removed produces 44% more TAN than in tanks, where they are removed. These studies call for regular removal of sludge from the pond bottom to improve water quality as any increase in pH or temperature of the pond water can result in increased ammonia production which is lethal to shrimps. In closed systems, rapid accumulation of ammonia occurs and reduces growth even at low concentrations [21]. This could be the reason why the relatively huge perennial ponds were not affected though they had higher levels of TAN, whereas the smaller seasonal ponds 1 and 2 with low water exchange were both affected by WSSV and had higher TAN levels at that time.

High levels of TAN were recorded in the perennial ponds and the seasonal ponds. It was 2–180 times more than the safe level (0.01 mg/L NH_3-N) calculated by Chin and Chen [33] for P. monodon larvae. The proportion of NH_3 to NH_4^+ in water increases with increase in water temperature, pH and decrease in salinity. Unionized ammonia is toxic, as it has high lipid solubility and readily diffuses across cell membranes [34]. Even low levels can spell doom for shrimps if the pH increases as was the case seen in our study. Ammonia in water can suppress the immune system of Littopenaeus vannamei and increases mortality by Vibrio alginolyticus [35].

The nitrite levels recorded in our study were always lower than the safe limit (1.28 mg/L) recommended by Law [20]. This is in tune with a study in Taiwan between 1989 and 1990, where only few ponds (<0.5%) recorded nitrite concentrations higher than the safe level. These low levels are because they are intermediate products and unstable nature in the aquatic environment which limits their accumulation and consequently low concentrations [36].

Nitrate levels varied greatly (0.12–0.19 mg/L) and were found to be more than that recorded by Rao et al. [37]. The nitrate levels in their study were between 0.002–0.213 mg/L, whereas the lowest nitrate value (0.028 mg/L) itself was higher than that encountered in their study. The average nitrate levels recorded were also higher than what was recorded by [38] in the shrimp ponds of Thailand (0.08–0.16 mg/L).

Bacterial count fluctuated during the study period, and there was no increase in the bacterial number, as the crop age increased. This trend is supported in a study by Burford et al. [39]. Total heterotrophic bacterial count in water was between 0.1×10^4–9×10^4 cfu/mL. They were lower compared to the levels recorded by Anand et al. [40] which was of the order of 10^6–10^7 cfu/mL. But the TVC counts of both water and sediment were found to be higher than that recorded by Janakiram et al. [41] in extensive and semi-intensive shrimp ponds in Andhra, Pradesh, India.

TVC levels in the sediment (1.3×10^4–1.79×10^7 cfu/g) were higher than that of the water. Water and sediment of the feeder canal showed relatively higher levels of TVC, TVLO, and TC count compared to the other ponds. Owing to the constant changes in the environment of a tidal estuary, it is difficult to study the influence of selected parameters on microbial population [42]. Janakiram et al. [41] also reported that a correlation between the physicochemical parameters and bacterial load could not be found.

The high bacterial count in the sediment may be attributed to the presence of high organic matter and nutrients in the sediments than in the water column, as the THB count is dependent on the availability of nutrient sources and the sediment bacteria play a major role in the remineralisation of nutrients that accumulate at the pond bottom [43]. The heterotrophic bacteria oxidize the organic matter consuming oxygen and release carbon dioxide. Thus, the water quality in an aquaculture pond is greatly influenced by microbial degradation of organic matter. In all aquatic system including ponds, microbes are an integral part of the food web, thus directly influencing the productivity of the system. The heterotrophic bacterial community of an aquatic ecosystem is involved in the breakdown of complex organic compounds to simple molecules, hence providing an important food source [44, 45]. They are beneficial to shrimp ponds by their involvement in mineralization and regeneration of nutrients [46]. But the bacterial flora of the shrimp environment also has several facultative and opportunistic pathogens which turn virulent under poor environmental conditions leading to mass mortality of cultured shrimps [47, 48]. Species level characterization of Vibrio and Aeromonas revealed the presence of pathogenic strains. The prevalence of Vibrio sp. in the pond environment poses a threat to both humans and shrimps as Vibrio are pathogenic to humans and shrimps as well. V. harveyi, V. vulnificus, V. parahemolyticus, V. anguillarum, and V. splendidus are some of the Vibrio sp. reported as shrimp pathogens. They are common inhabitants of shrimp hatcheries, pond water, and sediment, and they turn pathogenic under poor environmental conditions [49]. In experimental studies, shrimp exposed to ammonium stress prior to challenge showed higher susceptibility to vibrios [35]. It has also been indicated that a primary WSSV infection may weaken shrimp, increasing their susceptibility to bacterial infections [50].

Higher coliform count in sediment than water observed in our study is supported by a similar work by Harish et al. [51]. Presence of fecal coliforms is an indicator of pollution, and it is a common factor in environments closer to human existence. Most of the bacteria encountered in a shrimp pond are opportunistic pathogens, as they are part of the natural ecosystem as well as the normal flora of the shrimp. We were able to isolate A. hydrophila from the water, sediment, and the shrimps. They are part of the normal flora of the shrimp environment and turn pathogenic under stressful conditions which are in most cases poor environmental and physiological parameters [52]. The pond water, diseased fish, frogs, and convalescent frogs may harbour motile aeromonads, which can cause septicemia under stress conditions such as low dissolved oxygen level, elevated temperature [53], elevated levels of ammonia, and carbon dioxide [54]. The ponds under study are extensive and traditional with other fish and crabs which would also play a role in disease outbreak, as they might harbor harmful bacteria.

Unlike terrestrial animals, which live in a fairly constant environment (stable oxygen and carbon dioxide content), shrimps live in an environment which changes often abruptly. Manifestation of diseases and stress are closely related, and stress is induced by sudden changes in the temperature, oxygen, salinity, and ammonium. Farmers often do not do anything to check the variability in the aquatic environment [55]. In a pond ecosystem, it is difficult to separate shrimps from other crustaceans like crabs and copepods, as they may be alternate hosts of shrimp diseases. Disease avoidance is the only way out for avoiding mortality and increasing production in these ponds. In South and South East Asia, shrimp farmers have some level of success in controlling WSSV by limiting water exchange which probably increases water temperature and prevents pathogens from entering the ponds [9]. Sengupta et al. [56] show how antibiotic resistant Vibrio sp. are transmitted through the feeder canal to grow-out ponds.

Though there are studies supporting that ammonia, nitrite, and hydrogen sulphide is toxic to shrimps, a single metabolite may not be held responsible for retarded growth and mortality of shrimps. Pai et al. [57] have reported that cumulative factors like sudden and steep decline in salinity, lowering of DO levels, increase in iron content in the hypolimnion following monsoon and subsequent increase in ammonia and hydrogen sulphide triggers the WSSV from latency to virulence, thus showing that more than one factor is always involved in disease outbreak. The amplification of viral loads and onset of disease can be induced by environmental or physiological stress or at ambient temperatures below 30°C.

Studies have shown the inability of WSSV to replicate in higher temperatures. Temperature reduction due to monsoon outpours in June and cold water conditions in December could be attributed to the outbreak of WSSV and species of the marine environment spread easily and are more devastating [58]. WSSV seems to be triggered or aggravated by changes in seawater quality including hardness, temperature, and dissolved oxygen [59]. A study by Jiang et al. [60] shows that both ammonia-N and WSSV decrease the plasma proteins and total hemocyte count of P. japonicus significantly which can lead to increased mortality. Plasma proteins play a major role in crustacean immune system and high level of ammonia-N in water is the main reason for decrease in plasma proteins. Due to high ambient ammonia-N, accumulation of haemolymph ammonia and urea occurs, leading to the catabolism of proteins to amino acids [61–63]. Fluctuations in temperature, pH, and salinity are important risk factors, which can lead to WSSV infection [64]. They have also found out that shrimps in less transparent deep ponds are more susceptible to WSSV infection than those in less transparent shallow ponds.

Abrupt fluctuations in temperature and salinity due to heavy rain have been found to contribute to increase in viral loads in the shrimps and have caused 80% mortality in shrimps in a study in Mexico [65]. The same could be said of our study, as there was an outbreak of WSSV as soon as the monsoon started. Presence of the epicommensal protozoan, Zoothamnium, was a consistent occurrence among the WSSV-infected shrimp in the field outbreaks of India and Korea [66]. This is consistent with our observations too, as there was Zoothamnium infection in WSSV-positive P. monodon collected from the traditional ponds.

Along with low temperature, high pH and low dissolved oxygen caused by algal blooms (unpublished data) must have been a very stressful environment for the shrimps, aiding the rapid multiplication of WSSV and disease manifestation in the shrimps. It is very difficult to avoid diseases in an open-air culture system with close connection to the sea, by eliminating the causal agent or by using disease-free stocks. Rather, disease management based on providing conditions that will not help the development of the organism should be followed [9]. In our study period, there was no WSSV outbreak in perennial ponds, where feed input was either nil or minimal with almost stable dissolved oxygen and pH values throughout the culture period. In December, perennial pond-2 was treated with turmeric powder in an attempt to avoid disease outbreak which could have worked.

Bioaugmentation of the ponds with bacteria that can oxidize ammonia and nitrite could be used to improve water quality, as Rao et al. [37] have suggested that though these bacteria (involved in nitrogen and sulphur cycle) are present as natural flora of shrimp ponds, their numbers are low, particularly at the later part of the culture period. Survey conducted by Balasubramaniam et al. [67] in shrimp farms along the Chilka lagoon, Orissa, India, show that the production of zero-water exchange ponds has remained the same for years in that area, thus indicating the sustainability of the practice. Their production rate is also almost similar to the farms with regular water exchange. To replicate that sort of success in the seasonal farms along the Cochin backwaters, it is advised that farmers pay attention to water quality issues that plague them.

Acknowledgment

The authors would like to acknowledge the whole-hearted cooperation rendered by the local shrimp farmers.

References

[1] M. N. Kutty, "Towards sustainable freshwater prawn aquaculture—lessons from shrimp farming, with special reference to India," *Aquaculture Research*, vol. 36, no. 3, pp. 255–263, 2005.

[2] FAO, "Fishery and aquaculture country profiles. India. Fishery and aquaculture country profiles," in *FAO Fisheries and Aquaculture Department*, Rome, 2004.

[3] J. Bostock, B. McAndrew, R. Richards et al., "Aquaculture: global status and trends," *Philosophical Transactions of the Royal Society B*, vol. 365, no. 1554, pp. 2897–2912, 2010.

[4] E. Bachere, "Shrimp immunity and disease control," *Aquaculture*, vol. 191, no. 1–3, pp. 3–11, 2000.

[5] H. Y. Chou, C. Y. Huang, C. H. Wang, H. C. Chiang, and C. F. Lo, "Pathogenicity of a baculovirus infection causing white spot syndrome in cultured penaeid shrimp in Taiwan," *Diseases of Aquatic Organisms*, vol. 23, no. 3, pp. 165–173, 1995.

[6] Y. Takahashi, T. Itami, M. Kondo et al., "Electron microscopic evidence of bacilliform virus infection in kuruma shrimp (Penaeus japonicus)," *Fish Pathology*, vol. 29, pp. 121–125, 1994.

[7] S. N. Chen, "Current status of shrimp aquaculture in Taiwan," in *Proceedings of the Special Session of Shrimp Farming (Aquaculture' 95)*, C. L. Browdy and J. S. Hopkins, Eds., Swimming Through Troubled Water, pp. 29–34, World Aquaculture Society, Baton Rouge, La, USA, 1995.

[8] K. Inouye, K. Yamano, N. Ikeda et al., "The penaeid rod-shaped DNA virus (PRDV), which causes penaeid acute viremia (PAV)," *Fish Pathology*, vol. 31, no. 1, pp. 39–45, 1996.

[9] J. Cock, T. Gitterle, M. Salazar, and M. Rye, "Breeding for disease resistance of Penaeid shrimps," *Aquaculture*, vol. 286, no. 1-2, pp. 1–11, 2009.

[10] T. W. Flegel and V. Alday-Sanz, "The crisis in Asian shrimp aquaculture: current status and future needs," *Journal of Applied Ichthyology*, vol. 14, no. 3-4, pp. 269–273, 1998.

[11] M. S. Hossain, A. Chakraborty, B. Joseph, S. K. Otta, I. Karunasagar, and I. Karunasagar, "Detection of new hosts for white spot syndrome virus of shrimp using nested polymerase chain reaction," *Aquaculture*, vol. 198, no. 1-2, pp. 1–11, 2001.

[12] Y. Takahashi, T. Itami, M. Kondo et al., "Immunodefense system of crustacea," *Fish Pathology*, vol. 30, pp. 141–150, 1995.

[13] J. Z. Zhang and C. J. Fischer, "A simplified resorcinol method for direct spectrophotometric determination of nitrate in seawater," *Marine Chemistry*, vol. 99, no. 1–4, pp. 220–226, 2006.

[14] K. Muroga, M. Higashi, and H. Keitoku, "The isolation of intestinal microflora of farmed red seabream (Pagrus major) and black seabream (Acanthopagrus schlegeli) at larval and juvenile stages," *Aquaculture*, vol. 65, no. 1, pp. 79–88, 1987.

[15] G. I. Barrow and R. K. A. Feltham, *Cowan and Steel's Manual for the Identification of Medical Bacteria*, Cambridge University Press, Cambridge, UK, 3rd edition, 1993.

[16] J. G. Holt, N. R. Krieg, P. H. A. Sneath, J. T. Staley, and S.T. Willliams, *Bergy's Manual of Determinative Bacteriology*, Lipponcott Williams and Wilkins, 9th edition, 2000.

[17] J. P. Harley and L. M. Prescott, *Laboratory Exercise in Microbiology*, McGraw-Hill, New York, NY, USA, 5th edition, 2002.

[18] W. M. Strauss, "Preparation of genomic DNA from mammalian tissue," in *Current Protocols in Molecular Biology*, F. M. Ausubel, R. Brent, R. E. Kingston et al., Eds., vol. 3, pp. 2.2.1–2.2.3, Greene Publishing Associates and John Wiley and Sons, New York, NY, USA, 1994.

[19] C. F. Lo, J. H. Leu, C. H. Ho et al., "Detection of baculovirus associated with white spot syndrome (WSBV) in penaeid shrimps using polymerase chain reaction," *Diseases of Aquatic Organisms*, vol. 25, no. 1-2, pp. 133–141, 1996.

[20] A. T. Law, "Water quality requirements for *Penaeus monodon* culture," in *Proceedings of the Seminar on Marine Prawn Farming in Malaysia*, Malaysia Fisheries Society, Serdang, Malaysia, 1988.

[21] J. C. Chen and T. C. Wang, "Culture of tiger shrimp and red-tailed shrimp in a semistatic system," in *Proceedings of the 2nd Asian Fisheries Forum*, R. Hirano and I. Hanyu, Eds., pp. 77–80, Asian Fishery Society, Manila, Philippines, 1990.

[22] C. E. Boyd, "Water quality management and aeration in shrimp farming," Fisheries and Allied Aquacultures Departmental Series No. 2, Auburn, Ala, USA, 1989.

[23] H. Lemonnier, E. Bernard, E. Boglio, C. Goarant, and J. C. Cochard, "Influence of sediment characteristics on shrimp physiology: pH as principal effect," *Aquaculture*, vol. 240, no. 1–4, pp. 297–312, 2004.

[24] D. F. Cawthorne, T. Beard, J. Davenport, and J. F. Wickins, "Responses of juvenile Penaeus monodon Fabricius to natural and artificial sea waters of low salinity," *Aquaculture*, vol. 32, no. 1-2, pp. 165–174, 1983.

[25] R. K. Chakraborti, M. L. Bhowmik, and D. D. Halder, "Effect of change in salinity on the survival of Penaeus monodon (Fabricius) postlarvae," *Indian Journal of Fisheries*, vol. 33, pp. 484–487, 1986.

[26] S. Athithan, T. Francis, N. Ramanathan, and V. Ramadhas, *A Note on Monoculture of Penaeus Monodon in a Hard Water Seasonal Pond*, vol. 24, Naga, The ICLARM Quarterly, 2001.

[27] Y. Hirono, "Current practices of water quality management in shrimp farming and their limitations," in *Proceedings of the Special Session on Shrimp Farming*, J. Wyban, Ed., World Aquaculture Society, 1992.

[28] N. Wajsbrot, A. Gasith, M. D. Krom, and T. M. Samocha, "Effect of dissolved oxygen and the molt stage on the acute toxicity of ammonia to juvenile green tiger prawn penaeus semisulcatus," *Environmental Toxicology and Chemistry*, vol. 9, no. 4, pp. 497–504, 1990.

[29] L. E. Burnett and K. G. Burnett, "The effects of hypoxia and hypercampnia on cellular defenses of oysters, shrimp and fish," *Comparative Biochemistry and Physiology B*, vol. 126, p. S20, 2000.

[30] S. R. Guerrero-Galván, F. Páez-Osuna, A. C. Ruiz-Fernández, and R. Espinoza-Angulo, "Seasonal variation in the water quality and chlorophyll *a* of semi-intensive shrimp ponds in a subtropical environment," *Hydrobiologia*, vol. 391, pp. 33–45, 1998.

[31] M. A. Burford and A. R. Longmore, "High ammonium production from sediments in hypereutrophic shrimp ponds," *Marine Ecology Progress Series*, vol. 224, pp. 187–195, 2001.

[32] R. D. Yearsley, C. L. W. Jones, and P. J. Britz, "Effect of settled sludge on dissolved ammonia concentration in tanks used

to grow abalone (Haliotis midae L.) fed a formulated diet," *Aquaculture Research*, vol. 40, no. 2, pp. 166–171, 2009.

[33] T. S. Chin and J. C. Chen, "Acute toxicity of ammonia to larvae of the tiger prawn, Penaeus monodon," *Aquaculture*, vol. 66, no. 3-4, pp. 247–253, 1987.

[34] R. P. Trussell, "The percent un-ionized ammonia in aqueous ammonia solutions at different pH levels and temperatures," *Journal of Fish Resources Board Canada*, vol. 29, pp. 1505–1507, 1972.

[35] C. H. Liu and J. C. Chen, "Effect of ammonia on the immune response of white shrimp Litopenaeus vannamei and its susceptibility to Vibrio alginolyticus," *Fish and Shellfish Immunology*, vol. 16, no. 3, pp. 321–334, 2004.

[36] Y. H. Chien, "Water quality requirements and management for marine shrimp culture," in *Proceedings of the Special Session on Shrimp Farming*, J. Wyban, Ed., pp. 144–156, World Aquaculture Society, Baton Rouge, La, USA, 1992.

[37] P. S. S. Rao, I. Karunasagar, S. K. Otta, and I. Karunasagar, "Incidence of bacteria involved in nitrogen and sulphur cycles in tropical shrimp culture ponds," *Aquaculture International*, vol. 8, no. 5, pp. 463–472, 2000.

[38] M. R. P. Briggs and S. J. Funge-Smith, "A nutrient budget of some intensive marine shrimp ponds in Thailand," *Aquaculture & Fisheries Management*, vol. 25, no. 8, pp. 789–811, 1994.

[39] M. A. Burford, P. J. Thompson, R. P. McIntosh, R. H. Bauman, and D. C. Pearson, "Nutrient and microbial dynamics in high-intensity, zero-exchange shrimp ponds in Belize," *Aquaculture*, vol. 219, no. 1–4, pp. 393–411, 2003.

[40] T. P. Anand, J. K. P. Edward, and K. Ayyakkannu, "Monitoring of a shrimp culture system with special reference to Vibrio and fungi," *Indian Journal of Marine Sciences*, vol. 25, no. 3, pp. 253–258, 1996.

[41] P. Janakiram, L. Jayasree, and R. Madhavi, "Bacterial abundance in modified extensive and semi-intensive shrimp culture ponds of Penaeus monodon," *Indian Journal of Marine Sciences*, vol. 29, no. 4, pp. 319–323, 2000.

[42] F. L. Singleton, R. Attwell, S. Jangi, and R. R. Colwell, "Effects of temperature and salinity on Vibrio cholerae growth," *Applied and Environmental Microbiology*, vol. 44, no. 5, pp. 1047–1058, 1982.

[43] M. A. Burford, E. L. Peterson, J. C. F. Baiano, and N. P. Preston, "Bacteria in shrimp pond sediments: their role in mineralizing nutrients and some suggested sampling strategies," *Aquaculture Research*, vol. 29, no. 11, pp. 843–849, 1998.

[44] D. J. W. Moriarty, "Bacterial productivity in ponds used for culture of penaeid prawns," *Microbial Ecology*, vol. 12, no. 3, pp. 259–269, 1986.

[45] P. K. Anderson, "Production and decomposition in aquatic ecosystems and implications for aquaculture," in *Proceedings of the 14th ICLARM Conference*, D. J. W. Moriarty and R. S. V. Pullin, Eds., Detritus and Microbial Ecology in Aquaculture, pp. 123–145, ICLARM, Manila, Philippines, 1987.

[46] E. R. Leadbetter and J. S. Pointexter, *Bacteria in Nature*, vol. 1 of *Bacteriological Activities Perspectives*, Plenum Press, New York, NY, USA, 1985.

[47] L. Ruangpan and T. Kitao, "Vibrio bacteria isolated from black tiger shrimp, Penaeus monodon Fabricius," *Journal of Fish Disease*, vol. 14, pp. 383–388, 1991.

[48] L. D. de la Peña, T. Tamaki, K. Momoyama, T. Nakai, and K. Muroga, "Characteristics of the causative bacterium of vibriosis in the kuruma prawn, Penaeus japonicus," *Aquaculture*, vol. 115, no. 1-2, pp. 1–12, 1993.

[49] L. Jayasree, P. Janakiram, and R. Madhavi, "Characterization of Vibrio spp. associated with diseased shrimp from culture ponds of Andhra Pradesh (India)," *Journal of the World Aquaculture Society*, vol. 37, no. 4, pp. 523–532, 2006.

[50] J. Selvin and A. P. Lipton, "Vibrio alginolyticus associated with white spot disease of Penaeus monodon," *Diseases of Aquatic Organisms*, vol. 57, no. 1-2, pp. 147–150, 2003.

[51] R. Harish, K. S. Nisha, A. A. M. Hatha et al., "Prevalence of opportunistic pathogens in paddy cum shrimp farms adjoining Vembanadu Lake, Kerala, India," *Asian Fisheries Science*, vol. 16, pp. 185–194, 2003.

[52] R. C. Cipriano, "Aeromonas hydrophila and motile Aeromonad septicemias of fish," Fish Disease Leaflet 68. United states Department of the Interior. Fish and Wildlife Service Division of Fishery Research, Washington, DC, USA, 2001.

[53] G. W. Esch and T. C. Hazen, "Stress and body condition in a population of largemouth bass: implications for red-sore disease," *Transactions of the American Fisheries Society*, vol. 109, pp. 532–536, 1980.

[54] G. R. Walters and J. A. Plumb, "Environmental stress and bacterial infection in channel catfish, Ictalurus punctatus Rafinesque," *Journal of Fish Biology*, vol. 17, no. 2, pp. 177–185, 1980.

[55] P. M. Biggs, "Infectious animal disease and its control," *Philosophical Transactions of the Royal Society of London B*, vol. 310, no. 1144, pp. 259–274, 1985.

[56] T. Sengupta, D. Sasmal, and T. J. Abraham, "Antibiotic susceptibility of luminous bacteria from shrimp farm environs of West Bengal," *Indian Journal of Marine Sciences*, vol. 32, no. 4, pp. 334–336, 2003.

[57] S. S. Pai, M. Manjusha, S. Ranjit et al., "Evaluation of environmental factors imfluencing the outbreak of white spot syndrome in latent P. monodon," in *Proceedings of the Conference on Microbiology of the Tropical Seas*, National Institute of Oceanography, Dona Paula, Goa, India, 2004.

[58] H. I. McCallum, A. Kuris, C. D. Harvell, K. D. Lafferty, G. W. Smith, and J. Porter, "Does terrestrial epidemiology apply to marine systems?" *Trends in Ecology and Evolution*, vol. 19, no. 11, pp. 585–591, 2004.

[59] N. Kautsky, P. Rönnbäck, M. Tedengren, and M. Troell, "Ecosystem perspectives on management of disease in shrimp pond farming," *Aquaculture*, vol. 191, no. 1–3, pp. 145–161, 2000.

[60] G. Jiang, R. Yu, and M. Zhou, "Modulatory effects of ammonia-N on the immune system of Penaeus japonicus to virulence of white spot syndrome virus," *Aquaculture*, vol. 241, no. 1–4, pp. 61–75, 2004.

[61] J. C. Chen, F. H. Nan, and S. Y. Cheng, "Effects of ambient ammonia on ammonia-N and protein concentration in haemolymph an ammonia-N excretion of Penaeus chinensis," *Marine Ecology Progress Series*, vol. 98, pp. 203–208, 1993.

[62] J. C. Chen, C. T. Chen, and S. Y. Cheng, "Nitrogen excretion and changes of hemocyanin, protein and free amino acid levels in the hemolymph of Penaeus monodon exposed to different concentrations of ambient ammonia-N at different salinity levels," *Marine Ecology Progress Series*, vol. 110, no. 1, pp. 85–94, 1994.

[63] J. C. Chen, S. Y. Cheng, and C. T. Chen, "Changes of haemocyanin, protein and free amino acid levels in the haemolymph of Penaeus japonicus exposed to ambient ammonia," *Comparative Biochemistry and Physiology*, vol. 109, no. 2, pp. 339–347, 1994.

[64] E. A. Tendencia, R. H. Bosma, and J. A. J. Verreth, "WSSV risk factors related to water physico-chemical properties and microflora in semi-intensive Penaeus monodon culture ponds in the Philippines," *Aquaculture*, vol. 302, no. 3-4, pp. 164–168, 2010.

[65] L. I. Peinado-Guevara and M. López-Meyer, "Detailed monitoring of white spot syndrome virus (WSSV) in shrimp commercial ponds in Sinaloa, Mexico by nested PCR," *Aquaculture*, vol. 251, no. 1, pp. 33–45, 2006.

[66] K. V. Rajendran, S. C. Mukherjee, K. K. Vijayan, S. J. Jung, Y. J. Kim, and M. J. Oh, "A comparative study of white spot syndrome virus infection in shrimp from India and Korea," *Journal of Invertebrate Pathology*, vol. 84, no. 3, pp. 173–176, 2003.

[67] C. P. Balasubramaniam, S. M. Pillai, and P. Ravichandran, "Zero- water exchange shrimp farming systems (extensive) in the periphery of Chilka lagoon, Orissa, India," *Aquaculture International*, vol. 12, no. 6, pp. 555–572, 2004.

15

A Refined Methodology for Defining Plant Communities Using Postagricultural Data from the Neotropics

Randall W. Myster

Biology Department, Oklahoma State University, Oklahoma City, OK 73107, USA

Correspondence should be addressed to Randall W. Myster, rwmyster@gmail.com

Academic Editor: Qinfeng Guo

How best to define and quantify plant communities was investigated using long-term plot data sampled from a recovering pasture in Puerto Rico and abandoned sugarcane and banana plantations in Ecuador. Significant positive associations between pairs of old field species were first computed and then clustered together into larger and larger species groups. I found that (1) no pasture or plantation had more than 5% of the possible significant positive associations, (2) clustering metrics showed groups of species participating in similar clusters among the five pasture/plantations over a gradient of decreasing association strength, and (3) there was evidence for repeatable communities—especially after banana cultivation—suggesting that past crops not only persist after abandonment but also form significant associations with invading plants. I then showed how the clustering hierarchy could be used to decide if any two pasture/plantation plots were in the same community, that is, to define old field communities. Finally, I suggested a similar procedure could be used for any plant community where the mechanisms and tolerances of species form the "cohesion" that produces clustering, making plant communities different than random assemblages of species.

1. Introduction

The study of plant communities has been problematic, in part not only because of the various ways available to define them, but also due to the continuing debate about whether they even exist [1, 2]. Opinion has varied from a belief in strongly interacting plant communities [3], to the more commonly accepted individualistic view of plant assemblages [4]. But even if species behave individualistically, they may still form communities due to, for example, similar responses to mechanisms and overlapping tolerances [5]. These issues continue to influence how plant ecologists think of such constructs as biomes, ecotones, and ecoclines [6].

Criteria for a plant community may be that it only has a nonrandom subset of the regional pool of available species [7]. Alternatively, a plant community may contain properties such as (1) assembly rules that filter out species and traits until a community is left with only the most well-adapted species [8], (2) niche limitation, (3) stability, (4) resilience, (5) discontinuity/discreteness, (6) self-organization, (7) emergence, (8) coevolution [1, 9–11], or (9) "integratedness" such as linkage between processes [12]. Indeed, different communities may be (1) areas with different physiognomies [2, 9], (2) areas that contain species with different C, S, or R affinities [13], or (3) areas that have different functional groups. Plant communities may even be made up simply of complementary guilds of plants that share resources, such as light, water, and soil nutrients [14].

A common approach has been to first find broad structural characteristics that all plant communities must have—such as species composition, species richness, species evenness, and biomass [14]—and then measure those characteristics in field plots. An "index" based on these characteristics may also be computed [15]. If the variation of selected traits, or said index, within a subset of plots is small compared to the variation among all plots then those plots are considered to be in the same community [16]. This methodology is implicit in multivariate ordinations which group vegetation quadrats into community types according to how far apart they are in an ordination "space" defined on axes that are correlated with specific plant species, soil factors, or other parameters measured in those quadrats [17]. The problem with this approach [18–20] is that it does not necessarily include whether or not the plant species

TABLE 1: Positive association summary.

Country/field type	Total no. of positive associations	Largest no. for any year and plot
Puerto Rico USA/pasture	342	7
Ecuador/left banana	159	4
Ecuador/right banana	133	4
Ecuador/right sugarcane	132	4
Ecuador/left sugarcane	158	4

common to the plots actually occur together over the larger spatial and temporal scales of their distributions, where they associate naturally, which is fundamental to what makes a plant community.

In this paper I (1) start with the observation that positive plant associations among species are central to defining plant communities regardless of the mechanisms and/or tolerances that produce them, (2) suggest that the key question of whether or not two plots are in the same plant community is not answerable as "yes or no" but only in terms of degree, and hence (3) compare recently sampled plots to a hierarchy, built from positive plant associations taken from many plots sampled over time from the same community type [21]. Such long-term plots are needed to observe the natural "affinity" that these plant species have for each other.

Here I show how to define a common plant community (old fields) using postagricultural data sampled in the Neotropics by first computing all significant pairwise plant species associations in plots from five abandoned pastures and plantations sampled annually for a decade and then clustering those associations into a hierarchy using an association metric of decreasing strength [22, 23]. Finally I show how the key question of whether or not two plots are in the same community can be answered using that hierarchy. Such an approach can thus be used to define any plant community because it contains degrees of integration [11] and also captures the individualistic, overlapping distributions of plants found over space on gradients [24] and over time after a disturbance [25].

2. Methods and Materials

All five study pastures and plantations are located within tropical lower montane wet forests [26] of similar plant taxa [21]. All study areas receive between 3 m and 5 m of rain annually with small seasonally variation [27], and their temperatures range between 15°C and 25°C. All soils are fertile andisols and volcaniclastic in origin [28, 29]. The study pasture was never seeded with grasses and grazed for decades before abandonment. The pasture borders the Luquillo Experimental Forest (LEF) of northeastern Puerto Rico USA, close to the town of Sabana (18°20′N, 65°45′W: [30–34]) where the LEF is a long-term ecological research (LTER) site of the National Science Foundation (http://luq.lternet.edu/). The two study banana (*Musa* sp.) plantations (named left and right for convenience) and the two study sugarcane (*Saccharum officinarum*) plantations

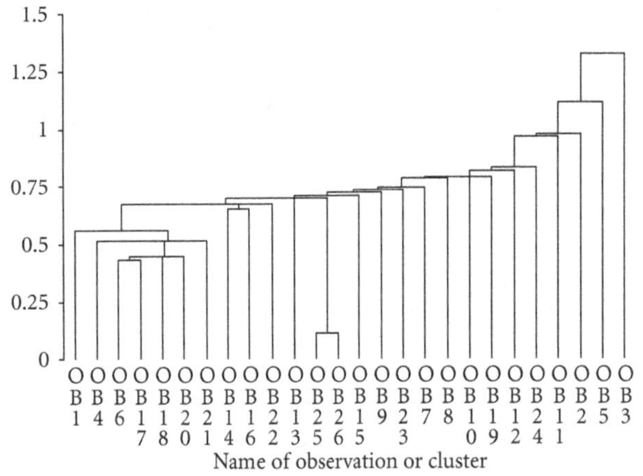

FIGURE 1: Dendrogram for pasture species (OB1, OB2, ..., OB26) which are numbered as in Table 2.

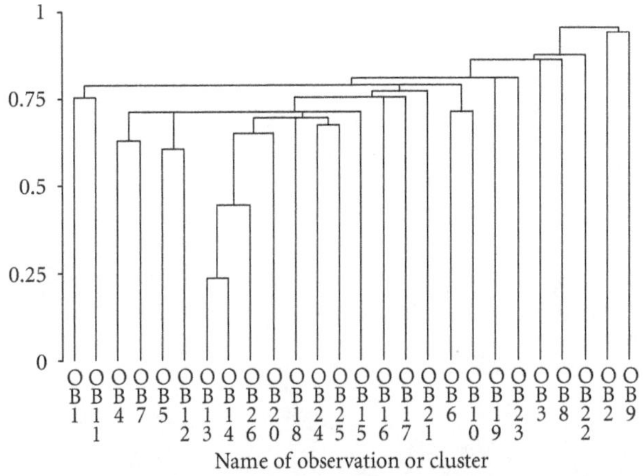

FIGURE 2: Dendrogram for left banana species (OB1, OB2, ..., OB26) which are numbered as in Table 3.

(also left and right) are located in the Maquipucuna Reserve, Ecuador (0°05′N, 78°37′W; http://www.maqui.org; [35–40]).

Within each study pasture and plantation, twenty-five 5 m × 2 m contiguous plots were laid out in 1996 [33] with the long side parallel to and bordering the forest in order to maximize any edge effects. Past analysis of this plot data [21, 32, 39] has shown them to be of sufficient size to capture community structure. No plots had any remnant trees or sprouting tree roots at the beginning of the study, and their tree seed bank was very small [34]. Starting in May of 1997, and continuing annually in May since then, each plot has been sampled for percent cover of each plant species. Percent cover—an indication of a species' ability to capture light and, therefore, to dominate these areas which are in the process of becoming forested communities [32]— was estimated visually in relation to each plot's area. Trained on-site LTER plant taxonomists were employed to identify plant species in Puerto Rico and plant taxonomists, trained at the University of Georgia where voucher specimens are kept on file [41, 42], assisted in the identification of species by using specimens located on site in Ecuador [30, 38]. The

TABLE 2: Half-matrix containing the number of significant positive spearman rank correlation coefficients among all plant species in a Puerto Rican pasture for each year over the first 10 years of succession, with a maximum of ten. Plant species and families are (1) *Bromelia* spp., (2) *Guarea guidonia*, (3) *Ocotea leucoxylon*, (4) *Citrus frutius*, (5) *Syzygium jambos*, (6) *Desmodium* spp., (7) *Gleichenia bifida*, (8) *Inga laurina*, (9) *Citrus limon*, (10) *Casearia sylvestris*, (11) *Prestoea montana*, (12) *Calophyllum calaba*, (13) *Miconia prasina*, (14) *Eugenia pseudopsidium*, (15) *Tabebuia heterophylla*, (16) *Eugenia malaccensis*, (17) *Piper hispidum*, (18) *Andira inermis*, (19) *Psychotria brachiata*, (20) *Miconia racemosa*, (21) *Psychotria berteriana*, (22) *Xanthosoma* spp., (23) *Clidemia hirta*, (24) *Panicum* spp., (25) *Myrcia splendens*, and (26) *Ocotea sintenisii*.

	2	3	4	5	6	7	8	9	10	11	12	13	14	15	16	17	18	19	20	21	22	23	24	25	26
1	1	1		2		1		2		1	3			2			1								2
2		6	2	7	2	3		2		5	2	1		3				1	1	3	1	2	2	4	2
3			1	3		1	1	2	4	2	3	1	3	2	3							2	4	4	2
4				1	2	2			1	1	1		1	1		2	1		1						
5					1	1	4	2		5	7	2	1	2	2		1		2	2	3	1	4	3	2
6						2		1		1							1					1	1		
7							1	2	2	1	4			1	1			3		1		1	2		1
8										5		1	3	4				1		1	1	1			2
9										2	1	1		2	1	3	1	1	1			1	1	1	5
10											2		2	2	1			5				2	2	2	
11														2	2	2	1	1	1	2		3	1	2	3
12												2	1	1	2	1		3				2	3	1	
13														4	2		1	4	1			2	1		
14														2	1					1		2	2		
15																			1		1				1
16																			1			2	1	3	1
17																									
18																						2			
19																				1		1			
20																				1	4			2	1
21																									
22																								2	2
23																									1
24																								4	1
25																									1
26																									

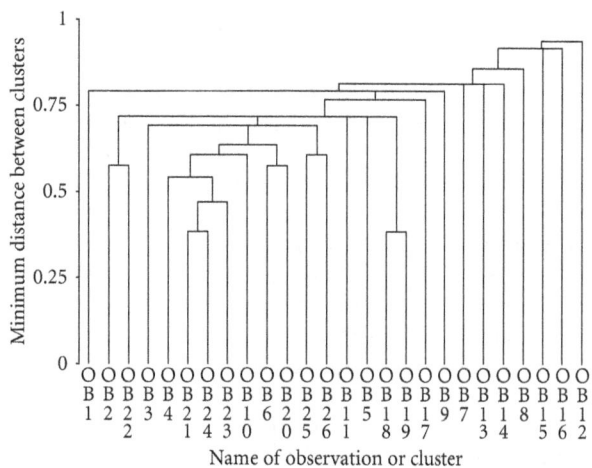

FIGURE 3: Dendrogram for right banana species (OB1, OB2,..., OB26) which are numbered as in Table 4.

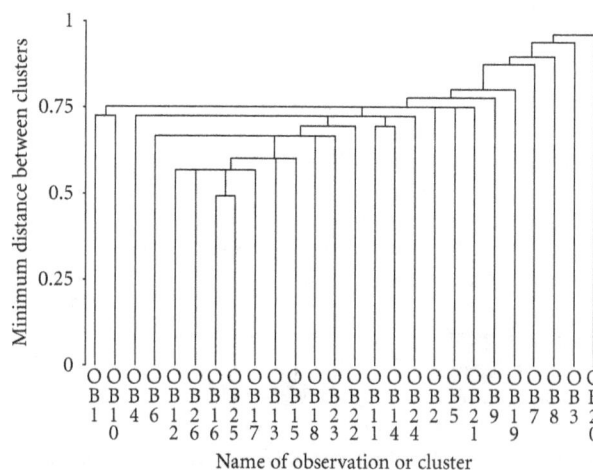

FIGURE 4: Dendrogram for right sugarcane species (OB1, OB2,..., OB26) which are numbered as in Table 5.

TABLE 3: Half-matrix containing the number of significant positive spearman rank correlation coefficients among all plant species in the Ecuador left banana plantation for each year over the first 10 years of succession. Plant species and families are (1) *Acalypha plalycephatus*, (2) *Begonia* spp., (3) *Geonoma undata*, (4) *Cyathea* spp., (5) *Musa* spp., (6) *Pilea* spp., (7) *Anthurium* spp., (8) *Trichipterix pilosissima*, (9) *Nectandra* spp., (10) *Ochroma* spp., (11) *Baccharis* spp., (12) *Anthurium* spp., (13) *Setaria* spp., (14) *Bocconia frutescens*, (15) *Piper aduncum*, (16) *Erythrina megistophyllta*, (17) *Vernonia patens*, (18) *Hedyosmum* spp., (19) *Commelina diffusa*, (20) *Alternantcera* spp., (21) *Siparuna piloso-lepidota*, (22) *Solanum* spp., (23) *Vernonia* spp., (24) *Digitaria sanguinalis*, (25) *Inga* spp., and (26) *Passiflora* spp.

	2	3	4	5	6	7	8	9	10	11	12	13	14	15	16	17	18	19	20	21	22	23	24	25	26
1	1				1	1	2		1	1		2	2						2	1					2
2		2	2	2	4	1	1		2	1	2									1	1	1			
3			3	1	1	3	1	3			1														
4					1	1	1	3				1	1												
5					2	1						1		1											1
6							2	1		2	1								1			1			
7							3	3		1												1			
8								3	2	1									1		1	2		1	
9									2									1							
10										3	2								1	1	1	1	1	1	1
11												2	2	2			1					1			1
12													1				1		1	1	1		1		
13														1											1
14																									1
15															3	1									
16																2					1				
17																	2	2					2		
18																		3	1						
19																			2	1					
20																				1	1				
21																					5				
22																							2		
23																							2	3	
24																								1	1
25																									
26																									

data from the plots in Puerto Rico (LTERDATB no. 97) and Ecuador (LTERDATB no. 101) are housed in the archives of the LEF LTER site.

First quantitative percent cover data, not presence/absence data, were used to generate pairwise Spearman coefficients of rank association [21, 26, 43, 44]. For each sampling year and field, the percent cover of any two species in each of the 25 plots (containing very few zeros) was used to compute a pair-wise association coefficient. Only the statistically significant (alpha < 0.05) positive associations are reported here but all associations, both negative and positive, can be found in [21]. Because only the first ten years of sampling data were used for each pasture and field, there is a maximum of 10 significant positive associations possible between any two plant species in Tables 1–5. This matrix of associations were then used to generate dendrograms for each separate pasture and field, after subtracting each cell value from a possible maximum of 10, using Ward agglomerative clustering [19, 43, 45] shown best for ecological data [46]. Clusters begin as single species and then form association clusters of more and more species (a hierarchy using species cooccurrence over large areas: [47]) based on a metric that becomes weaker as species form groups, eventually leading to all species clustered in one large group. Finally it should be remembered that any results given here may hold only for the original plot size.

3. Results

All pastures and fields showed a low amount of positive association in the context of the 6760 positive associations possible given all 26 species over 10 years. While no pasture/field showed more than 5% of the possible total (Table 1), the pasture in Puerto Rico had the greatest number of associations. In the Puerto Rican pasture, species that formed many positive associations included the trees *Syzygium*, *jambos*, *Guarea guidonia*, *Ocotea leucoxylon*, and *Prestoea montana* (Table 2) and for the left banana plantation of Ecuador key species with many positive associations included *Begonia* spp., *Trichipterix pilosissima*, and *Ochroma* spp. (Table 3).

TABLE 4: Half-matrix containing the number of significant positive spearman rank correlation coefficients among all plant species in the Ecuador right banana plantation for each year over the first 10 years of succession. Plant species and families are (1) *Acalypha plalycephaluss*, (2) *Costus* spp., (3) *Musa* spp., (4) *Solanum muricatum*, (5) Piperaceae, (6) *Setaria* spp., (7) *Tagetes terniflora*, (8) *Begonia* spp., (9) *Cuphea carthlagenensis*, (10) Polypodiaceae, (11) *Vernonia patens*, (12) *Brugmansia* spp., (13) *Digitaria sanguinalis*, (14) Urticaceae, (15) *Chusquea* spp., (16) *Nectandra* spp., (17) Piperaceae, (18) *Commelina diffusa*, (19) *Erythrina megistophyllta*, (20) *Heliotropium* spp., (21) *Inga* spp., (22) *Musa acuminate*, (23) *Chenopodium album*, (24) *Crataegus monogyna*, (25) *Bocconia frutescens*, and (26) *Cecropia monostachyta*.

	2	3	4	5	6	7	8	9	10	11	12	13	14	15	16	17	18	19	20	21	22	23	24	25	26
1	1					1	3	1	1			1			1	1									
2		1	1	1	2			2	1										1						
3						1		2			2	1	1	1						1					1
4								1	1			1	1		1									2	1
5						3		1	1	1				1							1				
6															1	1					2				
7							4	2	1	1						1									
8								2	1	1			1	2	1										
9									1	1	1	1			1					1	2		1		
10										1		1		1										2	1
11												3						1							
12												2	1		1	1	1	1	1	1					1
13													3	3					1		1				
14														3										1	1
15															3										
16																3	3								
17																			2	1				1	1
18																			1						
19																				1					
20																					3				
21																								1	2
22																									
23																								1	
24																									2
25																									
26																									

In the right banana plantation, *Begonia* spp., *Cuphea* spp., and *Brugmansia* spp. formed many associations (Table 4). In the left sugarcane plantation, key species included members of the families Asteraceae, Verbenaceae, and Papilionacea (Table 5), and in the other sugarcane plantation, *Cuphea* spp. and *Piper aduncum* were important (Table 6).

Clustering of the data in Table 2 (Puerto Rican pasture) showed that *Myrcia* and *Ocotea* clustered first, followed at a longer metric by *Desmodium* and *Piper*, which quickly formed a cluster with *Andira* and *Miconia*. That cluster fused then with *Citrus* and *Psychotria* and then with *Bromelia*. Then the rest of the species fused with all the previously mentioned species and clusters, except for *Prestoea*, *Syzygium* and *Ocotea* (Figure 1). Clustering of the data in Table 3 (left banana) showed that *Setaria* and *Bocconia* clustered first, united with *Passiflora* next, which clustered with *Alternanthera* sp. at about the same level as clusters form between *Musa* and *Anthurium* and between *Cyathea* and other *Anthurium*. The other species then clustered quickly, with *Begonia* and *Nectandra* forming a cluster last (Figure 2).

Clustering of the data in Table 4 (right banana) showed that *Inga* and *Crataegus* form a cluster at the same level as *Commelina* and *Erythrina*. The *Inga* cluster then fused with *Chenopodium* and later *Musa*. After that there were three clusters that formed between two species each: *Costus/Musa*, *Setaria/Heliotropium*, and *Bocconia/Cecropia*. A large cluster then formed which included all of the previously mentioned species plus *Vernonia* and *Piperaceae*. The rest of the species were added with *Chusquea*, *Nectandra*, and *Begonia* clustering last (Figure 3). Clustering of the data in Table 5 (right sugarcane) showed that *Costus* and *Columnea* formed the first cluster and it then united with *Rubus*, Orchidaceae, and *Miconia*. This cluster then united with *Commelina* and *Cecropia*, making a larger cluster that then joined with Passifloraceae and *Hieracium*. After this clustering, levels were similar among species until the end when *Piper*, *Lantana*, *Digitaria*, and *Chusquea* clustered last (Figure 4). Finally, clustering of the data in Table 6 (left sugarcane) showed that *Nectandra* and Polypodiaceae clustered first, then with Asteraceae and *Baccharis*, followed by *Sida* and *Commelina*.

TABLE 5: Half-matrix containing the number of significant positive spearman rank correlation coefficients among all plant species in the Ecuador right sugarcane plantation for each year over the first 10 years of succession. Plant species and families are (1) *Acalypha pladichephalus*, (2) Asteraceae, (3) *Digitaria sanguinalis*, (4) Polypodiaceae, (5) *Nectandra* spp., (6) *Stachys micheliana*, (7) Piperaceae, (8) *Lantana camara*, (9) Verbenaceae, (10) *Erythrina megistophyllta*, (11) *Piper aduncum*, (12) *Rubus* spp., (13) *Commelina diffusa*, (14) *Elephantopus mollis*, (15) *Cecropia* spp., (16) *Costus* spp., (17) *Miconia* spp., (18) Passifloraceae, (19) Fabaceae, (20) *Chusquea* spp., (21) Marantaceae, (22) *Pilea* spp., (23) *Hieracium* spp., (24) *Sabicea* spp., (25) *Columnea* spp., and (26) Orchidaceae.

	2	3	4	5	6	7	8	9	10	11	12	13	14	15	16	17	18	19	20	21	22	23	24	25	26
1	1	1	1	1	2		2			1	1									1	1		1		
2			1	1	1	4		2														1	2	1	
3														1							2	2	3		
4				1						1		1	2	2		1			1				1		
5					1	2		2																	1
6						1		2					1				1	1	1				1		
7							1			1							1					1			
8							1	3									1	1	2	2					
9											1		2	1					1				1		
10										1	1		2						1	1					
11											2						1						1		1
12													2									1		1	
13														1	1				3	1					
14														1									1	1	1
15																			1				1		
16																		1	1						
17																		3					1		
18																		2	1						1
19																			2	2	1	1	1	2	
20																				3	1			1	
21																						1	1		
22																									
23																							1		
24																									
25																									1
26																									

At the same level *Miconia* and *Vernonia* cluster and all of these species then join to be added with *Piper*, *Rubus*, *Polpyodiaceae*, and *Saccharum*. Finally the last species to cluster were *Piper*, *Orchidaceae*, and *Cuphea* (Figure 5).

4. Discussion

Because most of the significant associations in the study plot data were positive—unlike the mainly negative associations that were computed from plots in temperate old fields (using the same plot grid layout, sampling protocol, and analysis [26])—facilitation in these stressful, early successional fields may be more important than competition (also see [48]) which would challenge ecological paradigms regarding the pervasiveness of competition [49, 50].

The Puerto Rican pasture is different from the other fields with both different species and a different clustering pattern, although *Miconia* does cluster early here and in both sugarcane fields. Unfortunately without replication the cause of this difference—for example, it is because it is a pasture, because it revegetated naturally rather than was seeded with grass, because it is an island, or because it is Puerto Rico—cannot be determined. However both replicate banana fields in Ecuador show (1) *Musa* (their past crop) clustering in the middle of the pack of species and (2) that *Begonia* and *Nectandra* are the last two species to cluster. This suggests that recovering banana fields have distinct communities. In the sugarcane fields, (1) the past crop *Saccharum* is not as persistent as *Musa* was in the banana fields, (2) *Miconia* and *Commelina* clustered in both fields but at different levels, (3) *Acalypha* and *Erythrina* clustered in the middle of the pack, and (4) *Piper* clustered last in both fields. Consequently evidence for repeatable communities occurs in recovering sugarcane and banana plantations, but it is stronger in the banana plantations. In general, species groupings do not suggest taxonomic or obvious ecological (e.g., dispersal vector, seed size, shade tolerance) similarities, but there is a suggestion that the past crop not only persists after abandonment but also forms associations with invading plants [26].

TABLE 6: Half-matrix containing the number of significant positive spearman rank correlation coefficients among all plant species in the Ecuador left sugarcane plantation for each year over the first 10 years of succession. Plant species and families are: (1) *Musa* spp., (2) *Costus* spp., (3) *Cuphea carthagenensis*, (4) *Digitaria sanguinalis*, (5) *Miconia* spp., (6) *Piper* spp., (7) *Rubus* spp., (8) *Sida rhombifolia*, (9) Asteraceae, (10) *Baccharis* spp., (11) Polypodiaceae, (12) *Lantana camara*, (13) *Vernonia patens*, (14) *Acalypha pladichephalus*, (15) *Solanum* spp., (16) *Saccharum officinarum*, (17) *Piper aduncum*, (18) Verbenaceae, (19) *Commelina diffusa*, (20) *Erythrina megistophyllta*, (21) *Nectandra* spp., (22) *Altus* spp., (23) Orchidaceae, (24) *Polybotrya* spp., (25) *Vernonia* spp., and (26) Polypodiaceae.

	2	3	4	5	6	7	8	9	10	11	12	13	14	15	16	17	18	19	20	21	22	23	24	25	26
1		1				1			1		1			1	2	3	1		1				1		1
2	2	2	2											1			1				1	3	1	1	
3		3	1	4	2	1				3											1			1	
4			1	3	2	1	1				1														
5				1	1			1			1												1	1	
6					1					1			1				1						1		
7						1		2	1							1	1	1					1	1	
8							1						1		1		1						1	1	
9								1							2					1		1	1	1	1
10									1	1		1									1				
11														1							1				
12											4	1				1					1		1	1	
13											1	2				2		1							2
14											2	1	1	1						2	1				
15												1	2	2	1				1						
16													1	2					1						
17														1						4		1			
18																	2		1		1				
19																			1	1	1				
20																			4	2			1	2	
21																				1					
22																									
23																						3			
24																							3		
25																									
26																									

Results suggest that whether two plots are in the same plant community is not a "yes/no" proposition but rather a level of a hierarchy derived from the significant positive associations among the constituent plants themselves. This reflects the known individualistic, overlapping plant distribution patterns over both space (on gradients: [25]) and time (after a disturbance: [51]) where tolerances and mechanisms [21] produce the "cohesion" that makes plant communities something different than random assemblages of species.

How then should someone decide if two plots are in the same plant community, that is, how should we define a plant community? I suggest first deciding which plot species to focus on in the association analysis and then consulting the association hierarchies derived from long-term, repeated sampling of the same kind of communities (here, postagricultural) to find the level of association needed to cluster those species together. Defining what level defines a community is, of course, a basic issue in clustering methodologies [19] where taking all species as defining their own individual communities, or defining

only one community which contains all species, is not ecologically meaningful. One possible way to choose this level of association intensity could be based on the ecology and biology of the species themselves and/or the ecosystem they are found in. Alternatively, one may look for "cleavage points" in the clustering pattern where the break between groups is most clear, thereby making communities most distinct from each other. Or one may simply say that the association clustering pattern "is" the plant community. Finally it should be noted that association hierarchies may define communities that do not currently exist but may exist, or could have existed, at some other place and time.

In this paper I began with the observation that positive plant associations among species are central to defining plant communities regardless of the mechanisms and/or tolerances that produce them. I then suggested that the key question of whether or not two plots are in the same plant community is not answerable with "yes or no" but only in terms of degree. This leads to the construction of a hierarchy, built from positive plant associations after decades of plot sampling in

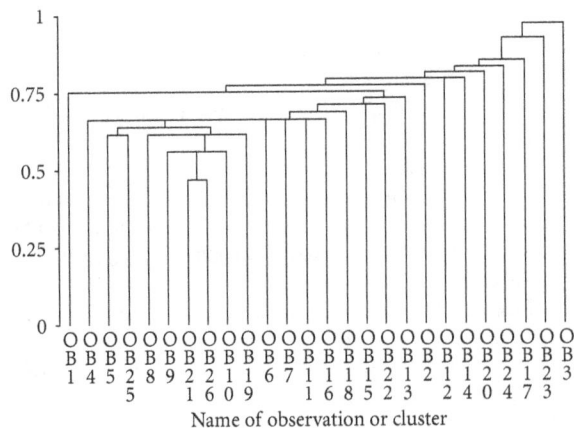

FIGURE 5: Dendrogram for left sugarcane species (OB1, OB2,..., OB26) which are numbered as in Table 6.

the same kind of community (old fields) using many plots located in different areas and using an association metric of decreasing strength. Only then can recently sampled plots be compared to the hierarchy to decide whether or not they are in the same community.

The association hierarchy and its clustering metric can be interpreted as showing (1) assembly rules defining a colonization process of permissible or forbidden species combinations [52], (2) functional groups where positive association means that species respond similarly to environmental factors and gradients [26], or (3) intrinsic "guilds" built up from community data [10]. Using this postagricultural data set we may also be able to address whether similar species group together regardless of whether they are in pasture, banana, or sugarcane (i.e., community convergence or divergence: [10]). Finally I suggest that future community investigations follow the sampling protocol of this data set with a hierarchy containing enough plots and a long enough sampling time to allow for significant individual plant-plant associations to develop as plants replace each other over time [21, 53]. Such an approach makes it much more likely that the species groupings that define actual plant communities will be found.

Acknowledgments

The authors thanks B. Witteveen for the help at Maquipucuna Reserve, N. Brokaw for help at LEF, and E. Melendez for the help with the LTER data sets. he also thanks M. Palmer, H. Bruelheide, and P. Burton for commenting on a previous draft of the paper. The author received support from Grant DEB-0218039 from the National Science Foundation to the Institute of Tropical Ecosystem Studies, University of Puerto Rico, and the USDA Forest Service, International Institute of Tropical Forestry as part of the Long-Term Ecological Research Program in the Luquillo Experimental Forest. Additional support was provided by the Forest Service (US Department of Agriculture) and the University of Puerto Rico.

References

[1] J. B. Wilson, "Does vegetation science exist?" *Journal of Vegetation Science*, vol. 2, pp. 289–290, 1991.

[2] M. W. Palmer and P. S. White, "On the existence of ecological communities," *Journal of Vegetation Science*, vol. 5, no. 2, pp. 279–282, 1994.

[3] F. E. Clements, *Plant Succession*, Carnegie Institute Washington Publication, Washington, DC, USA, 1916.

[4] H. A. Gleason, "The individualistic concept of the plant association," *Torrey Botanic Society Journal*, vol. 53, pp. 1–20, 1926.

[5] N. E. Zimmermann and F. Kienast, "Predictive mapping of alpine grasslands in Switzerland: species versus community approach," *Journal of Vegetation Science*, vol. 10, no. 4, pp. 469–482, 1999.

[6] R. W. Myster, *Ecotones between Forest and Grassland*, Springer, Berlin, Germany, 2012.

[7] S. L. Pimm, *The Balance of Nature? Ecological Issues in the Conservation of Species and Communities*, University of Chicago Press, Chicago, Ill, USA, 1991.

[8] P. A. Keddy, "Assembly and response rules: two goals for predictive community ecology," *Journal of Vegetation Science*, vol. 3, pp. 157–164, 1992.

[9] H. B. Johnson and H. S. Mayeux, "Viewpoint: a view on species additions and deletions and the balance of nature," *Journal of Range Management*, vol. 45, no. 4, pp. 322–333, 1992.

[10] C. L. Samuels and J. A. Drake, "Divergent perspectives on community convergence," *Trends in Ecology and Evolution*, vol. 12, no. 11, pp. 427–432, 1997.

[11] C. J. Lortie, R. W. Brooker, P. Choler et al., "Rethinking plant community theory," *Oikos*, vol. 107, no. 2, pp. 433–438, 2004.

[12] S. M. Scheiner, "An epistemology for ecology," *Bulletin of the Ecological Society of America*, vol. 74, pp. 17–21, 1993.

[13] J. P. Grime, *Plant Strategies and Vegetation Processes*, John Wiley & Sons, New York, NY, USA, 1979.

[14] A. J. Underwood, "What is a community?" in *Patterns and Processes in the History of Life*, D. M. Raup and D. Jablonski, Eds., pp. 351–368, Springer, New York, NY, USA, 1986.

[15] L. Tichý, "New similarity indices for the assignment of relevés to the vegetation units of an existing phytosociological classification," *Plant Ecology*, vol. 179, no. 1, pp. 67–72, 2005.

[16] N. W. H. Mason, M. Kit, J. B. Steel, and J. B. Wilson, "Do plant modules describe community structure better than biomass? A comparison of three abundance measures," *Journal of Vegetation Science*, vol. 13, no. 2, pp. 185–190, 2002.

[17] J. F. Duivenvoorden, "Tree species composition and rain forest-environment relationships in the middle Caqueta area, Colombia, NW Amazonia," *Vegetatio*, vol. 120, no. 2, pp. 91–113, 1995.

[18] R. H. Whittaker, *Ordination and Classification of Communities*, Junk, The Hague, The Netherlands, 1973.

[19] J. A. Ludwig and J. F. Reynolds, *Statistical Ecology: A Primer on Methods and Computing*, John Wiley & Sons, New York, NY, USA, 1988.

[20] P. Legendre and L. Legendre, *Numerical Ecology*, Elsevier, New York, NY, USA, 1998.

[21] R. W. Myster, *Post-Agricultural Succession in the Neotropics*, Springer, Berlin, Germany, 2007.

[22] R. V. O'Neill, R. H. Gardner, B. T. Milne, M. G. Turner, and B. Jackson, "Heterogeneity and spatial hierarchies," in *Ecological Heterogeneity*, J. Kolasa and S. T. A. Pickett, Eds., pp. 85–96, Springer, Berlin, Germany, 1991.

[23] R. R. Cook and J. F. Quinn, "An evaluation of randomization models for nested species subsets analysis," *Oecologia*, vol. 113, no. 4, pp. 584–592, 1998.

[24] R. H. Whittaker, *Communities and Ecosystems*, Macmillan, New York, NY, USA, 1975.

[25] R. W. Myster and S. T. A. Pickett, "Dynamics of associations between plants in ten old fields during 31 years of succession," *Journal of Ecology*, vol. 80, no. 2, pp. 291–302, 1992.

[26] J. Edmisten, "Some autoecological studies of *Ormosia krugii*," in *A Tropical Rainforest*, H. T. Odum and F. A. Pigeon, Eds., pp. 24–35, Oak Ridge, Tenn, USA, Atomic Energy Commission, 1970.

[27] R. W. Myster, "Regeneration filters in post-agricultural fields of Puerto Rico and Ecuador," *Plant Ecology*, vol. 172, no. 2, pp. 199–209, 2004.

[28] J. R. Thomlinson, M. I. Serrano, T. D. M. López, T. M. Aide, and J. K. Zimmerman, "Land-use dynamics in a post-agricultural Puerto Rican landscape (1936–1988)," *Biotropica*, vol. 28, no. 4, pp. 525–536, 1996.

[29] R. W. Myster and F. O. Sarmiento, "Seed inputs to microsite patch recovery on two tropandean landslides in Ecuador," *Restoration Ecology*, vol. 6, no. 1, pp. 35–43, 1998.

[30] T. M. Aide, J. K. Zimmerman, L. Herrera, M. Rosario, and M. Serrano, "Forest recovery in abandoned tropical pastures in Puerto Rico," *Forest Ecology and Management*, vol. 77, no. 1–3, pp. 77–86, 1995.

[31] R. W. Myster, "Seed regeneration mechanisms over fine spatial scales on recovering Coffee plantation and pasture in Puerto Rico," *Plant Ecology*, vol. 166, no. 2, pp. 199–205, 2003.

[32] R. W. Myster, "Vegetation dynamics of a permanent pasture plot in Puerto Rico," *Biotropica*, vol. 35, no. 3, pp. 422–428, 2003.

[33] R. W. Myster, "Effects of species, density, patch-type, and season on post-dispersal seed predation in a Puerto Rican pasture," *Biotropica*, vol. 35, no. 4, pp. 542–546, 2003.

[34] R. W. Myster, "Shrub vs. grass patch effects on the seed rain and the seed bank of a five-year pasture in Puerto Rico," *Ecotropica*, vol. 12, pp. 12–19, 2006.

[35] F. O. Sarmiento, "Arrested succession in pastures hinders regeneration of Tropandean forests and shreds mountain landscapes," *Environmental Conservation*, vol. 24, no. 1, pp. 14–23, 1997.

[36] C. C. Rhoades, G. E. Eckert, and D. C. Coleman, "Effect of pasture trees on soil nitrogen and organic matter: implications for tropical montane forest restoration," *Restoration Ecology*, vol. 6, no. 3, pp. 262–270, 1998.

[37] C. C. Rhoades and D. C. Coleman, "Nitrogen mineralization and nitrification following land conversion in montane Ecuador," *Soil Biology and Biochemistry*, vol. 31, no. 10, pp. 1347–1354, 1999.

[38] R. A. Zahawi and C. K. Augspurger, "Early plant succession in abandoned pastures in Ecuador," *Biotropica*, vol. 31, pp. 123–129, 1999.

[39] R. W. Myster, "Post-agricultural invasion, establishment, and growth of neotropical trees," *Botanical Review*, vol. 70, no. 4, pp. 381–402, 2004.

[40] R. W. Myster, "Early successional pattern and process after sugarcane, banana, and pasture cultivation in Ecuador," *New Zealand Journal of Botany*, vol. 45, no. 1, pp. 35–44, 2007.

[41] A. H. Gentry, *A Field Guide to the Families and Genera of Woody Plants of Northwest South America*, University of Chicago Press, Chicago, Ill, USA, 1996.

[42] A. H. Liogier and L. F. Martorell, *Flora of Puerto Rico and Adjacent Islands: A Systematic Synopsis*, University of Puerto Rico, Río Piedras, Puerto Rico, 1999.

[43] *SAS User's Guide: Statistics*, SAS Institute, Cary, NC, USA, 5th edition, 1985.

[44] A. Milbau, I. Nijs, I. F. De Raedemaecker, D. Reheul, and B. De Cauwer, "Invasion in grass-land gaps: the role of neighborhood richness, light availability and species complementarily during two successive years," *Functional Ecology*, vol. 19, pp. 27–37, 2005.

[45] M. De Cáceres, X. Font, P. Vicente, and F. Oliva, "Numerical reproduction of traditional classifications and automatic vegetation identification," *Journal of Vegetation Science*, vol. 20, no. 4, pp. 620–628, 2009.

[46] W. Singh, *Robustness of three hierarchical agglomerative clustering techniques for ecological data*, M.S. thesis, in Environment and Natural Resources, University of Iceland, Reykjavik, Iceland, 2008.

[47] E. T. Azeria, D. Fortin, C. Hébert, P. Peres-Neto, D. Pothier, and J. C. Ruel, "Using null model analysis of species co-occurrences to deconstruct biodiversity patterns and select indicator species," *Diversity and Distributions*, vol. 15, no. 6, pp. 958–971, 2009.

[48] E. Ruprecht, S. Bartha, Z. Botta-Dukát, and A. Szabó, "Assembly rules during old-field succession in two contrasting environments," *Community Ecology*, vol. 8, no. 1, pp. 31–40, 2007.

[49] D. Tilman, *Plant Strategies and the Dynamics and Structure of Plant Communities*, Princeton University Press, Princeton, NJ, USA, 1988.

[50] J. F. Bruno, J. J. Stachowicz, and M. D. Bertness, "Inclusion of facilitation into ecological theory," *Trends in Ecology and Evolution*, vol. 18, no. 3, pp. 119–125, 2003.

[51] R. W. Myster and S. T. A. Pickett, "Individualistic patterns of annuals and biennials in early successional oldfields," *Vegetatio*, vol. 78, no. 1-2, pp. 53–60, 1988.

[52] L. Stone, T. Dayan, and D. Simberloff, "Community-wide assembly patterns unmasked: the importance of species' differing geographical ranges," *American Naturalist*, vol. 148, no. 6, pp. 997–1015, 1996.

[53] R. W. Myster, "Plants replacing plants: the future of community modeling and research," *Botanical Review*. In press.

16

World Aquaculture: Environmental Impacts and Troubleshooting Alternatives

Marcel Martinez-Porchas[1] and Luis R. Martinez-Cordova[2]

[1] Departamento de Tecnología de Alimentos de Origen Animal, Centro de Investigación en Alimentación y Desarrollo, Km. 0.7 Carretera a La Victoria, Hermosillo, SON, Mexico
[2] Departamento de Investigaciones Científicas y Tecnológicas de la Universidad de Sonora, Boulevard Luis Donaldo Colosio s/n, 83000 Hermosillo, SON, Mexico

Correspondence should be addressed to Luis R. Martinez-Cordova, lmtz@guaymas.uson.mx

Academic Editors: E. Gilman and J. Kotta

Aquaculture has been considered as an option to cope with the world food demand. However, criticisms have arisen around aquaculture, most of them related to the destruction of ecosystems such as mangrove forest to construct aquaculture farms, as well as the environmental impacts of the effluents on the receiving ecosystems. The inherent benefits of aquaculture such as massive food production and economical profits have led the scientific community to seek for diverse strategies to minimize the negative impacts, rather than just prohibiting the activity. Aquaculture is a possible panacea, but at present is also responsible for diverse problems related with the environmental health; however the new strategies proposed during the last decade have proven that it is possible to achieve a sustainable aquaculture, but such strategies should be supported and proclaimed by the different federal environmental agencies from all countries. Additionally there is an urgent need to improve legislation and regulation for aquaculture. Only under such scenario, aquaculture will be a sustainable practice.

1. Introduction

Aquaculture, the farming of aquatic organisms, has been the agroindustrial activity with the highest growth rate worldwide in the last four decades. From 1970 to 2008 the production of aquaculture organisms grew at a rate of 8.3% per year, compared to less than 2% of fisheries, and 2.9% of livestock [1]. The annual aquaculture production is at present over 60 million tons (including marine plants), with an approximate value of 85 billion dollars [2]. The last FAO report revealed that the world population increased by 6.3% from 2004 to 2009, whereas the production of aquatic organisms by aquaculture increased by 31.5% in the same period (Figure 1) [2].

Despite the undeniable benefits of aquaculture such as the provision of good quality and accessible food for population and the generation of millions of jobs and billion dollars in budget for the developing countries, the activity is one of the most criticized worldwide, mainly because of the environmental impacts that have been and can be caused. Thus, the predominant and unavoidable question is: could aquaculture be a truly sustainable activity?

Understanding sustainability as "the ability to meet the needs of the present without compromising the ability of the future generations to meet their own needs" [3], many researchers, aquaculturists, and governmental instances have considered that a sustainable aquaculture is possible, but it depends on the way that the activity will be managed [4]. Additionally, other authors [5] have argued that "the sustainability of aquaculture not only requires neutral or benign effects on the environment, but also economic feasibility."

The present paper is a review of the world aquaculture and its environmental impacts. It analyzes the situation of aquaculture production up to date and summarizes the main problems faced by the activity, as well as the strategies suggested, evaluated, and proven to contribute to achieve a sustainable activity.

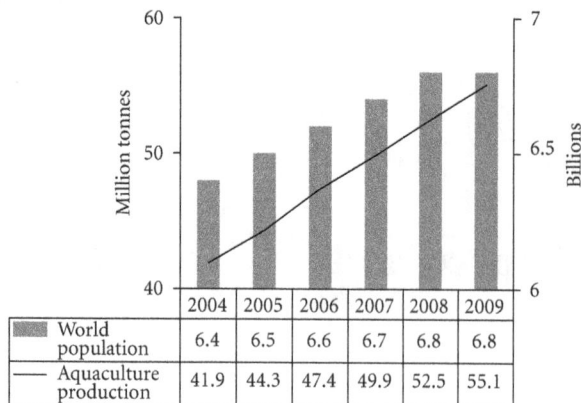

(a)

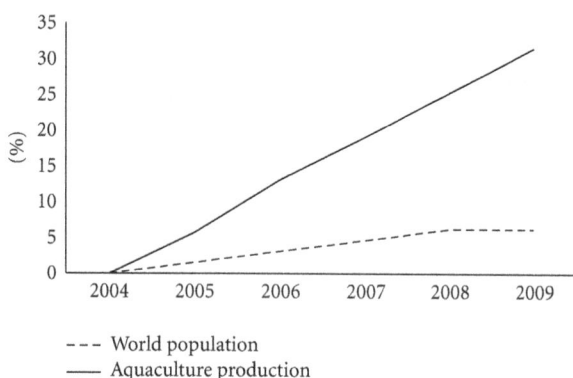

--- World population
—— Aquaculture production

(b)

FIGURE 1: Growth behavior of world population and aquaculture production during the last six years. (a) illustrates the total world population by year (billions) and the total production of aquatic organisms by aquaculture (million tonnes). (b) compares the percentage of annual increase of world population and aquaculture production, considering year 2004 as the basepoint. Data obtained from FAO Report 2010: World Review of Fisheries and Aquaculture.

Aquaculture Is an Essential Activity for World's Welfare. Aquaculture is considered as a double-edged sword, because it has not only tremendous benefits for the humanity, but also great repercussions to the environment. Considering the benefits, seafood produced by fisheries and aquaculture contributes with 15 to 20% of average animal protein consumption to 2.9 billion people worldwide [6] without considering the contribution of freshwater or brackish water species. The nutritional quality of aquatic products has a high standard and represents an important source of macro- and micronutrients for people from developing countries [7]. Additionally, aquaculture and fisheries are recognized as a source of employment; for instance, near to 43.5 million people were employed in 2006, and 520 million people relied on income from seafood production [6, 8].

Aquaculture products have also high trade potential as food commodities in the international market; fish and shellfish exports from developing countries have a greater value than the combination of important products such as coffee, tea, tobacco, meat, cocoa, rubber, and rice [6, 8]. In many

cases, the incomes generated by aquaculture exceed those from other agricultural activities, due to the high price market that some products can achieve and due to the most effective bioenergetics of some aquatic species. However, despite all these benefits, aquaculture is actually not considered a sustainable activity in the perception of the scientific community and the average population.

2. Why Aquaculture Is Considered a Nonsustainable Activity?

With or without valid arguments, aquaculture has been accused to be the cause of many environmental, social, economic, and inclusively esthetic problems. Ecosystems are not always as fragile as could be considered, instead, they have remarkable capacity of resiliency, and as long as basic processes are not irretrievably upset, ecosystems will continue to recycle and distribute energy [9]. However, irreversible damages have been already caused due to inadequate management of the activity. The main negative impacts attributed to the activity are as follows.

(1) Destruction of Natural Ecosystems, In Particular Mangrove Forests to Construct Aquaculture Farms [4, 10, 11]. The mangrove forests are important ecosystems considered as the main source of organic matter to the coastal zone [12, 13]; they are also nursery areas for many aquatic species ecologically and/or economically important, as well as refuge or nesting areas for bird, reptiles, crustaceans, and other taxonomic groups [14]. Mangroves are additionally accumulation sites for sediments, contaminants, nitrogen, carbon and offer protection against coastal erosion [15]. According to environmentalists [16], mangroves support diverse local fisheries and also provide critical nursery habitat and marine productivity which support wider commercial fisheries. These forests also provide valuable ecosystem services that benefit coastal communities, including coastal land stabilization and storm protection.

The cover of mangrove forest has decreased worldwide from 19.8 million hectares in 1980 to less than 15 millions in 2000. The annual deforestation rate was 1.7% from 1980 to 1990 and 1.0% from 1990 to 2000 [17], and the problem continues up today. Some authors have documented that aquaculture has been responsible for the deforestation of millions hectares of mangrove forest in Thailand, Indonesia, Ecuador, Madagascar, and other countries [18, 19]. From 1975 to 1993, the construction of shrimp farms in Thailand diminished the mangrove cover from 312,700 to 168,683 ha [20]. Philippines has reconverted 205,523 ha of mangrove and wetlands into aquaculture farms, Indonesia 211,000 ha, Vietnam 102,000 ha, Bangladesh 65,000 ha, and Ecuador 21,600 ha [21].

(2) Salinization/Acidification of Soils. Aquaculture farms are sometimes abandoned by multiple problems (operative, economic, sanitary, and etc.), and the soil from those former farms remain hypersaline, acid and eroded [22]. Therefore, those soils cannot be used for agricultural purposes and are

unusable for long periods. In addition, the application of lime and other chemicals used in aquaculture to treat the soil can also modify its physicochemical characteristics, which could aggravate the problem [23].

(3) Pollution of Water for Human Consumption. Although few studies have been conducted in relation with such topic, there are some signs indicating that inland aquaculture has been responsible for the deterioration of water bodies used for human consumption [21]. For instance, preliminary calculations revealed that an intensive aquaculture system farming three tons of freshwater fish can be compared, in respect to waste generation, to a community of around 240 inhabitants [24].

Although most of the aquaculture farms produce marine species, there is a growing sector of aquaculture farms producing freshwater species, which is a point of concern considering the above information.

(4) Eutrophication and Nitrification of Effluent Receiving Ecosystems. The eutrophication or organic enrichment of water column is mainly produced by nonconsumed feed (especially due to overfeeding), lixiviation of aquaculture feedstuffs [25, 26], decomposition of died organisms, and overfertilization [27–30]. It is well documented that from the total nitrogen supplemented to the cultured organisms, only 20 to 50% is retained as biomass by the farmed organisms, while the rest is incorporated into the water column or sediment [31, 32], and eventually discharged in the effluents toward the receiving ecosystems, causing diverse impacts such as phytoplankton blooms (sometimes of toxic microalgaes, such as red tides) [33], burring, and death of benthic organisms, as well as undesirable odors and the presence of pathogens in the discharge sites [34]. The impact may be more or less severe depending on some factors such as the intensification of the system (density of organisms), which is directly related to the amount of feed supplied [26, 35]. The feed conversion ratio (FCR) is a well indicator of the effectiveness of feeding and, consequently, of the retention of nitrogen and carbon as biomass of the farmed organisms. For instance, farms culturing the tiger shrimp *Penaeus monodon* usually report FCRs ranging from 1 to more than 2.5; such huge difference is later reflected in the amount of organic matter, nitrogen, and phosphorous discharged in the effluents, which may range from 500 to 1625 kg, 26 to 117 kg, and 13 to 38 kg, respectively, for each ton of shrimp harvested [28]. The estimated mean FCR worldwide for shrimp aquaculture is 1.8, which means that, for a world annual shrimp production around 5 million tons, 5.5 million tons of organic matter, 360,000 tons of nitrogen, and 125,000 tons of phosphorous are annually discharged to the environment. Unfortunately, these data considers only shrimp production, which represents around 8% of the total aquaculture production; if we assume that the FCRs are similar for the other farmed organisms and the diet formulations have some similitude [36], the total discharge of wastes may be multiplied by 12.5 from a very preliminary perspective. The nutrification is considered as the nutrient (N, P, C) enrichment of water column, mainly due to

fertilization, mineralization of organic matter, resuspension of sediments, and excretion of organisms into the ponds. The greatest concern in this aspect is the increasing production of nitrogenous metabolites especially ammonia, which is highly toxic in its unionized form (NH_3) for many aquatic organisms [37].

(5) Ecological Impacts in Natural Ecosystems because of the Introduction of Exotic Species. The negative impacts of the "biological contamination" for the introduction of exotic aquacultural species on the native populations have been well documented [18, 38, 39]. The main reported problems are the displacement of native species, competition for space and food, and pathogens spread. To cite an example, recent reports have revealed a parasite transmission of sea lice from captive to wild salmon [40]. The authors of such study have hypothesized that "if outbreaks continue, then local extinction is certain, and a 99% collapse in pink salmon abundance is expected in four salmon generations."

(6) Ecological Impacts Caused by Inadequate Medication Practices. Farmers usually expose their cultured organisms to medication regimes, for different purposes such as avoiding disease outbreaks and improving growth performance. However, monitoring studies have detected low or high levels of a wide range of pharmaceuticals, including hormones, steroids, antibiotics, and parasiticides, in soils, surface waters, and groundwaters [41]. These chemicals have caused imbalances in the different ecosystems. In particular, the use of hormones in aquaculture and its environmental implications have been scarcely studied.

(7) Changes on Landscape and Hydrological Patterns. The agricultural and aquacultural activities have contributed to the degradation of ecosystems including important modification on landscape [10, 18, 22, 42]. The construction of shrimp farms in the river beds has modified the hydrological patterns in many regions of the world with the consequent impacts on the regional ecosystems and the local weather.

(8) Trapping and Killing of Eggs, Larvae, Juveniles, and Adults of Diverse Organisms. It has been estimated that, for each million of shrimp postlarvae farmed, four to seven millions of other organisms are killed by trapping in the nets of farms inlet [18, 43].

(9) Negative Effect on Fisheries. Although aquaculture has been proclaimed as a solution to avoid overfishing, it has contributed in more or less proportion to the fisheries collapse. Fishermen who work in places near to aquaculture farms argue that the contamination produced by farms has decreased the population of aquatic organisms and in consequence their volume captures. Additionally, another problem of similar magnitude is the extremely high aquaculture's dependence of fishmeal and fish oil, which could be another nonsustainable practice in aquaculture. The proportion of fishmeal supplies used for fish production have increased from 10% in 1988 to more than 30% in the last years, which

classifies aquaculture as a potential promoter of the collapse of fisheries stocks worldwide [24].

(10) Some Other Accusations. Some other accusations for aquaculture include the production of fish and shellfish with high concentrations of toxins and/or heavy metals; genetic pollution and infestation of nondesirable phytoplankton and/or zooplankton species [44–47].

(11) In Its Role as Food Producer, Aquaculture Is Far from Complying an Adequate Distribution of Food. Overlaying net exports, governance, and undernourishment suggest that "seafood's contribution as a source of protein and livelihood is precarious" [6]. Moreover, it has been revealed that some countries with undernourishment and weak governance usually play a role as baler of seafood from countries well nourished and with strong economic capacity [6].

For these above reasons remains a generalized perception that the sustainability of aquaculture is at present being threatened or, in some cases, far from being reached.

3. What to Do for a Sustainable Aquaculture?

Many strategies have been suggested, evaluated, and/or proven in order to advance in the sustainability of aquaculture. Basically, all of them respond to the criticisms and are possible solutions to the problems attributed to the activity. The main aspects that have to be performed to advance toward such goal are the correct selection of the farming sites and species; the implementation of the most adequate culture system; use of the best feed and feeding practices; the use of bioremediation systems; decreasing the dependence of fishmeal and fish oil; adequate management of the effluents; achieving certification of compliance with sustainability; improving research and legislation related to evaluation and solutions for aquaculture impacts.

(1) In the context of the site selection it is necessary to consider the following.

(a) The vocation of the selected site. It would be absurd to select a site for aquaculture purposes if it is excellent for agriculture or livestock. Unfortunately, this is the case in many regions of the world, where agricultural lands have been reconverted to aquaculture farms. The vocation of a selected site is determined for many aspects (which can change from region to region) such as physical and chemical soil characteristics; water availability, soil fertility, topography, wild vegetal and animal communities, proximity to cities, towns, tourism zones, and so forth; priorities of the region or country (food, fuels, tourism budget, aquaculture budget, and etc.).

(b) The carrying capacity of the water bodies from the sites considered to supply the farms or used as effluent discharge places. It is very important to evaluate how much water can be taken from a particular water body or how much effluents it can receive without important alterations on its ecological equilibrium

[48]. The use of advanced technologies such as remote sensing could be an excellent auxiliary in this field [11].

(2) For the selection of species it is crucial to consider the following.

(a) It is always better to select native instead of exotic species. The introduction of exotic species causes many and diverse problems as mentioned in the previous section. Additionally, the obtaining and maintenance of broodstock of exotic species could be difficult and expensive.

(b) It is necessary to have the most possible knowledge about the biology and ecology of the organism that is pretended to be farmed (life cycle, feeding habits and nutritional requirements, tolerance to environmental parameters, and etc.).

(c) It is important to select organisms with a good market and price when farmed for commercial purposes.

(3) Regarding implementation of the best culture system, the main aspects to consider include the following.

(a) The type and size of farming structure [49]. Depending on the species, intensity, land and water availability, and economic investment, it is possible to use different types of farming structures for the culture of the same species or group. Some of them are more adequate and sustainable. For the case of shrimp farming, for instance, it has been suggested that floating or submerged cages could have a lower impact on the environment than earthen ponds. The same suggestion is applicable for culture of fishes or mollusks. Regarding size of production units, small ponds or farming structure is easier to manage in aspects such as feeding, monitoring, cleaning, pond bottom management, and harvesting. Such considerations usually lead to lower environmental impacts.

(b) Intensity. The stocking density and the consequent biomass harvested are absolutely related to the sustainability of aquaculture. The increase of the intensity implies an increase in the supplemental feed and in consequence, in the organic matter, nitrogen, and phosphorous in the effluents. Additionally, intensive or super intensive systems require the use of diverse chemicals (antibiotics, algaecides, parasiticides, and etc.), which also contribute to increasing the pollution [50]. The most adequate intensity depends on the land and water availability, as well as the carrying capacity of the water body or terrestrial ecosystems which will receive the effluents. However, recalculating and zero water exchange systems can eliminate the environmental impact while maintaining extremely high densities of aquatic organisms. Promising results have been achieved in the culture of fish and crustaceans using biofloc systems with zero water exchange [51].

(c) An adequate design of the water inlet and outlet systems, considering the water quality, weather conditions, marine currents and tide patterns (for sea water), and hydrological patterns (for continental waters) [52]. The modifications of oceanic currents patterns may have implications on the sediment transport and consequently on the beaches conformation.

(d) The possibility of farming simultaneously two or more species (polycultures or integrated multitrophic aquaculture (IMTA)). This strategy has proven to be one of the most effective ways to recuperate the carbon, nitrogen, and phosphorous supplied to the system as biomass of the farmed organisms and to diminish the environmental impacts caused by the effluents [53–55]. Polyculture is commonly referred to organisms of the same environment (marine, brackish wáter, or continental waters) and trophic level, while IMTA is mostly referred to organisms from different trophic levels and inclusively different environments. The implementation of such alternative systems improves the nutrient cycling within the culture units. In short, while in a traditional aquaculture system, 25 to 35% of the nitrogen supplied is recuperated as biomass of the farmed organisms, in a polyculture or IMTA, the recuperation could be increased by more than 50%. A pilot project made aware and informed a group of participants about the benefits of IMTA; the authors revealed that 50% of the participants were willing to pay an extra 10% of products labeled as of "IMTA products." Moreover, the authors were optimistic regarding the social impacts caused by the implementation of IMTA as a sustainable practice [5].

(4) Since supplemental feed is considered the main source of contamination of aquaculture systems and effluent receiving ecosystems [56], the improvement of these feed, as well as the feeding, strategies could be considered as an important part of the solution for a sustainable aquaculture [28, 57]. The main aspects in which the feedstuffs must be improved include the following.

(a) Better and more precise formulations for the particular species to be farmed, which consider the best concentration and quality of the nutrients. A common practice of world aquaculture is the use of diets with protein contents higher than those required, thus affecting not only the price of the feed but also increasing the pollution potential, considering that protein catabolism produces ammonium nitrogen as the main metabolite. Regarding nutrient quality, it is important to use ingredients with high digestibility; the low digestibility of ingredients (protein, lipid, carbohydrate) is partially the responsible for a low retention of those nutrients in the farmed organisms and their increase in the water column and sediment, augmenting the polluting potential [58].

(b) Higher hydrostability. One of the most important causes of nutrient losses of aquafeeds is the low hydrostability, which provoke fast disintegration and lixiviation, decreasing the nutrient incorporation efficiency by the farmed organisms and increasing the concentration in the water column. Fishes are faster swimmers and can consume a formulated feed within minutes, but crustaceans are usually less active and can consume the formulated feed within minutes or even hours. The hydrostability of feedstuffs can be improved by incorporation of effective binders and/or for the use of special fabrication processes [59].

(c) Better attractability and palatability. It is necessary to produce feeds which can be consumed as soon as possible to avoid nutrient losses. This is possible with the incorporation of effective attractants and improving the palatability with ingredients such as fish oils and others. Many of these ingredients have been sufficiently proven [60].

(5) Regarding to the feeding strategies some important advances have been achieved but there are yet much more to advance in aspects such as forms to supply the feed, adjustment of the ration, and frequency of feeding.

(a) The use of feeding trays and the increase of feeding frequency have been demonstrated to diminish the pollution potential of the effluents in shrimp farms [37]; however these strategies are suitable only for high-intensity systems (intensive or superintensive), but not economically feasible for extensive, semiextensive of semiintensive systems.

(b) The promotion, management, and rational utilization of natural feed, including microorganisms (biofilm, biofloc), are considered as a promising strategy for the culture of shrimp, fishes, and mollusks. Some authors [61–63] have successfully enhanced the production of zooplankton and benthos in shrimp ponds and demonstrated their great contribution not only in the production response, but also in the nutritional, sanitary, and immune condition of the farmed organisms. Additionally, the use and contribution of microorganisms associated to biofilms and bioflocs for the nutrition of farmed organisms have been also documented [24, 64–66]. Such practice may also decrease the dependence of fishmeal and fish oil; however other strategies such as the use of plant ingredients and the use of bioflocflour have been tested and proposed to substitute at different rates the fishmeal in formulated feeds [51].

(c) The practice of subfeeding or intermittent-feeding regimes is a strategy aimed to achieve average growth performances in aquatic organisms, but supplying significantly lower amounts of formulated feed. Such alternative takes advantage of the compensatory growth process of shrimp and crustaceans [67].

(6) The adequate management of effluents is indubitably one of the central aspects to consider for a sustainable aquaculture. Diverse strategies have been proven or suggested to minimize the environmental impacts of effluents. The most

promising are settling lagoons [34], treatments with septic tanks [68], the implementation of systems with low or zero water exchange [69], the utilization of recirculation systems [70, 71], the use of mangrove forests as sinks for nutrients, organic matter, and contaminants [72], the polyculture or integrated multitrophic aquaculture systems [55, 73], and the bioremediation [54, 74].

(a) It is considered as bioremediation the use of individual or combined organisms (including animal, vegetal, and bacteria) to minimize the contaminating charge of effluents from any activity (including aquaculture). This practice takes advantage of the natural or modified abilities of those organisms to reduce and/or transform waste products [75].

(b) There are different ways to conduct bioremediation: in situ, ex situ, biostimulation, bioaugmentation, and others. Many successfully examples of bioremediation practices can be mentioned: the use of plants (phytoremediation), macroalgae, microalgae, filter feeders, biofilters (polymer spheres with immobilized microorganisms), biofilms, and bioflocs [76, 77]. There are also combined systems which use two or more of these practices. Many studies have been conducted to use individual or combined organisms for bioremediation [78–80]. However, the ideal strategy would be the decreasing or complete halting of effluent discharge and using zero water exchange systems.

(7) Achieve certification of compliance with sustainability.

A combination of analyses has been suggested to evaluate the sustainability of commercial aquaculture farms [81]. For example, the authors of such contribution suggest the calculation of mass balances and undesirable outputs of shrimp farms; calculation of the input distance function approach which provides a complete characterization of the structure of multiinput, multioutput efficient production technology and provides a measure of the distance from each producer to that efficient-sustainable technology; finally a productivity measurement with and without undesirable outputs. However, the analysis of the socioeconomic impacts caused by farms ought to be included in the list.

Additionally, certification processes can be followed to assure the sustainability of aquaculture or to compare the standards established by the different agencies and check if the practices of any farm cope with those standards. The certification of aquaculture is performed by the International Standards Organization (ISO), the WTO Technical Barriers to Trade (TBT), the FAO Guidelines for the Ecolabelling of Fish and Fishery Products from Marine Capture Fisheries, and the Network of Aquaculture Centres in Asia-Pacific (NACA) and others [82].

According to the FAO criterion [82], certification "is a procedure through which written or equivalent assurance states that a product, process, or service conforms to specified requirements. Within the aquaculture sector certification can be applied to a process followed by a production unit (pond, cage, farm, processing plant), a specific product or commodity or to the inputs being applied to the system before or during production."

(8) Finally, there is an unavoidable need to improve research and legislation regarding evaluation and solutions for aquaculture impacts.

(a) One of the reasons of the severe environmental impacts of aquaculture is that scientific research in some developing countries is firstly focused on increasing biomass production (improvement of formulated feeds, production systems, genetically improved organisms, etc.) and later on the environmental impacts; however, it is desirable to evaluate the potential impacts of any farm that is pretended to be installed, rather than monitoring the pollution that is already being caused by any farm constructed without considering its environmental impact.

(b) In addition, there is a great heterogeneity regarding policies and legislation of aquaculture impacts among different countries; while some developed countries have complete and concrete legislation for aquaculture in order to avoid environmental impacts, others have weak policies that do not protect their environment from aquaculture wastes; under such scenario ecological imbalances and disasters have been caused, with some of them being irreversible. Herein, Smith et al. asserted that "some developing countries often lack the institutions necessary to prevent deleterious ecosystem impacts of seafood production and to sustain trade benefits" [6]; they also argued that "the developed countries have a history of these problems as well, but with less-obvious consequences." In the same report, the authors revealed that with base in the World Bank indicators, more than 60% of the countries had inefficient governance regarding the regulation of aquaculture and fisheries activities; a possible cause of such result is related to corruption and regulatory quality. Thus, it is absolutely essential for the future of aquaculture that the governments and the producers be attuned with each other to reach agreements that resolve the problems of this activity.

(c) Finally, contrasting actions have been observed by governmental instances, while some instances try to protect the environment and achieve a sustainable aquaculture, others have directly or indirectly supported the unsustainable aquacultural practices; for instance, according to a recent report [16] "the conversion of mangroves to aquaculture ponds has been fuelled by governmental support, private sector investment and external assistance from multilateral development agencies such as the World Bank and Asian Development Bank" [83, 84].

In conclusion, aquaculture is a possible panacea, but at present is also responsible for diverse problems related with the environmental sanity; however the new strategies proposed during the last decade have proven that it is possible to reach a sustainable aquaculture, but such strategies should be supported and proclaimed by the different federal environmental agencies from all countries. Only under such scenario, aquaculture will be a sustainable practice. The implementation of the different alternatives stated above would depend on particular circumstances of any farm. Fortunately, there are reports of some aquaculture farms along the world on sustainable practices.

References

[1] L. Luchini and S. Panné-Huidobro, *Perspectivas en Acuicultura: Nivel Mundial, Regional y Local*, Dirección de Acuicultura. Subsecretaría de Pesca y Acuicultura, Buenos Aíres, Argentina, 2008.

[2] FAO, *World Review of Fisheries and Aquaculture*, 2010.

[3] World Commission on the Environment and Development (WCED), *Our Common Future*, Oxford University Press, New York, NY, USA, 1987.

[4] R. R. Stickney and J. P. McVey, *Responsible Marine Aquaculture*, World Aquaculture Society, New York, NY, USA, 2002.

[5] K. Barrington, N. Ridler, T. Chopin, S. Robinson, and B. Robinson, "Social aspects of the sustainability of integrated multi-trophic aquaculture," *Aquaculture International*, vol. 18, no. 2, pp. 201–211, 2010.

[6] M. D. Smith, C. A. Roheim, L. B. Crowder et al., "Sustainability and global seafood," *Science*, vol. 327, no. 5967, pp. 784–786, 2010.

[7] N. Roos, M. A. Wahab, C. Chamnan, and S. H. Thilsted, "The role of fish in food-based strategies to combat vitamin A and mineral deficiencies in developing countries," *Journal of Nutrition*, vol. 137, no. 4, pp. 1106–1109, 2007.

[8] FAO, *The State of the World Fisheries and Aquaculture 2008*, FAO, Rome, Italy, 2009.

[9] A. Frankic and C. Hershner, "Sustainable aquaculture: developing the promise of aquaculture," *Aquaculture International*, vol. 11, no. 6, pp. 517–530, 2003.

[10] B. R. DeWalt, J. R. Ramirez-Zavala, L. Noriega, and E. González, "Shrimp aquaculture, the people and the environment in coastal Mexico," Tech. Rep., World Bank, NACA, WWF y FAO Consortium Program on Shrimp Farming and the Environment, 2002.

[11] K. Rajitha, C. K. Mukherjee, and R. Vinu Chandran, "Applications of remote sensing and GIS for sustainable management of shrimp culture in India," *Aquacultural Engineering*, vol. 36, no. 1, pp. 1–17, 2007.

[12] J. H. Tidwell and G. L. Allan, "Fish as food: aquaculture's contribution. Ecological and economic impacts and contributions of fish farming and capture fisheries," *EMBO Reports*, vol. 2, no. 11, pp. 958–963, 2001.

[13] E. J. Olguín, M. E. Hernández, and G. Sánchez-Galván, "Contaminación de manglares por hidrocarburos y estrategias de biorremediación, fitorremediación y restauración," *Revista Internacional de Contaminacion Ambiental*, vol. 23, no. 3, pp. 139–154, 2007.

[14] F. Páez-Osuna, "Retos y perspectivas de la camaronicultura en la zona costera," *Revista Latinoamericana de Recursos Naturales*, vol. 1, pp. 21–31, 2005.

[15] D. M. Alongi, "Present state and future of the world's mangrove forests," *Environmental Conservation*, vol. 29, no. 3, pp. 331–349, 2002.

[16] B. B. Walters, P. Rönnbäck, J. M. Kovacs et al., "Ethnobiology, socio-economics and management of mangrove forests: a review," *Aquatic Botany*, vol. 89, no. 2, pp. 220–236, 2008.

[17] FAO, *El Estado Mundial de la Pesca y Acuicultura 2006*, FAO, Rome, Italy, 2007.

[18] R. L. Naylor, R. J. Goldburg, J. H. Primavera et al., "Effect of aquaculture on world fish supplies," *Nature*, vol. 405, no. 6790, pp. 1017–1024, 2000.

[19] G. J. Harper, M. K. Steininger, C. J. Tucker, D. Juhn, and F. Hawkins, "Fifty years of deforestation and forest fragmentation in Madagascar," *Environmental Conservation*, vol. 34, no. 4, pp. 325–333, 2007.

[20] E. Barbier and S. Sathirathai, *Shrimp Farming and Mangrove Loss in Thailand*, Edward Elgar, 2003.

[21] F. Páez-Osuna, "The environmental impact of shrimp aquaculture: causes, effects, and mitigating alternatives," *Environmental Management*, vol. 28, no. 1, pp. 131–140, 2001.

[22] J. A. Rodríguez-Valencia, D. Crespo, and M. López-Camacho, "La camaronicultura y la sustentabilidad del Golfo de California," 2010, http://www.wwf.org.mx.

[23] L. R. Martínez-Córdova, M. Martínez-Porchas, and S. Pedrín-Avilés, "Selección de sitios, construcción y preparación de estanques," in *Camaronicultura Sustentable*, L. R. Martínez-Córdova, Ed., chapter I, p. 179, Trillas, D.F., Mexico, 2009.

[24] Y. Avnimelech, *Biofloc Technology. A Practical Guide Book*, The World Aquaculture Society, Baton Rouge, La, USA, 2009.

[25] S. Focardi, I. Corsi, and E. Franchi, "Safety issues and sustainable development of European aquaculture: new tools for environmentally sound aquaculture," *Aquaculture International*, vol. 13, no. 1-2, pp. 3–17, 2005.

[26] R. Crab, Y. Avnimelech, T. Defoirdt, P. Bossier, and W. Verstraete, "Nitrogen removal techniques in aquaculture for a sustainable production," *Aquaculture*, vol. 270, no. 1–4, pp. 1–14, 2007.

[27] M. A. Burford and K. C. Williams, "The fate of nitrogenous waste from shrimp feeding," *Aquaculture*, vol. 198, no. 1-2, pp. 79–93, 2001.

[28] A. G. J. Tacon and I. P. Forster, "Aquafeeds and the environment: policy implications," *Aquaculture*, vol. 226, no. 1–4, pp. 181–189, 2003.

[29] Y. Y. Feng, L. C. Hou, N. X. Ping, T. D. Ling, and C. I. Kyo, "Development of mariculture and its impacts in Chinese coastal waters," *Reviews in Fish Biology and Fisheries*, vol. 14, no. 1, pp. 1–10, 2004.

[30] A. Gyllenhammar and L. Håkanson, "Environmental consequence analyses of fish farm emissions related to different scales and exemplified by data from the Baltic—a review," *Marine Environmental Research*, vol. 60, no. 2, pp. 211–243, 2005.

[31] C. Jackson, N. Preston, P. J. Thompson, and M. Burford, "Nitrogen budget and effluent nitrogen components at an intensive shrimp farm," *Aquaculture*, vol. 218, no. 1–4, pp. 397–411, 2003.

[32] O. Schneider, V. Sereti, E. H. Eding, and J. A. J. Verreth, "Analysis of nutrient flows in integrated intensive aquaculture systems," *Aquacultural Engineering*, vol. 32, no. 3-4, pp. 379–401, 2005.

[33] R. Alonso-Rodríguez and F. Páez-Osuna, "Nutrients, phytoplankton and harmful algal blooms in shrimp ponds: a review with special reference to the situation in the Gulf of California," *Aquaculture*, vol. 219, no. 1–4, pp. 317–336, 2003.

[34] L. R. Martínez-Córdova and F. Enriquez-Ocaña, "Study of the benthic fauna in a discharge lagoon of a shrimp faro with special emphasis on polychaeta," *Journal of Biological Sciences*, vol. 7, pp. 12–17, 2007.

[35] L. Deutsch, S. Gräslund, C. Folke et al., "Feeding aquaculture growth through globalization: exploitation of marine ecosystems for fishmeal," *Global Environmental Change*, vol. 17, no. 2, pp. 238–249, 2007.

[36] E. A. H. Priya and S. J. Davies, "Growth and feed conversion ratio of juvenile *Oreochromis niloticus* fed with replacement of fishmeal diets by animal by-products," *Indian Journal of Fisheries*, vol. 54, pp. 51–58, 2007.

[37] R. Casillas-Hernández, H. Nolasco-Soria, T. García-Galano, O. Carrillo-Farnes, and F. Páez-Osuna, "Water quality, chemical fluxes and production in semi-intensive Pacific white shrimp (*Litopenaeus vannamei*) culture ponds utilizing two different feeding strategies," *Aquacultural Engineering*, vol. 36, no. 2, pp. 105–114, 2007.

[38] H. A. González-Ocampo, L. F. Beltrán Morales, C. Cáceres-Martínez et al., "Shrimp aquaculture environmental diagnosis in the semiarid coastal zone in Mexico," *Fresenius Environmental Bulletin*, vol. 15, no. 7, pp. 659–669, 2006.

[39] W. L. Shelton and S. Rothbard, "Exotic species in global aquaculture—a review," *Israeli Journal of Aquaculture-Bamidgeh*, vol. 58, no. 1, pp. 3–28, 2006.

[40] M. Krkošek, J. S. Ford, A. Morton, S. Lele, R. A. Myers, and M. A. Lewis, "Declining wild salmon populations in relation to parasites from farm salmon," *Science*, vol. 318, no. 5857, pp. 1772–1775, 2007.

[41] A. B. A. Boxall, "The environmental side effects of medication," *EMBO Reports*, vol. 5, no. 12, pp. 1110–1116, 2004.

[42] C. A. Berlanga-Robles and A. Ruiz-Luna, "Assessment of landscape changes and their effects on the San Blas estuarine system, Nayarit (Mexico), through Landsat imagery analysis," *Ciencias Marinas*, vol. 32, no. 3, pp. 523–538, 2006.

[43] F. Páez-Osuna, "The environmental impact of shrimp aquaculture: a global perspective," *Environmental Pollution*, vol. 112, no. 2, pp. 229–231, 2001.

[44] S. C. Johnson, R. B. Blaylock, J. Elphick, and K. D. Hyatt, "Disease induced by the sea louse (*Lepeophtheirus salmonis*) (Copepoda: Caligidae) in wild sockeye salmon (*Oncorhynchus nerka*) stocks of Alberni Inlet, British Columbia," *Canadian Journal of Fisheries and Aquatic Sciences*, vol. 53, no. 12, pp. 2888–2897, 1996.

[45] G. M. Kruzynski, "Cadmium in oysters and scallops: the BC experience," *Toxicology Letters*, vol. 148, no. 3, pp. 159–169, 2004.

[46] C. E. Boyd, "Guidelines for aquaculture effluent management at the farm-level," *Aquaculture*, vol. 226, no. 1–4, pp. 101–112, 2003.

[47] P. Read and T. Fernandes, "Management of environmental impacts of marine aquaculture in Europe," *Aquaculture*, vol. 226, no. 1–4, pp. 139–163, 2003.

[48] F. J. Magallón-Barajas, A. Arreola, G. Portillo-Clark et al., "Capacidad de carga y capacidad ambiental en la camaronicultura," in *Camaronicultura Sustentable*, L. M. Córdova, Ed., pp. 37–80, Trillas, D.F., Mexico, 2009.

[49] L. R. Martínez-Córdova, M. M. Porchas, and E. Cortés-Jacinto, "Camaronicultura mexicana y mundial: Actividad sustentable o industria contaminante?" *Revista Internacional de Contaminacion Ambiental*, vol. 25, no. 3, pp. 181–196, 2009.

[50] M. J. Costello, A. Grant, I. M. Davies et al., "The control of chemicals used in aquaculture in Europe," *Journal of Applied Ichthyology*, vol. 17, no. 4, pp. 173–180, 2001.

[51] M. Emerenciano, E. L. C. Ballester, R. O. Cavalli, and W. Wasielesky, "Biofloc technology application as a food source in a limited water exchange nursery system for pink shrimp *Farfantepenaeus brasiliensis* (Latreille, 1817)," *Aquaculture Research*. In press.

[52] C. E. Boyd, J. A. Hargreaves, and J. W. Clay, "Codes of conduct for marine shrimp aquaculture," in *The New Wave, Proceedings of Special Session on Sustainable Shrimp Culture. Aquaculture*, L.C. Browdy and D. E. Jory, Eds., pp. 303–321, The World Aquaculture Society, Baton Rouge, La, USA, 2001.

[53] T. Chopin, A. H. Buschmann, C. Halling et al., "Integrating seaweeds into marine aquaculture systems: a key toward sustainability," *Journal of Phycology*, vol. 37, no. 6, pp. 975–986, 2001.

[54] L. R. Martínez-Córdova, J. A. López-Elías, G. Leyva-Miranda, L. Armenta-Ayón, and M. Martínez-Porchas, "Bioremediation and reuse of shrimp aquaculture effluents to farm whiteleg shrimp, *Litopenaeus vannamei*: a first approach," *Aquaculture Research*, vol. 42, no. 10, pp. 1415–1423, 2011.

[55] M. Martínez-Porchas, L. R. Martínez-Córdova, M. A. Porchas-Cornejo, and J. A. López-Elías, "Shrimp polyculture: a potentially profitable, sustainable, but uncommon aquacultural practice," *Reviews in Aquaculture*, vol. 2, no. 2, pp. 73–85, 2010.

[56] A. Sapkota, A. R. Sapkota, M. Kucharski et al., "Aquaculture practices and potential human health risks: current knowledge and future priorities," *Environment International*, vol. 34, no. 8, pp. 1215–1226, 2008.

[57] L. Martínez-Córdova, M. Porchas, and H. Villarreal, "Efecto de tres diferentes estrategias de alimentaciön sobre el fitoplancton, zooplancton y bentos en estanques de cultivo de camarön café penaeus californiens' (holmes 1900)," *Ciencias Marinas*, vol. 24, no. 3, pp. 267–281, 1998.

[58] L. E. Cruz-Suárez, M. Nieto-López, C. Guajardo-Barbosa, M. Tapia-Salazar, U. Scholz, and D. Ricque-Marie, "Replacement of fish meal with poultry by-product meal in practical diets for *Litopenaeus vannamei*, and digestibility of the tested ingredients and diets," *Aquaculture*, vol. 272, no. 1–4, pp. 466–476, 2007.

[59] C. J. Simon, "The effect of carbohydrate source, inclusion level of gelatinised starch, feed binder and fishmeal particle size on the apparent digestibility of formulated diets for spiny lobster juveniles, *Jasus edwardsii*," *Aquaculture*, vol. 296, no. 3-4, pp. 329–336, 2009.

[60] T. Ho, *Feed attractants for juvenile Chinook salmon (Oncorhynchus tshawytscha) prepared from hydrolisates of Pacific hake (Merluccius productus)*, M.S. thesis, University of British Columbia, 2009.

[61] L. R. Martínez-Córdova and M. A. Porchas-Cornejo, "Manejo del alimento y la alimentación en estanques camaronícolas: una estrategia de acuicultura sustentable," in *Camaronicultura Sustentable*, L. R. Martínez-Córdova, Ed., chapter III, p. 176, Trillas, D.F., Mexico, 2009.

[62] A. Campaña-Torres, L. R. Martínez-Córdova, H. Villarreal-Colmenares, and E. Cortés-Jacinto, "Evaluation of different concentrations of adult live Artemia (*Artemia franciscana*, Kellogs 1906) as natural exogenous feed on the water quality and production parameters of *Litopenaeus vannamei* (Boone 1931) pre-grown intensively," *Aquaculture Research*, vol. 42, no. 1, pp. 40–46, 2010.

[63] M. A. Porchas-Cornejo, L. R. Martínez-Córdova, L. Ramos-Trujillo, J. Hernández-López, M. Martínez-Porchas, and F. Mendoza-Cano, "Effect of promoted natural feed on the production, nutritional, and immunological parameters of *Litopenaeus vannamei* (Boone, 1931) semi-intensively farmed," *Aquaculture Nutrition*, vol. 17, no. 2, pp. e622–e628, 2011.

[64] M. E. Azim, D. C. Little, and J. E. Bron, "Microbial protein production in activated suspension tanks manipulating C:N ratio in feed and the implications for fish culture," *Bioresource Technology*, vol. 99, no. 9, pp. 3590–3599, 2008.

[65] E. L. C. Ballester, P. C. Abreu, R. O. Cavalli, M. Emerenciano, L. de Abreu, and W. Wasielesky Jr., "Effect of practical diets with different protein levels on the performance of *Farfantepenaeus paulensis* juveniles nursed in a zero exchange suspended microbial flocs intensive system," *Aquaculture Nutrition*, vol. 16, no. 2, pp. 163–172, 2010.

[66] M. J. Becerra-Dorame, L. R. Martínez-Córdova, P. Martínez-Porchas, and J. A. López-Elías, "Evaluation of zero water exchange autotrophic and heterotrophic microcosm-based systems on the production response of *Litopenaeus vannamei* intensively nursed without Artemia," *Israeli Journal of Aquaculture-Bamidgeh*, vol. 63, pp. 1–7, 2011.

[67] L. Stumpf, N. S. Calvo, F. C. Díaz, W. C. Valenti, and L. S. L. Greco, "Effect of intermittent feeding on growth in early juveniles of the crayfish *Cherax quadricarinatus*," *Aquaculture*, vol. 319, no. 1-2, pp. 98–104, 2011.

[68] R. C. Summerfelt and C. R. Penne, "Septic tank treatment of the effluent from a small-scale commercial recycle aquaculture system," *North American Journal of Aquaculture*, vol. 69, no. 1, pp. 59–68, 2007.

[69] C. P. Balasubramanian, S. M. Pillai, and P. Ravichandran, "Zero-water exchange shrimp farming systems (extensive) in the periphery of Chilka lagoon, Orissa, India," *Aquaculture International*, vol. 12, pp. 555–572, 2005.

[70] C. Lezama-Cervantes, J. J. Paniagua-Michel, and J. Zamora-Castro, "Bioremediacion of effluents ones of the culture of *Litopenaeus vannamei* (Boone, 1931) using microbial mats in a recirculating system," *Latin American Journal of Aquatic Research*, vol. 38, no. 1, pp. 129–142, 2010.

[71] M. B. Timmons, J. M. Ebeling, F. W. Wheaton, S. T. Summerfelt, and B. J. Vinci, *Sistemas de Recirculación para la Acuicultura*, Fundación Chile,Vitacura, Santiago de Chile, Chile, 2002.

[72] V. H. Rivera-Monroy, L. A. Torres, N. Bahamon, F. Newmark, and R. R. Twilley, "The potential use of mangrove forests as nitrogen sinks of shrimp aquaculture pond effluents: the role of denitrification," *Journal of the World Aquaculture Society*, vol. 30, no. 1, pp. 12–25, 1999.

[73] L. R. Martinez-Cordova and M. Martinez-Porchas, "Polyculture of Pacific white shrimp, *Litopenaeus vannamei*, giant oyster, *Crassostrea gigas* and black clam, *Chione fluctifraga* in ponds in Sonora, Mexico," *Aquaculture*, vol. 258, no. 1–4, pp. 321–326, 2006.

[74] J. Paniagua-Michel and O. Garcia, "Ex-situ bioremediation of shrimp culture effluent using constructed microbial mats," *Aquacultural Engineering*, vol. 28, no. 3-4, pp. 131–139, 2003.

[75] P. Chávez-Crooker and J. Obreque-Contreras, "Bioremediation of aquaculture wastes," *Current Opinion in Biotechnology*, vol. 21, no. 3, pp. 313–317, 2010.

[76] P. De Schryver, R. Crab, T. Defoirdt, N. Boon, and W. Verstraete, "The basics of bio-flocs technology: the added value for aquaculture," *Aquaculture*, vol. 277, no. 3-4, pp. 125–137, 2008.

[77] D. D. Kuhn, G. D. Boardman, S. R. Craig, G. J. Flick, and E. Mc Lean, "Use of microbial flocs generated from tilapia effluent as a nutritional supplement for shrimp, *litopenaeus vannamei*, in recirculating aquaculture systems," *Journal of the World Aquaculture Society*, vol. 39, no. 1, pp. 72–82, 2008.

[78] J. M. E. Hussenot, "Emerging effluent management strategies in marine fish-culture farms located in European coastal wetlands," *Aquaculture*, vol. 226, no. 1–4, pp. 113–128, 2003.

[79] J. Liu, Z. Wang, and W. Lin, "De-eutrophication of effluent wastewater from fish aquaculture by using marine green alga *Ulva pertusa*," *Chinese Journal of Oceanology and Limnology*, vol. 28, no. 2, pp. 201–208, 2010.

[80] Y. Zhou, H. Yang, H. Hu et al., "Bioremediation potential of the macroalga *Gracilaria lemaneiformis* (Rhodophyta) integrated into fed fish culture in coastal waters of north China," *Aquaculture*, vol. 252, no. 2-4, pp. 264–276, 2006.

[81] F. J. Martinez-Cordero and P. Leung, "Sustainable aquaculture and producer performance: measurement of environmentally adjusted productivity and efficiency of a sample of shrimp farms in Mexico," *Aquaculture*, vol. 241, no. 1–4, pp. 249–268, 2004.

[82] F. Corsin, S. Funge-Smith, and J. Clausen, *A qualitative assessment of standards and certification schemes applicable to aquaculture in the Asia-Pacific region*, RAP Publication 2007/25. Food and Agriculture Organization for the United Nations. Regional Office for the Asia and the Pacific, Bangkok, Thailand, 2007.

[83] S. E. Siddall, J. A. Atchue III, and P. L. Murray Jr., "Mariculture development in mangroves: a case study of the Philippines, Panama and Ecuador," in *Coastal Resources Management: Development Case Studies. Renewable Resources Information Series, Coastal Management*, J. R. Clark, Ed., National Park Service, U.S. Dept. of the Interior, and the U.S. Agency for International Development. Research Planning Institute, Columbia, SC, USA, 1985.

[84] W. J. M. Verheugt, A. Purwoko, F. Danielsen, H. Skov, and R. Kadarisman, "Integrating mangrove and swamp forests conservation with coastal lowland development; the Banyuasin Sembilang swamps case study, South Sumatra Province, Indonesia," *Landscape and Urban Planning*, vol. 20, no. 1–3, pp. 85–94, 1991.

17

Diversity of Woodland Communities and Plant Species along an Altitudinal Gradient in the Guancen Mountains, China

Dongping Meng,[1] Jin-Tun Zhang,[2] and Min Li[1]

[1] Institute of Loess Plateau, Shanxi University, Taiyuan 030006, China
[2] College of Life Sciences, Beijing Normal University, Beijing 100875, China

Correspondence should be addressed to Jin-Tun Zhang, zhangjt@bnu.edu.cn

Academic Editors: B. B. Castro and H. Gjosaeter

Study on plant diversity is the base of woodland conservation. The Guancen Mountains are the northern end of Luliang mountain range in North China. Fifty-three quadrats of 10 m × 20 m of woodland communities were randomly established along an altitudinal gradient. Data for species composition and environmental variables were measured and recorded in each quadrat. To investigate the variation of woodland communities, a Two-Way Indicator Species Analysis (TWINSPAN) and a Canonical Correspondence Analysis (CCA) were conducted, while species diversity indices were used to analyse the relationships between species diversity and environmental variables in this study. The results showed that there were eight communities of woodland vegetation; each of them had their own characteristics in composition, structure, and environment. The variation of woodland communities was significantly related to elevation and also related to slope, slope aspect, and litter thickness. The cumulative percentage variance of species-environment relation for the first three CCA axes was 93.5%. Elevation was revealed as the factor which most influenced community distribution and species diversity. Species diversity was negatively correlated with elevation, slope aspect, and litter thickness, but positively with slope. Species richness and heterogeneity increased first and then decreased but evenness decreased significantly with increasing elevation. Species diversity was correlated with slope, slope aspect, and litter thickness.

1. Introduction

Variations of woodland communities and species diversity are important in conservation of natural areas and have been frequently studied in plant ecology [1–6]. In China, mountainous regions are more significant in the conservation practice because most woodland communities are centralized in mountains with limited area [7–9]. The variation of plant communities and species diversity can be linked to several ecological gradients [10, 11]. Altitudinal gradient is known to be one of the decisive factors shaping the spatial patterns of vegetation and species diversity [12–14]. The relationship of community structure, composition, and species diversity of woodland with elevation gradient and other environmental variables have emerged as a key issue in ecological and environmental sciences [6, 15–17].

The patterns of species and community diversity along elevation gradient have been frequently tested [10, 18, 19].

The most commonly observed pattern is a maximum diversity at the intermediate altitudinal range [10, 16]. However, there are still a number of exceptions to this pattern [2, 20]. Some authors argued that whether the species diversity will increase or decrease with increasing elevation or peak at intermediate elevation depends largely on specific patterns of interactions among plant communities, species, and environmental factors [13, 18, 21]. Thus further test of the hypothesis in different mountains should be carried out [22–24].

The Guancen Mountains, located at the north-eastern area of Luliang Mountain Range of the Loess Plateau, is the main distribution area of cold-temperate conifer woodland and is a famous ecological-tourism region in North China [14, 25]. Vegetation plays a significant role in local development and should be protected and utilized reasonably in the Guancen Mountains [9]. Some studies related to floristic characteristics and plant resources have been carried out

FIGURE 1: The location of the Guancen Mountains in Shanxi province of China (the coordinate system of this map was WGS 1984). The star is Beijing, the capital of China.

in this area [26–28]. However, no studies have examined the variations of vegetation and species diversity associated with the major environmental variables in the Guancen Mountains. Quantitative analysis of vegetation data, such as classification and ordination, is an important approach to generate and test hypotheses with respect to vegetation and environment [3, 29–33]. Therefore, the woodland plant species composition and diversity were analysed and their relationships with environmental variables were investigated in the present study. Our objectives were (1) to test the hypothesis of a maximum diversity at the intermediate altitudinal range, (2) to analyse the interdependencies among community characteristics and topographic variables, and (3) to identify the key environmental variable influencing plant community composition and species diversity.

2. Materials and Methods

2.1. Study Area. The Guancen Mountains is located at E111° 05′-120° 40′, N38° 31′-39° 8′, and is the northern end of Luliang mountain range in Shanxi Province, China (Figure 1). It lies on the eastern part of the Loess Plateau and is on the transitional area from forest-steppe zone to warm-temperate forest zone [7, 27]. The climate of this area is temperate and semihumid with continental characteristics and controlled by seasonal wind. The annual mean temperature is 6.2°C, and the monthly mean temperatures of January and July are −9.9°C and 20.1°C respectively. The annual mean precipitation varies from 470 mm to 770 mm in this mountain, and 70% precipitation falls from July to September. Several soil types, such as loess soil, mountain cinnamon soil, and brown forest soil, can be found in this area. The elevation varies from 800 m to 2 620 m, but the area between 800 and 1 600 m is covered by crop fields. Most area above 1 600 m is covered with woodlands. This study concerns all woodland communities distributed from 1 620 to 2 620 m. The total

area of woodland in this region is over 850,000 ha [27]. The woodlands form secondary natural vegetation with frequent disturbance connecting with grazing and logging of timber or firewood until the end of 1980s when a national park was found there [26].

2.2. Sampling Design. Along the altitudinal gradient between 1 620 and 2 620 m a. s. l., 20 sampling points separated by 50 meters in elevation were set up, and 2 or 3 quadrats around each sampling point were established randomly. Species data were recorded in each quadrat. The quadrat size was 10 m × 20 m, in which three 5 m × 5 m and three 2 m × 2 m small quadrats were used to record shrubs and herbs, respectively. The cover, height, and abundance of trees, shrubs and herbs, as well as the basal area of trees were measured in each quadrat. The plant height was measured by using a height-meter for trees and a ruler for shrubs and herbs. The basal diameter of trees was measured by using a caliper and was used to calculate the basal area. A total of 112 plant species were recorded in 53 quadrats. Elevation, slope, slope aspect, and the litter thickness for each quadrat were also recorded. The elevation in each quadrat was measured by using an altimeter, the slope and slope aspect were measured by using a compass meter, and the litter depth was measured by using a ruler directly [14, 26]. The elevation, slope, and litter thickness were reading values, while the aspect measurements were classified from 1 to 8 in the following way: 1 (337.6°–22.5°), 2 (22.6°–67.5°), 3 (292.6°–337.5°), 4 (67.6°–112.5°), 5 (247.6°–292.5°), 6 (112.6°–157.5°), 7 (202.6°–247.5°), and 8 (157.6°–202.5°).

2.3. Data Analysis. The Importance Value of each species was calculated and used as data in multivariate analysis of communities and species diversity. The importance value was calculated by the formula [14, 26]:

$$
IV_{Tree} = \frac{(Relative\ cover + Relative\ dominance + Relative\ height)}{300},
\tag{1}
$$

$$
IV_{Scrubs\ and\ Herbs} = \frac{(Relative\ cover + Relative\ height)}{200}.
\tag{2}
$$

The relative dominance referred to species basal area. The species data were importance values of 112 species in 53 quadrats. The environmental variables included elevation, slope, slope aspect, and litter thickness of each quadrat.

A Two-Way Indicator Species Analysis (TWINSPAN) [30] and a Canonical Correspondence Analysis (CCA) [33] were conducted to identify plant communities and analyse their relationship with environmental variables. The calculation of TWINSPAN and CCA was carried out by computer program of TWINSPAN [30] and CANOCO [33], respectively.

Six species diversity indices, two for species richness, two for species heterogeneity, and two for species evenness, were used to calculate diversity values [14, 34]. Different indices may be suitable to different ecological data, and therefore their results can be compared [35–38]. These indices were

Species number (as a richness index):

$$D = S. \tag{3}$$

Margalef richness index:

$$R2 = \frac{S - 1}{\ln(N)}. \tag{4}$$

Shannon-Wiener heterogeneity index:

$$H' = -\sum P_i \ln P_i. \tag{5}$$

Hill heterogeneity index:

$$N2 = \frac{1}{\sum_{i=1}^{S} N_i(N_i - 1)/N(N - 1)}. \tag{6}$$

Pielou evenness index:

$$E1 = \frac{H'}{\ln(S)}. \tag{7}$$

Sheldon evenness index:

$$E2 = \frac{e^{H'}}{S}, \tag{8}$$

where P_i is the relative importance value of species i, N_i the importance value of species i, N the sum of importance values for all species in a quadrat, and S the species number present in a quadrat [25–32].

The Spearman rank correlation and regression were used to analyse the relationships between species diversity and environmental variables.

3. Results

3.1. Variation of Communities. A prior DCA analysis provided a great gradient of 6.0 for the first DCA axis, which suggested that TWINSPAN and CCA were suitable for the analyses of these data [9].

TWINSPAN classified the 53 quadrats into 8 clusters, representing 8 woodland communities (Figure 2). The names and the main composition of the 8 communities are as follows. The community name was followed by the dominant species rules, that is, Dominant trees—dominant scrubs—dominant herbs [7].

I Comm: *Hippophae rhamnoides + Ostryopsis davidiana – Dendianthena chanetii.* The common species in this community are *Artemisia sacrorum, Artemisia sieversiana, Wikstroemia chamaedaphne, Cymbopogon nardus,* and *Carex lanceolata.*

II Comm: *Hippophae rhamnoides + Wikstroemia chamaedaphne – Artemisia sacrorum.* The common species are *Caragana intermedia, Larix principis-ruprechtii, Artemisia sacrorum, Populus davidiana, Wikstroemia chamaedaphne, Oxytropis caerulea,* and *Fragaria arientalis.*

III Comm: *Larix principis-ruprechtii – Caragana intermedia + Wikstroemia chamaedaphne – Artemisia sacrorum.* The common species are *Populus davidiana, Spiraea*

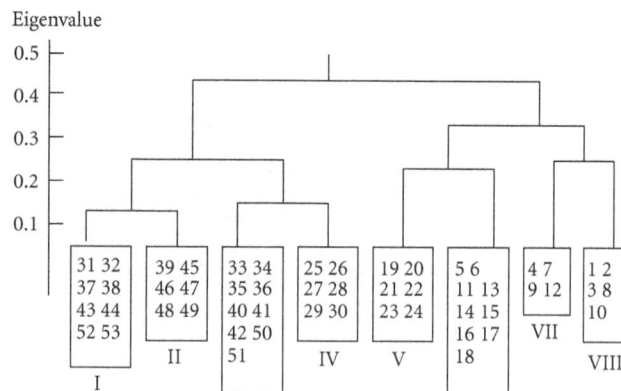

FIGURE 2: The dendrogram of TWINSPAN results for 53 samples of woodland communities in the Guancen Mountains, China. I–VIII refer to the 8 communities and Arabic numbers in rectangles refer to quadrat number.

pubescens, Oxytropis caerulea, Anemone raddeana, Scabiosa tschiliensis, Carex lanceolata, and *Patrinia heterophylla.*

IV Comm: *Spiraea pubescens – Artemisia sacrorum + Oxytropis caerulea.* The common species are *Abelia biflora, Rosa bella, Spiraea trilobata, Thalictrum petaloideum, Chamaenerion angustifolium,* and *Agtimonia pilosa.*

V Comm: *Picea wilsonii + Larix principis-ruprechtii + Betula platyphylla – Salix pseudotongii – Carex lanceolata + Roegneria kamoji.* The common species are *Tilia amurensis, Populus davidiana, Lonicera hispida, Geranium wilfordii, Carex lanceolata, Galium verum,* and *Cymbopogon* sp.

VI Comm: *Larix principis-ruprechtii + Picea wilsonii – Hippophae rhamnoides – Carex lanceolata.* The common species are *Betula platyphylla, Salix pseudotongii, Hippophae rhamnoides, Viburnum schensianum, Ribes burejense,* and *Sanguisorba officinalis.*

VII Comm: *Picea wilsonii + Larix principis-ruprechtii – Lonicera hispida – Carex lanceolata + Sanguisorba officinalis.* The common species are *Salix pseudotongii, Populus davidiana, Betula platyphylla, Hippophae rhamnoides, Cymbopogon* sp., *Lespedeza floribunda, Dendianthena chanetii, Saposhnikovia divaricata,* and *Taraxacum mongolicum.*

VIII Comm: *Larix principis-ruprechtii – Sanguisorba officinalis + Cymbopogon* sp.+ *Geranium wibfordii.* The common species are *Picea wilsonii, Carex lanceolata, Artemisia* spp., *Saussurea japonica, Anemone rivularis, Polygonum viviparum, Oxytropis caerulea,* and *Geranium wilfordii.*

The characteristics of communities' structure and environment above were listed in Table 1. The variation of communities was clear and related to ecological gradients (Figure 2, Table 2). The elevation decreased from left to right, whereby the temperature increased and the soil water content decreased from left to right of Figure 2 [29–38].

3.2. Community Variation Related to Environment. Figure 3 was the biplot of 53 quadrats and 4 environmental variables in CCA ordination space. In CCA ordination, the Monte Carlo permutation test indicated that the eigenvalues for the first four axes were all significant ($P < 0.05$). The

TABLE 1: Characteristics of environmental variables and community structure of woodland communities in the Guancen Mountains, China.

Communities	Elevation (m)	Slope (°)	Aspect (classes)	Litter thickness(cm)	Soils types	Plant cover (%)			
						Total	Trees	Shrubs	Herbs
I	1700–1800	15–35	1–3	1.0–3.0	Mt. cinnamon	80–95	5–10	70–85	40–55
II	1600–1700	15–40	1–3	0–2.0	Mt. Cinnamon	80–90	5–10	70–80	35–55
III	1700–1750	8–10	3	1.0–3.5	Mt. cinnamon	80–90	30–50	55–70	45–60
IV	2000–2050	20–40	2–4	2.0–5.0	Brown forest	85–90	5–10	80–90	40–60
V	2150–2350	20–25	1–5	3.0–6.5	Brown forest	90–98	85–95	30–45	50–65
VI	2150–2400	5–25	2–4	3.0–7.0	Brown forest	90–95	80–90	35–45	65–80
VII	2500–2600	2–20	2–6	6.0–9.0	Brown forest	90–95	85–90	30	70–80
VIII	2550–2600	1–2	4–5	6.0–10.0	Brown forest and meadow	100	10	1–5	95–100

Community type: I Comm: *Hippophae rhamnoides + Ostryopsis davidiana − Dendianthena chanetii*; II Comm: *Hippophae rhamnoides + Wikstroemia chamaedaphne − Artemisia sacrorum*; III Comm: *Larix principis-ruprechtii − Caragana intermedia + Wikstroemia chamaedaphne − Artemisia sacrorum*; IV Comm: *Spiraea pubescens − Artemisia sacrorum + Oxytropis caerulea*; V Comm: *Picea wilsonii + Larix principis-ruprechtii + Betula platyphylla − Salix pseudotongii − Carex lanceolata + Roegneria kamoji*; VI Comm: *Larix principis-ruprechtii + Picea wilsonii −Hippophae rhamnoides − Carex lanceolata*; VII Comm: *Picea wilsonii + Larix principis-ruprechtii − Lonicera hispida − Carex lanceolata + Sanguisorba officinalis*; VIII Comm: *Larix principis-ruprechtii − Sanguisorba officinalis + Cymbopogon sp.+ Geranium wibfordii*. Aspect classes: 1 (337.6°–22.5°), 2 (22.6°–67.5°), 3 (292.6°–337.5°), 4 (67.6°–112.5°), 5 (247.6°–292.5°), 6 (112.6°–157.5°), 7 (202.6°–247.5°), and 8 (157.6°–202.5°).

TABLE 2: Interset correlation coefficients of environmental variables with CCA axes in woodland communities in the Guancen Mountains, China.

Environmental variables	CCA axes		
	Axis 1	Axis 2	Axis 3
Elevation	−0.962***	0.035	−0.078
Slope	0.427**	0.526***	−0.392**
Aspect	−0.606***	0.384**	0.336**
Litter thickness	−0.804***	0.239**	0.230*

*$P < 0.05$, **$P < 0.01$, ***$P < 0.001$.

TABLE 3: Correlation coefficients between environmental variables in woodland communities in the Guancen Mountains, China.

Environmental variables	Elevation	Slope	Aspect	Litter thickness
Elevation	1			
Slope	−0.350**	1		
Aspect	0.651***	−0.396**	1	
Litter thickness	0.843***	−0.373**	0.699***	1

*$P < 0.05$, **$P < 0.01$, ***$P < 0.001$.

eigenvalues of the first three CCA axes were 0.605, 0.236, and 0.216, respectively; the species-environment correlations of the first three CCA axes were 0.968, 0.774, and 0.711; and the cumulative percentage variance of species-environment relation was 57.4%, 77.9%, and 93.5%; which showed that CCA performed well in describing relations between species, communities, and environmental gradients [33–35]. The Monte Carlo permutation test also indicated that the species-environment correlations with the CCA axes were significant. CCA result showed that the first CCA axis was significantly related to elevation, slope, slope aspect, and litter thickness, and elevation is the most significant factor related to the first CCA axis ($r = 0.962$, $P < 0.0010$; Figure 3, Table 2). The second and the third CCA axes are related to slope, slope aspect, and litter thickness. The altitudinal gradient from left to right was very clear in Figure 3, and along

this gradient the elevation was decreasing gradually. The communities on the left were usually distributed in the hills with high elevation, such as Assoc. *Larix principis-ruprechtii − Sanguisorba officinalis + Cymbopogon sp.+ Geranium wibfordii*, Assoc. *Larix principis-ruprechtii + Picea wilsonii − Hippophae rhamnoides − Carex lanceolata*, and Assoc. *Picea wilsonii + Larix principis-ruprechtii − Lonicera hispida − Carex lanceolata + Sanguisorba officinalis*. These communities were forests with high canopy density. The communities on the right were distributed in comparatively low hills, for example, Assoc. *Hippophae rhamnoides + Ostryopsis davidiana − Dendianthena chanetii* and Assoc. *Hippophae rhamnoides + Wikstroemia chamaedaphne − Artemisia sacrorum*.

The four environmental variables were significantly correlated with each other (Table 3).

TABLE 4: Spearman rank correlation coefficients between environmental variables and species diversity in woodland communities in the Guancen Mountains, China.

Environmental variables	Diversity indices					
	Species no.	R1	H'	N2	E1	E2
Elevation	−0.567***	−0.581***	−0.525***	−0.545***	−0.489***	−0.174
	($R^2 = 0.408***$)	($R^2 = 0.449***$)	($R^2 = 0.378***$)	($R^2 = 0.372***$)	($R^2 = 0.234***$)	($R^2 = 0.032$)
Slope	0.398 **	0.462 ***	0.458 ***	0.391 **	0.362 **	0.175
Slope aspect	−0.526 ***	−0.499 ***	−0.499 ***	−0.461 ***	−0.398 **	−0.110
Litter thickness	−0.471***	−0.512***	−0.577***	−0.597***	−0.523***	−0.211

$^*P < 0.05$, $^{**}P < 0.01$, $^{***}P < 0.001$; R^2 in brackets refers to the significance of unimodal regression; R1: Margalef richness index; H': Shannon-Wiener heterogeneity index; N2: Hill heterogeneity index; E1: Pielou evenness index; and E2: Sheldon evenness index.

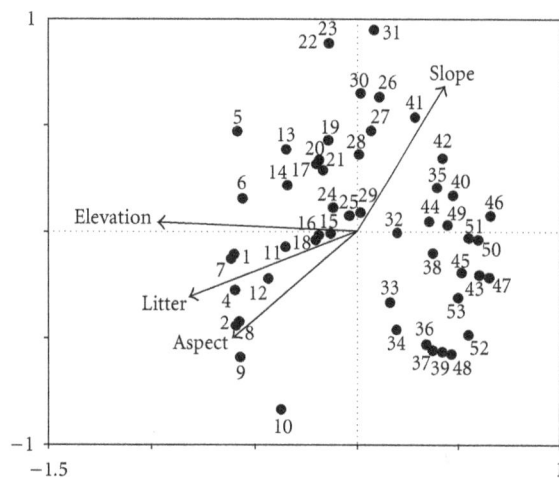

FIGURE 3: CCA ordination Bi-plot of 53 quadrats and four environmental variables of woodland communities in the Guancen Mountains, China. The numbers refer to quadrat number.

3.3. Species Diversity. Correlation analyses showed that species diversity was significantly correlated with all environmental variables, and positively correlated with slope but negatively correlated with elevation, slope aspect, and litter thickness (Table 4). We also analysed the relationships between species diversity indices and altitudinal gradient by nonlinear regression model (Table 4) because elevation was the most important variable in affecting the vegetation and species distribution in the Guancen Mountains based on the CCA analyses. Species richness, species heterogeneity, and species evenness showed almost all a significant relationship with elevation change (Table 4). Species richness and heterogeneity increased first and then decreased with increasing elevation in the Guancen Mountains, but species evenness decreased with increasing elevation. This suggests that elevation was an important factor to species diversity.

4. Discussion

The variation of woodland communities was apparent in the Guancen Mountains. TWINSPAN had successfully distinguished them as different vegetation communities. The eight communities were representative of the general vegetation in

the Guancen Mountains [7, 27] and conform to the Chinese vegetation classification system [7, 26]. They were all secondary vegetation, following destruction of the original cold-temperate coniferous forests [9]. The distribution of dominant species determined vegetation differentiation [7, 39]. This was also true in the Guancen Mountains. The distribution of dominant species, such as *Larix principis-ruprechtii, Picea wilsonii, Betula platyphylla, Hippophae rhamnoides,* and *Ostryopsis davidiana,* played important roles in vegetation patterning [14, 24].

The variation of woodland communities was closely related to the environmental variables, such as elevation, slope aspects, slope, and litter thickness, among which elevation was the most important factor affecting community variation in the Guancen Mountains. The change of woodland communities in CCA space clearly illustrated the relationships of plant communities and environmental variables. Each community had its own distribution area and was related to special combination of environmental variables [25, 37]. The first CCA axis was significantly correlated with the four environmental variables measured and was mainly an altitudinal gradient, that is, from left to right of CCA ordination diagram; elevation was decreasing gradually. Elevation change leads to the change of humidity, temperature, soil type, and so forth, which influence the variation of communities [15, 24, 40].

Community variation was also closely related to other environmental variables, such as slope aspect, slope, and the litter depth [11, 22]. These variables were significantly correlated with elevation in the Guancen Mountains. The altitude and the litter depth were positively correlated with each other and had similar effects on community changes [7]. The litter thickness decreased with increasing elevation, which may be due to the effects of mean temperature on the decomposition rate of litter with elevation increase [9]. The effects of slope and aspect on vegetation were also significant [17, 38].

Species diversity in communities was an important feature in community structure and its change was a part of community variation [16, 21, 22]. Five out of the six indices of species diversity used were significantly correlated with elevation and also related to litter thickness, slope aspect, and slope (Table 4). Species diversity was negatively correlated with elevation, slope aspect and litter but positively correlated with slope. All indices showed a nonlinear relationship with elevation change; that is, they were increased first and then decreased along the altitudinal gradient. These patterns were consistent with the hypothesis of maximum diversity at intermediate level of elevation [16, 17, 19]. The maximum richness and heterogeneity appeared at 1800–1900 m, but the maximum evenness at 1600 m. The curve peaks were not very obvious, which may be due to the fact that this altitudinal gradient (1620–2620 m) was not a whole but only a part of elevation gradient in the Guancen Mountains. The whole altitudinal range varied from 800 m to 2620 m for the Guancen Mountains, but crop fields occurred to all areas below 1600 m [27]. Therefore, the pattern of species diversity along altitudinal gradient in this study was, in fact, a typical pattern of maximum diversity at intermediate level of elevation [16, 17, 40].

Species diversity was also related to litter thickness, slope, and slope aspect in the Guancen Mountains. In fact, all the changes of species richness, heterogeneity, and evenness were significantly related to community variation and environmental diversity [9, 22]. Elevation was one of the most important variables controlling community change and species diversity in the Guancen Mountains, which was identical to that of many other studies [15, 40].

Five of the six indices of species diversity used in this work were very effective; they were Species number, Margalef richness index, Shannon-Wiener heterogeneity index, Hill heterogeneity index, and Pielou evenness index. These indices provide similar results because some of them were similar, correlated, or in one index family [5, 25, 41]. However, Sheldon evenness index was not sensitive to detect the changes of species diversity among communities and their relationships with environmental variables in this study. This suggests that species indices need to be compared and selected in different studies [12, 14]. More than one index was combined and compared in one study and was a common choice in species diversity research [22, 41–43].

Acknowledgments

The study was financially supported by the National Natural Science Foundation of China (Grants nos. 31170494, 30870399) and the Teachers' Foundation of Education Ministry of China.

References

[1] A. E. Magurran, Ecological Diversity and Its Measurement, Princeton University Press, London, UK, 1988.

[2] J. G. Pausas, "Species richness patterns in the understorey of Pyrenean Pinus sylvestris forest," Journal of Vegetation Science, vol. 5, pp. 517–524, 1994.

[3] D. Martins, E. Odd, F. Eli, E. L. Jonas, and A. Erik, "Beech forest communities in the Nordic countries—a multivariate analysis," Plant Ecology, vol. 140, no. 2, pp. 203–220, 1999.

[4] M. Loreau, S. Naeem, P. Inchausti et al., "Ecology: biodiversity and ecosystem functioning: current knowledge and future challenges," Science, vol. 294, no. 5543, pp. 804–808, 2001.

[5] B. Tóthmérész, "Comparison of different methods for diversity ordering," Journal of Vegetation Science, vol. 6, no. 2, pp. 283–290, 1995.

[6] M. Fetene, Y. Assefa, M. Gashaw, Z. Woldu, and E. Beck, "Diversity of afroalpine vegetation 16 and ecology of treeline species in the Bale Mountains, Ethiopia, and the influence of fire," in Land Use Change and Mountain Biodiversity, E. M. Spehn, M. Liberman, and C. Korner, Eds., pp. 25–38, CRC PRESS, New York, NY, USA, 2006.

[7] Z. Y. Wu, Vegetation of China, Science Press, Beijing, China, 1980.

[8] J.-T. Zhang, "Conservation of biodiversity and sustainable development," Economic Geography, vol. 19, no. 2, pp. 70–75, 1999.

[9] J.-T. Zhang, "Succession analysis of plant communities in abandoned croplands in the eastern Loess Plateau of China," Journal of Arid Environments, vol. 63, no. 2, pp. 458–474, 2005.

[10] M. Kessler, "Patterns of diversity and range size of selected plant groups along an elevational transect in the Bolivian Andes," Biodiversity and Conservation, vol. 10, no. 11, pp. 1897–1921, 2001.

[11] I. Schmidt, S. Zerbe, J. Betzin, and M. Weckesser, "An approach to the identification of indicators for forest biodiversity—the solling Mountains (NW Germany) as an example," Restoration Ecology, vol. 14, no. 1, pp. 123–136, 2006.

[12] T. J. Stohlgren, A. J. Owen, and M. Lee, "Monitoring shifts in plant diversity in response to climate change: a method for landscapes," Biodiversity and Conservation, vol. 9, no. 1, pp. 65–86, 2000.

[13] J. H. Brown, "Mammals on mountainsides: elevational patterns of diversity," Global Ecology and Biogeography, vol. 10, no. 1, pp. 101–109, 2001.

[14] J.-T. Zhang, W. Ru, and B. Li, "Relationships between vegetation and climate on the Loess Plateau in China," Folia Geobotanica, vol. 41, no. 2, pp. 151–163, 2006.

[15] C. Q. Tang and M. Ohsawa, "Zonal transition of evergreen, deciduous, and coniferous forests along the altitudinal gradient on a humid subtropical mountain, Mt. Emei, Sichuan, China," Plant Ecology, vol. 133, no. 1, pp. 63–78, 1997.

[16] F. Ojeda, T. Marañón, and J. Arroyo, "Plant diversity patterns in the Aljibe Mountains (S. Spain): a comprehensive account," Biodiversity and Conservation, vol. 9, no. 9, pp. 1323–1343, 2000.

[17] G. Austrheim, "Plant diversity patterns in semi-natural grasslands along an elevational gradient in southern Norway," Plant Ecology, vol. 161, no. 2, pp. 193–205, 2002.

[18] M. V. Lomolino, "Elevation gradients of species-density: historical and prospective views," Global Ecology and Biogeography, vol. 10, no. 1, pp. 3–13, 2001.

[19] J.-T. Zhang, Y. Xi, and J. Li, "The relationship between environment and plant communities in the middle part of Taihang Mountain Range, North China," Community Ecology, vol. 7, no. 2, pp. 155–163, 2006.

[20] G. C. Stevens, "The elevational gradient in altitudinal range: an extension of Rapoport's latitudinal rule to altitude," American Naturalist, vol. 140, no. 6, pp. 893–911, 1992.

[21] J. I. Olten, G. Paulsen, and W. C. Oechel, Eds., Impacts of Climate Change on Natural Ecosystems, NINA, Trondheim, Norway, 1993.

[22] J.-T. Zhang and T. G. Chen, "Variation of plant communities along an elevation gradient in the Guandi Mountains, North China," Community Ecology, vol. 5, no. 2, pp. 227–233, 2004.

[23] J.-T. Zhang and T. Chen, "Effects of mixed Hippophae rhamnoides on community and soil in planted forests in the Eastern Loess Plateau, China," Ecological Engineering, vol. 31, no. 2, pp. 115–121, 2007.

[24] Z. Kikvidze, L. Khetsuriani, D. Kikodze, and R. M. Callaway, "Seasonal shifts in competition and facilitation in subalpine plant communities of the central Caucasus," *Journal of Vegetation Science*, vol. 17, no. 1, pp. 77–82, 2006.

[25] J.-T. Zhang, *Quantitative Ecology*, Science Press, Beijing, China, 2004.

[26] Z. Q. Ma, *Vegetation of Shanxi Province*, China Science and Technology, Beijing, China, 2001.

[27] J.-T. Zhang, "The vertical vegetation zones of Luya mountains in Shanxi Province," *Scientia Geographica Sinica*, vol. 9, no. 4, pp. 346–353, 1989.

[28] Suriguga, J.-T. Zhang, B. Zhang et al., "Forest community analysis in the Songshan National Nature Reserve of China using self-organizing map," *Russian Journal of Ecology*, vol. 42, no. 3, pp. 216–222, 2011.

[29] H. Hillebrand and B. Matthiessen, "Biodiversity in a complex world: consolidation and progress in functional biodiversity research," *Ecology Letters*, vol. 12, no. 12, pp. 1405–1419, 2009.

[30] M. O. Hill, *TWINSPN-A Fortran Program for Arranging Multivariate Data in an Ordered Two-Way Table by Classification of the Individuals and Attributes*, Cornell University, Ithaca, NY, USA, 1979.

[31] J.-T. Zhang and F. Zhang, "Diversity and composition of plant functional groups in mountain forests of the Lishan Nature Reserve, North China," *Botanical Studies*, vol. 48, no. 3, pp. 339–348, 2007.

[32] J. Podani, *Introduction to the Exploration of Multivariate Biological Data*, Backhuys, Leiden, The Netherlands, 2000.

[33] C. J. F. ter Braak and P. Šmilauer, *CANOCO Reference Manual and User's Guide to Canoco for Windows. Software for Canonical Community Ordination (version 4.5)*, Centre for Biometry Wageningen, Wageningen, The Netherlands; Microcomputer Power, Ithaca, NY, USA, 2002.

[34] E. C. Pielou, *Ecological Diversity*, Wiley and Sons, London-John, UK, 1975.

[35] J.-T. Zhang and R. Oxley, "A comparison of three methods of multivariate analysis of upland grasslands in North Wales," *Journal of Vegetation Science*, vol. 5, no. 1, pp. 71–76, 1994.

[36] S. M. Wilson, D. G. Pyatt, D. C. Malcolm, and T. Connolly, "The use of ground vegetation and humus type as indicators of soil nutrient regime for an ecological site classification of British forests," *Forest Ecology and Management*, vol. 140, no. 2-3, pp. 101–116, 2001.

[37] J. Doležal and M. Šrůtek, "Altitudinal changes in composition and structure of mountain-temperate vegetation: a case study from the Western Carpathians," *Plant Ecology*, vol. 158, no. 2, pp. 201–221, 2002.

[38] J. C. Lovett, A. R. Marshall, and J. Carr, "Changes in tropical forest vegetation along an altitudinal gradient in the Udzungwa Mountains National Park, Tanzania," *African Journal of Ecology*, vol. 44, no. 4, pp. 478–490, 2006.

[39] J.-T. Zhang and F. Zhang, "Ecological relations between forest communities and environmental variables in the Lishan Mountain Nature Reserve, China," *African Journal of Agricultural Research*, vol. 6, no. 2, pp. 248–259, 2011.

[40] X. Z. Zhu and J.-T. Zhang, "Altitudinal patterns of plant diversity of China Mountains," *Acta Botanica Boreali-Occidentalia Sinica*, vol. 25, no. 7, pp. 1480–1486, 2005.

[41] B. Tóthmérész, "On the characterization of scale-dependent diversity," *Abstracta Botanica*, vol. 22, no. 1-2, pp. 149–156, 1998.

[42] Z. H. Cheng and J.-T. Zhang, "Difference between tourism vegetation landscapes of different distance," *Journal of Mountain Science*, vol. 21, no. 6, pp. 647–652, 2001.

[43] Z. H. Cheng and J.-T. Zhang, "Difference between tourism vegetation landscapes of different distance," *Journal of Mountain Science*, vol. 21, no. 6, pp. 647–652, 2003.

18

The Ecological Response of *Carex lasiocarpa* Community in the Riparian Wetlands to the Environmental Gradient of Water Depth in Sanjiang Plain, Northeast China

Zhaoqing Luan,[1] **Zhongxin Wang,**[1,2] **Dandan Yan,**[1,2] **Guihua Liu,**[1,2] **and Yingying Xu**[1]

[1] *Key Laboratory of Wetland Ecology and Environment Science, Northeast Institute of Geography and Agroecology,*
 Chinese Academy of Sciences, 4888 Shengbei Road, Changchun 130102, China
[2] *University of Chinese Academy of Sciences, 19A Yuquan Road, Beijing 100049, China*

Correspondence should be addressed to Zhaoqing Luan; luanzhaoqing@neigae.ac.cn

Academic Editors: J. Bai, H. Cao, B. Cui, and A. Li

The response of *Carex lasiocarpa* in riparian wetlands in Sanjiang Plain to the environmental gradient of water depth was analyzed by using the Gaussian Model based on the biomass and average height data, and the ecological water-depth amplitude of *Carex lasiocarpa* was derived. The results indicated that the optimum ecological water-depth amplitude of *Carex lasiocarpa* based on biomass was [13.45 cm, 29.78 cm], while the optimum ecological water-depth amplitude of *Carex lasiocarpa* based on average height was [2.31 cm, 40.11 cm]. The intersection of the ecological water-depth amplitudes based on biomass and height confirmed that the optimum ecological water-depth amplitude of *Carex lasiocarpa* was [13.45 cm, 29.78 cm] and the optimist growing water-depth of *Carex lasiocarpa* was 21.4 cm. The TWINSPAN, a polythetic and divisive classification tool, was used to classify the wetland ecological series into 6 associations. Result of TWINSPAN matrix classification reflected an obvious environmental gradient in these associations: water-depth gradient. The relation of biodiversity of *Carex lasiocarpa* community and water depth was determined by calculating the diversity index of each association.

1. Introduction

Water regime, as distinct from instantaneously measured water depth, has been implicated in affecting the composition, diversity, and distribution of macrophyte communities [1–6]. While the influence of water level fluctuation on the germination and establishment of wetland seed banks has been well documented, there is few research on the impact of water level on the growth of mature plants [2]. Water regime (depth, duration, and frequency of flooding) is the principal factor determining plant species distribution along the the land-water interface in wetlands [1, 6–17]. *Carex lasiocarpa* wetland is the main wetland type in the mire wetlands in Sanjiang Plain, Northeast China [18]. *Carex lasiocarpa*, a perennial Cyperaceae moss grass, is a clonal perennial which can form nearly monospecific stands on shorelines and lakesides [2, 19–23]. Where water conditions permit, such as in bays protected from waves, the species

sometimes forms thick, floating mats. These floating mats often support a rich array of other plant life adapted to wet infertile conditions. Hence, this particular species of *Carex* is important in producing distinctive plant communities along lakes and rivers [2, 19, 20]. In wetlands in Sanjiang Plain it is generally considered to be an indicator species for wetlands [18, 19, 21, 22].

Responses of diversity of assemblages and individual species to water depth are necessary to be considered in the management and restoration efforts of wetland ecosystems; therefore, understanding plant response to hydrologic conditions is important both to the maintenance of native biodiversity and to the design of management strategies appropriate to a specific wetland [8, 24–26]. At present, research on the vegetation in Sanjiang plain wetlands still mostly focuses on the traditional classification and description, while little attention paid on the ecological pattern and the process of quantitative research, especially the quantitative research on

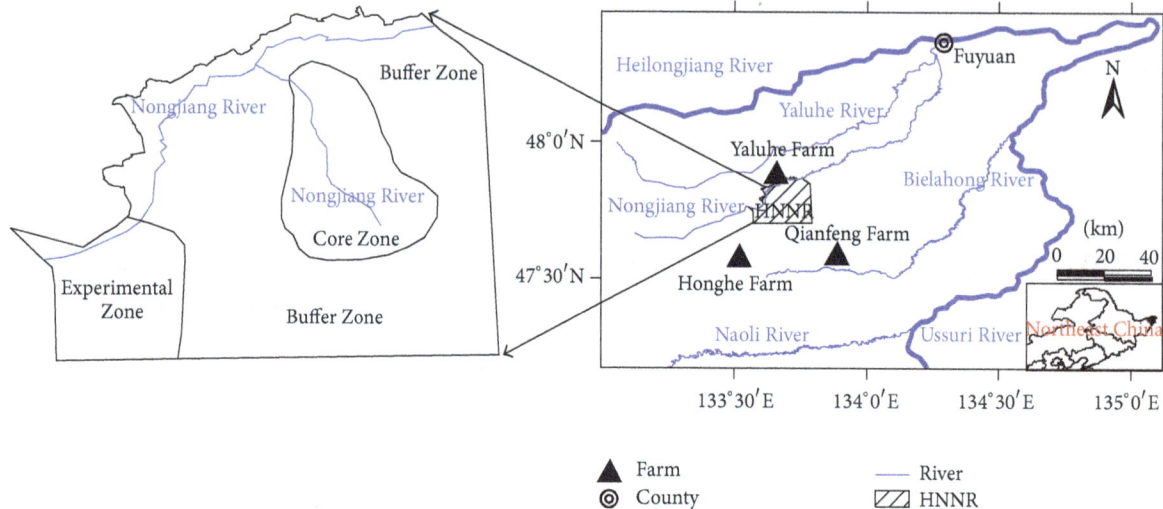

FIGURE 1: Location of the study area.

the relationship between wetland hydrological and vegetation [2, 20, 21, 27–31]. The present study sought to achieve the following two objectives: (1) analyze the response of *Carex lasiocarpa* populations to the environmental gradient of water depth using the biological characters of biomass and height based on Gaussian Model and figure out the ecological amplitude of reed populations to water depth; (2) explore the relationship between the ecological characteristics of *Carex lasiocarpa* communities and water-depth variables. Based on the relationship between the ecological characteristics of wetland vegetation and hydrological regime, the suggestion of ecohydrological management for wetland ecosystem is proposed.

2. Materials and Methods

2.1. Study Area.
The Honghe National Nature Reserve (HNNR) ($47°42'18''$N–$47°52'$N and $133°34'38''$E–$133°46'29''$E) is located in the northeast of Sanjiang Plain, Northeast China, with an area of 250.9 ha (Figure 1).

HNNR has been listed as the International Important Wetland (Ramsar wetland) since 2001 for being a typical inland wetland and fresh water ecosystem in the north temperate zone [32]. Presently, HNNR has 16 orders, 43 families, and 174 species of waterfowl, including ten species of nationally rare and endangered waterfowl. In addition, 1012 species of plants are founded in HNNR, including six species of nationally endangered plants. With a very low topographic gradient (average slope grade less than 1 : 10,000), this area is favorable to the formation of wetland ecosystems [33].

2.2. Methods

2.2.1. Sampling Method.
Samples were collected in 28 sampling spots (Figure 2) from May to September of 2011 and 47 sampling spots in 2012. Three quadrats (50×50 cm^2) of plants samples were collected with scissors at each sample point. Plants naturally growing in the quadrats were recorded

with their names, abundances, coverage, heights, and aboveground biomass (dry weight). Coordinate of each sampling quadrat was also recorded by using GPS.

2.2.2. Data Analysis.
(1) Gaussian Model was adopted to describe the species-environmental relations. The study on the response of reed to water depth based on the Gaussian Model has been achieved good effect [34, 35]. The Gaussian Model was shown as the following equation:

$$y = ce^{[-(1/2)(x-u)^2/t^2]}, \qquad (1)$$

where y represents an indicator of biological characteristics of plant species, which can be abundance, coverage, density or biomass, and so on; c is the maximum of y; x is the value of environmental factor, u is the optimum ecological amplitude of species to environmental factor; and t is species tolerance. Generally optimum ecological amplitude of species to environmental factor change within $2t$ range. The analysis was performed by Excel 2003.

(2) Two-Way Indicator Species Analysis (TWINSPAN) was used to classify the plant community in the study area based on the number of species in all quadrats [36]. The TWINSPAN analysis was performed by using winTWINS 2.3 [37].

(3) The ecological characteristics of plant community were reflected with the following indices [38–40].

Species richness: Margalef index (MA) was adopted, which is expressed as

$$MA = \frac{(S-1)}{\ln N}, \qquad (2)$$

where S is the number of species and N is the number of individuals of all species in a community.

Species diversity: Shannon-Weaver index (H) was used, which can be calculated as

$$H = -\Sigma Pi \ln Pi, \qquad (3)$$

The Ecological Response of Carex lasiocarpa Community in the Riparian Wetlands to the Environmental Gradient of
Water Depth in Sanjiang Plain, Northeast China

187

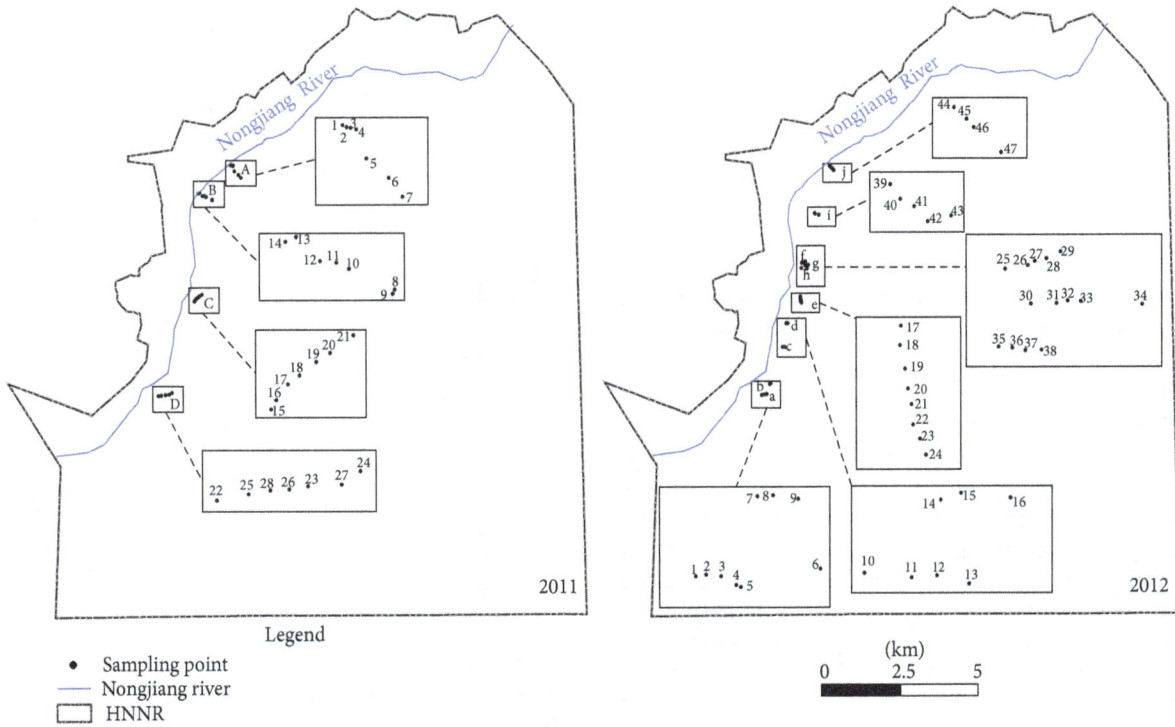

Legend

•   Sampling point
—   Nongjiang river
▭   HNNR

(km)

0   2.5   5

Figure 2: Location of sampling spots.

where $Pi = ni/N$, ni is the importance value of species i, and N is the sum of the importance value of all species in a community.

Species evenness: Pielou evenness index (E) was used, which is expressed as:

$$E = \frac{H}{\ln(S)}, \qquad (4)$$

where S is the number of species and H is the Shannon-Weaver index.

(4) Data was analyzed by using SPSS17.0 and Microsoft Excel.

3. Results and Discussion

3.1. Response of Carex lasiocarpa to Water Depth

3.1.1. Result of Statistical Analyses. 17 *Carex lasiocarpa* community sampling data (spots without *Carex lasiocarpa* removed) from May to September in 2011 were statistically analyzed (Table 1). The water depth ranged within 2.5–37.5 cm with a mean value of 17.18 cm. Average population height ranged within 38–73.25 cm, and the average value was 59.64 cm. Range of populations biomass was 1.3–47.68, and the mean value was 25.37. All the sampling data was of normal distribution.

3.1.2. Response of the Population Biomass of Carex lasiocarpa to Water Depth. Population biomass of *Carex lasiocarpa* was strongly correlated to water depth ($R^2 = 0.7229$,

Table 1: The statistical description of *Carex lasiocarpa* community in different sampling points.

Sampling points	Water depth (cm)	Population height (cm)	Population biomass (g/m²)
2	22.70	53.25	33.54
4	12.80	64.00	24.96
5	8.20	52.50	26.86
6	5.70	38.00	1.30
7	3.10	43.00	2.63
8	26.80	73.25	28.10
9	27.90	69.50	47.68
10	20.10	69.00	36.90
12	18.80	68.75	35.92
13	13.50	60.75	29.94
17	31.50	54.50	10.92
19	37.50	48.75	9.15
21	22.20	66.67	23.58
23	9.60	69.25	28.41
25	15.00	66.00	39.00
26	14.20	71.25	46.05
28	2.50	45.50	6.36
Mean	17.18	59.64	25.37
Max	37.50	73.25	47.68
Min	2.50	38.00	1.30

$P < 0.01$). Quadratic curve fitting was used to fit the relationship between population biomass data of *Carex lasiocarpa* (after natural logarithm transformation) and water-depth data, and the obtained quadratic curve was fit with gaussian regression (Figure 3). With regression analyses, the Gaussian regression equation and regression curve were obtained, which was expressed as the following equation:

$$y = 39.82 \exp\left[-\frac{(1/2)(x - 21.61)^2}{8.16^2}\right]. \qquad (5)$$

The results indicated that the optimum ecological amplitude of *Carex lasiocarpa* to water depth based on population biomass was [13.45 cm, 29.78 cm] and the optimist growing point is 21.6 cm.

3.1.3. Response of the Population Height of Carex lasiocarpa to Water Depth.
Population height of *Carex lasiocarpa* was strongly correlated to water depth ($R^2 = 0.6685$, $P < 0.01$). Quadratic curve fitting was used to fit the relationship between population height data of *Carex lasiocarpa* (after natural logarithm transformation) and water-depth data, and the obtained quadratic curve was fit with Gaussian regression (Figure 4). With regression analysis, the Gaussian regression equation and regression curve were obtained, which was expressed as the following equation:

$$y = 68.43 \exp\left[-\frac{(1/2)(x - 21.21)^2}{18.90^2}\right]. \qquad (6)$$

The results indicated that the optimum ecological amplitude of *Carex lasiocarpa* to water depth based on population average height was [2.31 cm, 40.11 cm] and the optimist growing point is 21.2 cm.

3.1.4. The Optimum Ecological Amplitude of Carex lasiocarpa to Water Depth.
An intersection of the ecological amplitudes based on biomass ([13.45 cm, 29.78 cm]) and height ([2.31 cm, 40.11 cm]) was carried out to figure out the ecological amplitude of *Carex lasiocarpa* to water depth in general. The final result confirmed that the optimum ecological amplitude of *Carex lasiocarpa* to water depth was [13.45 cm, 29.78 cm] and the optimist growing point of *Carex lasiocarpa* to water depth was 21.4 cm.

3.2. Response of Community Diversity of Carex lasiocarpa to Water Depth.
By using TWINSPAN, the 47 sampling spots in 2012 were classified into 6 groups at the end of division (Figure 5).

Association Group I: Association *Carex pseudocuraica* + *Carex lasiocarpa*. This was also a hygrophyte association group including 3 sampling spots (S19, S20, S24). *Carex pseudo-curaica* and *Carex lasiocarpa* were dominant species, while the others were companion species. *Carex pseudo-curaica* occurred in relatively deep water conditions.

Association Group II: Assoc. *Carex pseudo-curaica* + *Carex lasiocarpa* + *Glyceria spiculosa*. This was also a hygrophyte association group including 19 sampling spots (S2, S3, S4, S5, S7, S8, S9, S10, S12, S13, S14, S15, S16, S18, S21, S23, S25, S42, S43). The group occured in relatively moderate water-depth conditions.

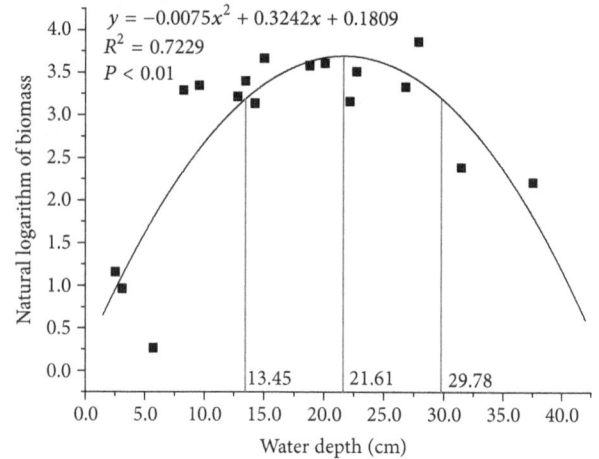

FIGURE 3: The secondary nonlinear regression based on Gaussian Model of *Carex lasiocarpa* population biomass and water depth.

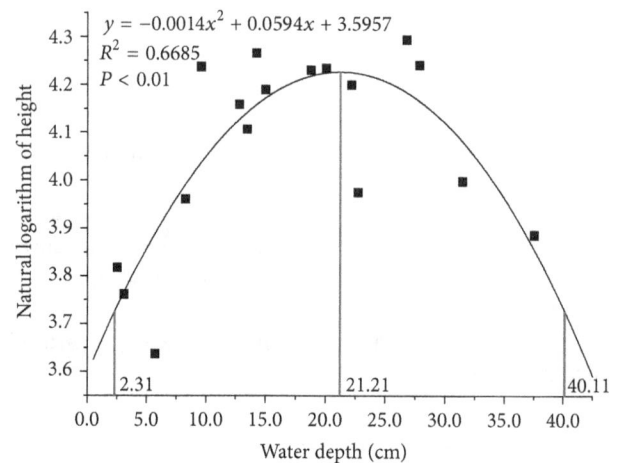

FIGURE 4: The secondary nonlinear regression based on Gaussian Mode of *Carex lasiocarpa* populations height and water depth.

Association Group III: Assoc. *Carex lasiocarpa* + *Carex pseudo-curaica* + *Glyceria spiculosa* + *Carex dispalata*. This was also a hygrophyte association group including 4 sampling spots (S31, S32, S33, S44). The group occured relatively in moderate water depth conditions.

Association Group IV: Assoc. *Glyceria spiculosa* + *Carex lasiocarpa* + *Carex pseudo-curaica* + *Calamagrostis angustifolia*. This was also a mesophyte association group including 12 sampling spots (S1, S11, S22, S26, S27, S34, S35, S36, S39, S40, S45, S47).

Association Group V: Assoc. *Carex lasiocarpa* + *Calamagrostis angustifolia* + *Carex pseudo-curaica*. This was also a mesophyte association group including 7 sampling spots (S6, S17, S30, S37, S38, S41, S46). *Carex lasiocarpa*, *Calamagrostis angustifolia* and *Carexpseudo-curaica* were dominant species, and the others were companion species.

Association Group VI: Assoc. *Calamagrostis angustifolia* + *Carex lasiocarpa*. This was also a mesophyte association

The Ecological Response of Carex lasiocarpa Community in the Riparian Wetlands to the Environmental Gradient of
Water Depth in Sanjiang Plain, Northeast China

189

```
               I                  II                              III       IV                      V            VI
            1 2 2    2 2                 1 1 1 1 1 1 2 4 4   3 3 3 4  1 2 3 4 3 2 2     3 4 4 3  1   3 4 4 3 3  2 2
            9 0 4  3 1 3 9 2 4 5 7 8 0 2 3 4 5 6 8 5 2 3    1 2 3 4  1 2 6 5 5 6 7 1 9 0 7 4   7 6 0 1 6 7 8  8 9

 2   2      - - -  5 4 5 5 5 4 4 5 4 - 5 4 5 5 3 - 5 5 4    5 3 4 5  5 2 5 5 5 5 5 5 5 5 5 5   - - 2 3 2 5 5  - -    0 0
12  12      2 3 3  - 3 3 - - - - - - - - - - - - - - - -    - - - -  - - - - - - - - - - - -   - - - - - - -  - -    0 0
13  13      - 3 -  - - - - - - - - - - - - - - - - - - -    - - - -  - - - - - - - - - - - -   - - - - - - -  - -    0 0
 1   1      5 5 5  5 5 5 5 5 5 5 5 5 5 5 4 5 5 5 5 5 5 5    5 5 5 5  5 5 4 5 5 3 5 5 5 5 5      5 5 5 5 5 5 5  3 4    0 1 0
 3   3      5 5 5  5 5 5 - 5 5 5 5 4 5 5 5 5 5 5 5 5 5 5    5 5 5 3  5 5 5 3 5 5 4 5 5 5 5 5    5 5 4 5 5 5 2  2 -    0 1 0
 6   6      - - -  - - - 4 - - - - - - - - - - - - - - -    - - - -  1 - - - - - - - - - - -   - - - 1 - - -  - -    0 1 0
 9   9      - - 2  - - - - - - 1 - 1 - - - 2 4 2 - - - 2    - - - 2  - - - 2 - 5 5 - - - - 2   - - - - - - 1  5 1    0 1 0
 5   5      - - -  - - - - - - - 1 - 1 - - - - - - - - -    3 - 2 -  3 2 5 4 5 - - - - - 1 1   3 - 1 - 1 5 1  - -    0 1 1
 7   7      - 3 1  2 1 2 - - - - - - - - - - - - - - - -    - - - -  1 - - - - - - - - - - -   3 2 - - - 2 -  - -    0 1 1
10  10      - - -  - - - - - - - - - - - - - - - - - - -    - - - -  - - - - - - - - - - - -   - 4 - - 2 2 -  - -    1 0 0 0
17  17      - - -  - - - - - - - - - - - - - - - - - - -    - - - -  - - 1 - - - - - - - - -   - - - - 1 3 -  - -    1 0 0 0
14  14      - - -  - - - - - - - - - - - - - - - 1 1       3 - - -  - - - - - - - - - - - -   - 3 1 - 2 2 -  - -    1 0 0 1
15  15      - - -  - - - - - - - - - - - - - - - - - -     3 4 4 3  - - - - - - - - - - - 2   - - 4 4 5 - 5  - -    1 0 0 1
16  16      - - -  - - - - - - - - - - - - - - - - - -     2 - 1 -  - - - 1 - - - - - - - 2   - - - - 2 2 -  - -    1 0 0 1
 8   8      2 4 -  1 1 - - 1 - - - - - - - 3 - - 1 - - -    - - 1 -  - - 2 - - 1 - - - - - 2   - - - - 5 5 -  - -    1 0 1
 4   4      - - -  - - - - - - - - - - - - - - - - - -     - - - -  - - - 2 2 5 3 2 5 4 5    5 5 5 4 5 5 5 5         1 1
11  11      - - -  - - - - - - - - - - - - - - - - - -     - - - -  - - - - - - - - - - - 2   - - - - - 2 -  - -    1 1


            0 0 0  0 0 0 0 0 0 0 0 0 0 0 0 0 0 0 0 0 0 0    0 0 0 0  0 0 0 0 0 0 0 0 0 0 0 0   0 1 1 1 1 1 1  1 1
            0 0 0  1 1 1 1 1 1 1 1 1 1 1 1 1 1 1 1 1 1 1    1 1 1 1  1 1 1 1 1 1 1 1 1 1 1 1   1 0 0 0 0 0 0  1 1
                   0 0 0 0 0 0 0 0 0 0 0 0 0 0 0 0 0 0 0    0 0 0 0  0 0 0 0 0 1 1 1 1 1 1 1   1 0 0 0 0 1 1
                   0 0 0 0 0 0 0 0 0 0 0 0 0 0 0 0 0 0 0    1 1 1 1  1 1 1 1 1 0 0 0 0 0 0 0   1
                     0 0 0 1 1 1 1 1 1 1 1 1 1 1 1 1 1 1 1  0 0 0 0  1 1 1 1 1 0 0 0 0 0 0 0 1
                       0 1 1 1 1 1 1 1 1 1 1 1 1 1 1 1      0 0 0 0 1 0 0 1 1 1 1
```

FIGURE 5: TWINSPAN analyses. Note: *1-Carex lasiocarpa, 2-Glyceria spiculosa, 3-Carex pseudo-curaica, 4-Calamagrostis angustifolia, 5-Galium manshuricum Kitag., 6-Galium dahuricum Turcz, 7-Comarum palustre L., 8-Equisetum fluviatile, 9-Carex humida, 10-Phragmites australis, 11-Anemone dichotoma, L. 12-Menyanthes trifoliate, 13-Achillea acuminate, 14-Lathyrus quinquenervius., 15-Carex dispalata, 16-Salix rosmarinifolia, L. 17-Caltha palustris var, sibirica.*

group including 2 sampling spots (S28, S29). *Calamagrostis angustifolia* occurred in relatively shallow water conditions.

TWINSPAN classification matrix results reflected an obvious environmental gradient: water depth. The weighted-average wetland indicator status for each community type reflected the distribution of community types along the hydrologic gradient. The matrix diagram reflected that from association 1 to association 6 the water depth was gradually reduced, which determined the distribution range of these species.

Based on the inquisitional data of sampling sites, the biodiversity of the *Carex lasiocarpa* community in HNNR was analyzed by adopting diversity index, richness index, and evenness index.

The characteristics of the plant community can reflect vegetation functions and ecological niche. In our research, the characteristics of plant community (species richness MA, species diversity index H, and species evenness index E) were supposed to be different among different water-depth areas. Therefore, 47 quadrats were divided into 5 groups based on different water depth, and then the biodiversities indices of plant community were analyzed (Figure 6). The

results demonstrated that the water depth was distinct in different vegetation types, and the optimal water depth for *Carexpseudo-curaica* was the deepest, followed by *Glyceria spiculosa, Carex lasiocarpa*, and *Calamagrostis angustifolia*. It was observed that the evenness index and diversity index of *Carex lasiocarpa* communities were low when water depth was too high or too low, while species evenness was poorer. When the water depth was moderate, *Carex lasiocarpa* species distributes evenly (Figure 5). Species richness of association I was high because of relatively more species, but when water depth was too high, *Carex pseudocuraica* and *Carex lasiocarpa* occured as dominant species, and other associated species were rare. *Carex pseudocuraica, Carex lasiocarpa*, and *Glyceria spiculosa* in association II were main dominant species, accompanying species was less, therefore species richness was low. The number of species in association III, IV, and V was more, and evenness and abundance of plant communities were higher. The growth of *Carex lasiocarpa* community vegetation was significantly correlated with water depth, which can be found from biodiversity of each association. With the increase or decrease of water depth, density of *Carex lasiocarpa* community decreased while biodiversity increased.

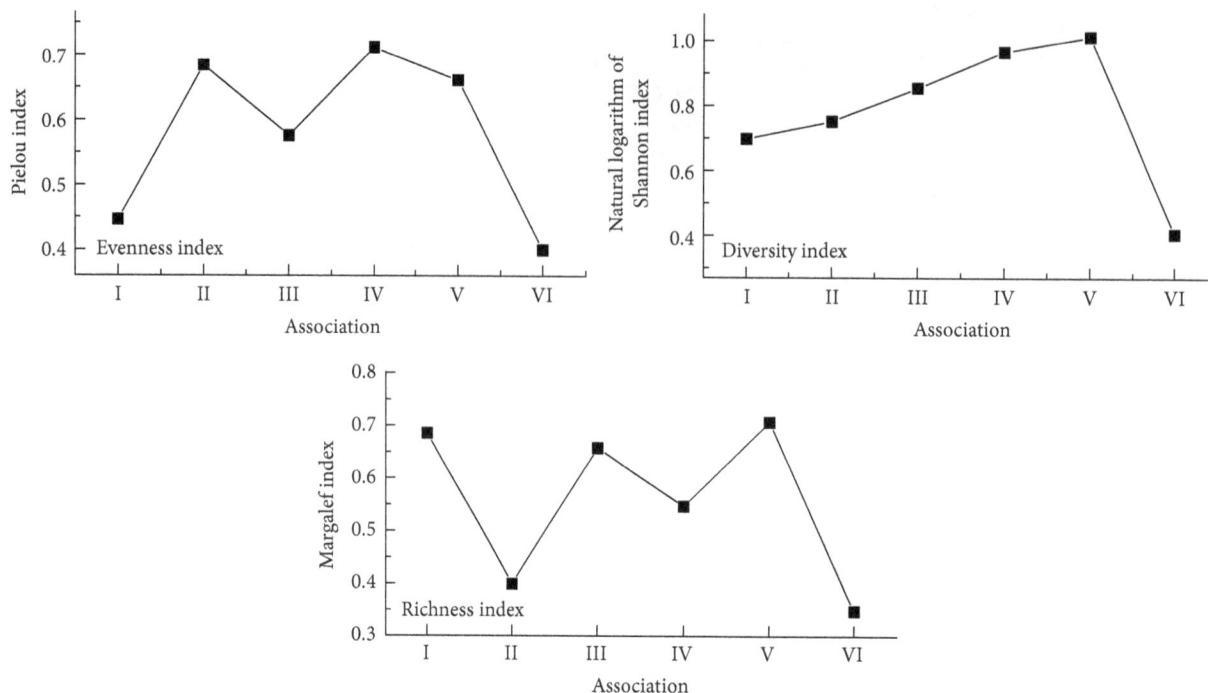

FIGURE 6: *Carex lasiocarpa* community biodiversity index.

4. Conclusions

The results indicated that the optimum ecological amplitude of *Carex lasiocarpa* to water depth based on population biomass was [13.45 cm, 29.78 cm], while the optimum ecological amplitude of *Carex lasiocarpa* to water depth based on average height was [2.31 cm, 40.11 cm]. The optimum ecological amplitude of *Carex lasiocarpa* to water depth was [13.45 cm, 29.78 cm] and the optimist growing point of *Carex lasiocarpa* to water depth was 21.4 cm.

TWINSPAN classification matrix results reflected an obvious environmental gradient for wetland plant species: water-depth gradient. *Carex lasiocarpa* maintains high cover across most water-depth gradients but requires high variation at the wettest conditions. Water depth for plant species in the freshwater marsh showed the order as *Carex pseudocuraica* > *Carex lasiocarpa* > *Glyceria spiculosa* > *Calamagrostis angustifolia*.

The growth of *Carex lasiocarpa* community was significantly correlated with water depth. With the increase or decrease of water depth, the densityof *Carex lasiocarpa* decreased, and the evenness index and diversity index of *Carex lasiocarpa* communities were low. In moderate water depth condition, the density of *Carex lasiocarpa* was the highest.

Acknowledgments

This study was funded by the National Natural Science Foundation of China (NSFC 41001050). The research also received support from the Projects of the National Basis Research Program of China (2009CB421103), and the Special S&T Project on Treatment and Control of Water Pollution (2012ZX07201004). The authors would like to thank the Sanjiang Marsh Wetland Experimental Station, Chinese Academy of Sciences, and Honghe National Natural Reserve for their help with our field work and ecohydrological monitoring. The authors are indebted to all editors and reviewers for their critical reading, kind remarks, and relevant comments.

References

[1] M. F. Carreño, M. A. Esteve, J. Martinez, J. A. Palazón, and M. T. Pardo, "Habitat changes in coastal wetlands associated to hydrological changes in the watershed," *Estuarine, Coastal and Shelf Science*, vol. 77, no. 3, pp. 475–483, 2008.

[2] Y. H. Ji, X. G. Lu, Q. Yang, and K. Y. Zhao, "The succession character of *Carex lasiocarpa* community in the Sanjiang Plain," *Wetland Science*, vol. 2, pp. 140–144, 2004.

[3] T. Nakayama, "Shrinkage of shrub forest and recovery of mire ecosystem by river restoration in northern Japan," *Forest Ecology and Management*, vol. 256, no. 11, pp. 1927–1938, 2008.

[4] M. C. Thoms, "Floodplain-river ecosystems: lateral connections and the implications of human interference," *Geomorphology*, vol. 56, no. 3-4, pp. 335–349, 2003.

[5] C. L. Yi, S. M. Cai, J. L. Huang, and R. R. Li, "Classification of wetlands and their distribution of the Jianghan-Dongting Plain, central China," *Journal of Basic Science and Engineering*, vol. 6, pp. 19–25, 1998.

[6] D. Zhou, H. Gong, Z. Luan, J. Hu, and F. Wu, "Spatial pattern of water controlled wetland communities on the Sanjiang Floodplain, Northeast China," *Community Ecology*, vol. 7, no. 2, pp. 223–234, 2006.

[7] S. C. L. Watt, E. García-Berthou, and L. Vilar, "The influence of water level and salinity on plant assemblages of a seasonally flooded Mediterranean wetland," *Plant Ecology*, vol. 189, no. 1, pp. 71–85, 2007.

[8] T. K. Magee and M. E. Kentula, "Response of wetland plant species to hydrologic conditions," *Wetlands Ecology and Management*, vol. 13, no. 2, pp. 163–181, 2005.

[9] K. A. Dwire, J. B. Kauffman, and J. E. Baham, "Plant species distribution in relation to water-table depth and soil redox potential in montane riparian meadows," *Wetlands*, vol. 26, no. 1, pp. 131–146, 2006.

[10] T. Riis and I. Hawes, "Relationships between water level fluctuations and vegetation diversity in shallow water of New Zealand lakes," *Aquatic Botany*, vol. 74, no. 2, pp. 133–148, 2002.

[11] B. D. Richter, "A method for assessing hydrologic alteration within ecosystems," *Conservation Biology*, vol. 10, no. 4, pp. 1163–1174, 1996.

[12] P. A. Keddy, *Wetland Ecology: Principles and Conservation Edition*, Cambridge University Press, Cambridge, UK, 2nd edition, 2010.

[13] K. S. Godwin, J. P. Shallenberger, D. J. Leopold, and B. L. Bedford, "Linking landscape properties to local hydrogeologic gradients and plant species occurrence in minerotrophic fens of New York State, USA: a hydrogeologic setting (HGS) framework," *Wetlands*, vol. 22, no. 4, pp. 722–737, 2002.

[14] H. Y. Zhang, Y. B. Qian, Z. N. Wu, and Z. C. Wang, "Vegetation-environment relationships between northern slope of Karlik Mountain and Naomaohu Basin, East Tianshan Mountains," *Chinese Geographical Science*, vol. 22, no. 3, pp. 288–301, 2012.

[15] J. Bai, Q. Wang, W. Deng, H. Gao, W. Tao, and R. Xiao, "Spatial and seasonal distribution of nitrogen in marsh soils of a typical floodplain wetland in Northeast China," *Environmental Monitoring and Assessment*, vol. 184, no. 3, pp. 1253–1263, 2012.

[16] J. H. Bai, H. F. Gao, R. Xiao, J. J. Wang, and C. Huang, "A review of Soil nitrogen mineralization in coastal wetlands: issues and methods," *Clean-Soil, Air, Water*, vol. 40, no. 10, pp. 1099–1105, 2012.

[17] R. Xiao, J. H. Bai, H. F. Gao, L. B. Huang, and W. Deng, "Spatial distribution of phosphorous in marsh soils from a typical land/inland water ectone along a hydrological gradient," *Catena*, vol. 98, pp. 96–103, 2012.

[18] K. Y. Zhao, *The Marsh of China*, Science Press, Beijing, China, 1999.

[19] Chinese wetland vegetation editing committee, *Wetland Vegetation in China*, Science Press, Beijing, China, 1999.

[20] Y. J. Lou and K. Y. Zhao, "Study of species diversity of *Carex lasiocarpa* commun ity in Sanjiang plain for 30 years," *Journal of Arid Land Resources and Environment*, vol. 5, no. 22, pp. 182–186, 2008.

[21] L. L. Wang, C. C. Song, J. M. Hu, and T. Yang, "Growth responses of *Carex lasiocarpa* to different water regimes at different growing stages," *ActaPratacul Turae Sinica*, vol. 18, pp. 17–24, 2009.

[22] X. T. Liu and X. H. Ma, *Natural Environment Change and Ecological Conservation of Sanjiang Plain*, Science Press, Beijing, China, 2002.

[23] Z. G. Liu, M. Wang, and X. H. Ma, "Estimation of storage and density of organic carbon in peatlands of China," *Chinese Geographical Science*, vol. 22, no. 6, pp. 637–646, 2012.

[24] B. Wen, X. Liu, X. Li, F. Yang, and X. Li, "Restoration and rational use of degraded saline reed wetlands: a case study in western Songnen Plain, China," *Chinese Geographical Science*, pp. 1–11, 2012.

[25] J. H. Bai, R. Xiao, K. J. Zhang, and H. F. Gao, "Arsenic and heavy metal pollution in wetland soils from tidal freshwater and salt marshes before and after the flow-sediment regulation regime in the Yellow River Delta, China," *Journal of Hydrology*, vol. 450-451, pp. 244–253, 2012.

[26] P. P. Liu, J. H. Bai, Q. Y. Ding, H. B. Shao, H. F. Gao, and R. Xiao, "Effects of water level and salinity on TN and TP contents in wetland soils of the Yellow River Delta, China," *Clean-Soil, Air, Water*, vol. 40, no. 10, pp. 1118–1124, 2012.

[27] C. He, "Dynamics of litter and under-ground biomass in Carex lasiocarpa wetland on Sanjiang Plain," *Chinese Journal of Applied Ecology*, vol. 14, no. 3, pp. 363–366, 2003.

[28] C. He and K. Zhao, "Fractal relationship between aboveground biomass and plant length or sheath height of *Carex lasiocarpa* population," *Chinese Journal of Applied Ecology*, vol. 14, no. 4, pp. 640–642, 2003.

[29] Q. He, B. S. Cui, X. S. Zhao, H. L. Fu, and X. L. Liao, "Relationships between salt marsh vegetation distribution/diversity and soil chemical factors in the Yellow River Estuary, China," *Acta Ecologica Sinica*, vol. 29, no. 2, pp. 676–687, 2009.

[30] X. Tan and X. Zhao, "Spatial distribution and ecological adaptability of wetland vegetation in Yellow River Delta along a water table depth gradient," *Chinese Journal of Ecology*, vol. 25, no. 12, pp. 1460–1464, 2006.

[31] D. L. Wu, T. L. ShangGuan, and J. T. Zhang, "Species diversity of wetland vegetation in Hutuo river valley," *Journal of Beijing Normal Univer Sity*, vol. 42, no. 2, pp. 195–199, 2006.

[32] The List of Wetlands of International Importance http://www .ramsar.org/pdf/sitelist_order.pdf.

[33] D. Zhou, Z. Luan, X. Guo, and Y. Lou, "Spatial distribution patterns of wetland plants in relation to environmental gradient in the Honghe National Nature Reserve, Northeast China," *Journal of Geographical Sciences*, vol. 22, no. 1, pp. 57–70, 2012.

[34] B. S. Cui, X. S. Zhao, Z. F. Yang, N. Tang, and X. J. Tan, "The response of reed community to the environment gradient of water depth in the Yellow River Delta," *Acta Ecologica Sinica*, vol. 26, no. 5, pp. 1533–1541, 2006.

[35] Z. L. Bi, X. Xiong, F. Lu, Q. He, and X. S. Zhao, "Studies on ecological amplitude of reed to the environmental gradient of water depth," *Shandong Forestry Science and Technology*, no. 4, pp. 1–3, 2007.

[36] M. O. Hill, *TWINSPAN: A FORTRAN Program for Arranging Multivariate Data in an Ordered Two-Way Table by Classification of the Individuals and Attributes, Ecology and Systematics*, Cornell University, Ithaca, NY, USA, 1979.

[37] M. O. Hill and P. Šmilauer, *TWINSPAN for Windows Version 2.3*, 2005.

[38] J. T. Zhang, *Quantitative Ecology*, Science Press, Beijing, China, 2010.

[39] F. Z. Kong, R. C. Yu, Z. J. Xu, and M. J. Zhou, "Application of excel in calculation of biodiversity indices," *Marine Sciences*, vol. 36, no. 4, pp. 57–62, 2012.

[40] J. Z. Ren, *Grassland Research Methods*, Chinese Agriculture Press, Beijing, China, 1998.

Impacts of Intensified Agriculture Developments on Marsh Wetlands

Zhaoqing Luan[1] and Demin Zhou[1,2]

[1] Key Laboratory of Wetland Ecology and Environment, Northeast Institute of Geography and Agricultural Ecology, Chinese Academy of Sciences, 4888 Shengbei Street, Changchun 130102, China
[2] College of Resources, Environment and Tourism, Capital Normal University, Beijing 100048, China

Correspondence should be addressed to Demin Zhou; zhoudemin@neigae.ac.cn

Academic Editors: J. Bai, H. Cao, and A. Li

A spatiotemporal analysis on the changes in the marsh landscape in the Honghe National Nature Reserve, a Ramsar reserve, and the surrounding farms in the core area of the Sanjiang Plain during the past 30 years was conducted by integrating field survey work with remote sensing techniques. The results indicated that intensified agricultural development had transformed a unique natural marsh landscape into an agricultural landscape during the past 30 years. Ninety percent of the natural marsh wetlands have been lost, and the areas of the other natural landscapes have decreased very rapidly. Most dry farmland had been replaced by paddy fields during the progressive change of the natural landscape to a farm landscape. Attempts of current Chinese institutions in preserving natural wetlands have achieved limited success. Few marsh wetlands have remained healthy, even after the establishment of the nature reserve. Their ecological qualities have been declining in response to the increasing threats to the remaining wetland habitats. Irrigation projects play a key role in such threats. Therefore, the sustainability of the natural wetland ecosystems is being threatened by increased regional agricultural development which reduced the number of wetland ecotypes and damaged the ecological quality.

1. Introduction

Natural ecosystems, especially freshwater ecosystems in the inland flood plain, are undergoing profound and extensive disturbances by humans worldwide [1–5]. A key indicator of these disturbances is that humans extensively reclaim natural wetlands to expand their economic benefits. Therefore, most habitats of natural ecosystems have been changed into farms or urban areas rapidly and continuously [6–8]. The disturbances have been representatively observed in China, the largest developing country in the world. A good example is the shrinking process of the marsh wetland landscapes on the Sanjiang Plain in Northeast China [9, 10].

With its rapid development, China can be regarded as a typical country of most other developing countries in the world. China has experienced high-speed development in the past 30 years. Scientifically assessing or even imagining the impact of urbanization and agricultural reclamation on natural ecosystems is difficult because few countries have comparably rapid and extensive development [11, 12]. During the past 30 years, a large number of natural habitats in China have been reclaimed into cropland, and numerous farmlands have been occupied and then urbanized into towns or cities [7]. With this progress, the Chinese population has rapidly increased and is currently 1.3 billion. The most natural habitats of the wetland ecosystems have been encroached upon during this progress [13]. Though food security is always the top priority for the massive Chinese population [10], the continuous reclamation of the few remaining natural habitats has difficulty meeting the demand of grain production.

Some developing countries, such as China, have published various administrative policies for natural resource protection during their rapid developmental stages. Many natural reserves have been established in the past few years. China has listed the most natural reserves in the world [14]. However, the institutional efficacy of these reserves remains questionable from a scientific perspective [15, 16]. In this paper, the Honghe National Nature Reserve (HNNR) was included within our study area as a wetland reserve. It is also an international wetland listed by the Ramsar Convention.

The institutional efficacy of this Chinese natural reserve was the topic of the present study. Researchers analyzed the spatiotemporal changes of the inner and outer landscapes of the reserve and reached some interesting scientific conclusions.

Chinese scholars have recently become concerned about the great changes in the natural marsh wetlands in China. Many papers have reported research results in this field [9, 13, 17–30]. In these studies, some researchers [9, 13, 31–33] analyzed the marsh landscape on the Sanjiang Plain over periods of 20 or even 50 years. Most research approaches were based on theories of landscape ecology. The integration of remote sensing techniques and geographical information systems was applied for the spatiotemporal analysis of marsh landscape segments. Landscape investigators obtain dynamic information on marsh landscapes with the support of remote sensing techniques [34]. However, these studies lack an analysis on the profound driving forces that impact the wetlands and especially lack a correlational analysis of the linkage between policy issues and regional characteristics that deal with the spatiotemporal dynamics of the marsh wetlands. These previous studies focused more on obtaining data and analyzing dynamic wetland landscapes on large regional scales (e.g., 10000 km^2), which is suitable for the application of remote sensing techniques [35]. Liu and Ma descriptively studied the changes in the natural environments on the entire Sanjiang Plain and its regional ecological response to such changes [9]. Rich survey data and historical statistics of wetlands were used in their study, but the spatiotemporal dynamics of the wetland landscapes were poorly assessed.

Many papers have studied the issue of land use and cover change caused by regional and international urbanization in the past few decades. An abundance of literature has addressed the impact of urbanization and regional development that have encroached on cropland or the reclamation of wild fields in China [10, 11, 36]. Most studies have focused on the spatiotemporal characteristics of changing land use or land cover or have analyzed the relative driving forces. Ecological impact issues related to agricultural activity have long been neglected [14]. Little research has focused on the impact on wetland ecology, linked the dynamics of the marsh landscape over the long term, and studied the driving forces of regional agriculture with a background analysis of historical national policies [7]. This paper provides a case study of the Sanjiang Plain in Northeast China and demonstrates the shrinking process of the typical marsh wetland and other natural landscapes driven by agricultural activity. The ecological impacts on the wetland ecosystems were also analyzed from a regional development perspective. This research will help better understand the gradual evolution of the disturbed natural ecosystems and elucidate the dependence of these natural ecosystems in developing countries. The goal is to help resource administrators determine the evolutionary direction of these ecosystems in the future [37, 38]. An identification of the common characteristics of these natural ecosystems will significantly impact decision making in the management of surviving natural ecosystems in developing countries [38, 39].

The present study sought to achieve three objectives: (1) present the spatiotemporal process of the encroachment of expanding farmland on wild marsh landscapes in the core area on the Sanjiang Plain since 1975, which is a microcosm of shrinking natural wetland ecosystems worldwide; (2) analyze the characteristics of the driving forces that continuously reduce the marsh wetland area in this region, with an emphasis on discussing Chinese policies related to intensified agricultural development on a local scale; and (3) study the negative impact of marsh reclamation on natural ecosystems. An international wetland is used as a typical example to show readers the ecological impact of agricultural activity on marsh wetlands and assess the functional efficacy of this natural reserve.

2. Materials and Methods

2.1. Study Area. The HNNR and its three surrounding farms (Yaluhe Farm, Honghe Farm, and Qianfeng Farm) were selected as our study area. The study area is located in the northeast region of Heilongjiang (47°25′N-48°1′N, 133°18′E-134°5′E), the core area of the Sanjiang Plain (Figure 1). It covers 2416.8 km^2 in the neighboring area of Tongjiang County and Fuyuan County. This area was a unique marsh wetland landscape 30 years ago. The establishment of local farms coincided with a gradual loss of the marsh wetlands. The establishment of the HNNR was useful for obtaining data on the later progression [40]. Therefore, our study area selection of both the HNNR and its surrounding farms was helpful for comparing and analyzing marsh wetland loss and the negative impacts of neighboring agricultural activity on the marsh landscape in the HNNR.

2.2. Methods. The database for this research derived mostly from LANDSAT satellite images. It included one MSS image from July 25, 1975, and two TM images from June 12, 1989, and August 30, 2006. Additional materials used for this research included a geographical map (1 : 100000 scale) and a QuickBird image with a high spatial resolution of 0.61 m from May 16, 2004. All of the landscape maps in raster format that were interpreted from the images were inputted into the ArcGIS 9.2 platform, in which a spatial resolution of less than 0.5 pixels was attained with the aid of a 1 : 10000 scale geographical map. The statistical analysis was complemented with the dynamics of local landscapes during the past 30 years using Excel 2003 software after careful topological examination in the ArcGIS. Data sources about current wetland plant survey and water fowl survey came from our field survey, and the comparable historic data source came from previous research publication (see details in Section 3.4).

A classification system of the landscapes needs to be based on the specific objectives of the research, and the hierarchical characteristics of a classification system need to match the corresponding spatial scale of the research. This research focused on the historical exchange between the natural landscapes and artificial landscapes according to the spatiotemporal information generated from the satellite

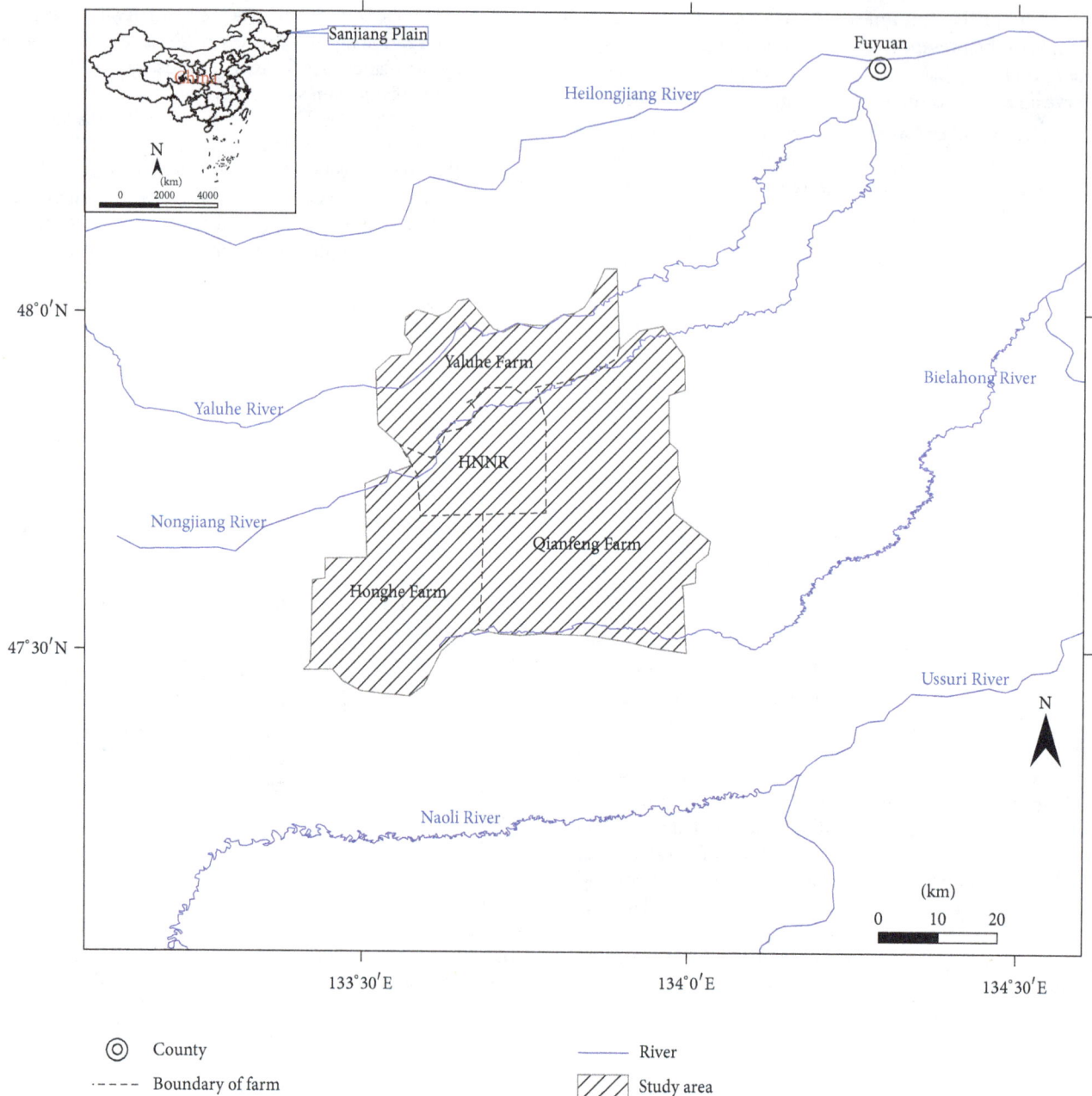

FIGURE 1: Location of the study area.

images on three different dates. The landscapes were classified into seven basic classifications that included three ecotypes to analyze the various landscape information in the images. The seven landscape classifications included marsh, river pond, meadow, forest, paddy field, dry farmland, and others. Among these, the river pond classification comprised natural rivers, ponds, and all other artificial water bodies. Few areas included residences in the study area between 1975 and 1989, although this increased in 2006. For an easier historical comparison of the different landscape classifications, residential areas, road areas, and other types of small landscapes were merged into one landscape classification termed "other."

The data processing method for this research included constructing a new multiple-band file for georeferenced remote sensing images and a mask for the boundary of the study area within the ENVI 4.0 platform. The mask was applied to the imagery data for the purpose of creating an image-based region of interest in the three specific dates. We utilized the layer stacking tool to construct a new file and then performed rapid filter enhancement on the images to meet the needs of image interpretation. The interpretation signs were then established, based on the images according to different colors, shapes, textures, and field investigation photographs. Manual interpretation was used to obtain the classification maps in raster format to describe the regional wetland landscapes in 1975, 1989, and 2006. The QuickBird image was used for reducing the uncertainty while manually delineating the similar landscapes, such as marsh and meadow. After resetting the digital boundaries of four inner units as the HNNR and three farms within the study area, the three thematic maps of the wetland landscapes were reproduced for dynamic analysis purposes (Figures 2(a), 2(b), and 2(c)). An accuracy estimation was made based on the confusion matrices generated from the database of ground truth and a variety of relevant maps (e.g., the previous land-use maps and a previous classification map of the wetlands) [16, 41]. The results of the accuracy assessment showed that the total classification accuracies reached 92.33%, 92.60%, and 90.41% in 1975, 1989, and 2006, respectively. The kappa coefficients (N = 365) were 86.66%, 89.47%, and 86.93%, respectively. Finally, a statistical analysis was performed to present the temporal and spatial changes of the regional dynamic landscapes using Excel 2003 software [41].

3. Results and Discussion

3.1. Basic Changes of the Landscapes in the Study Area.
The progression of gradual marsh landscape loss could be described quantitatively in the study area by comparing and analyzing the dynamic information from the three landscape maps in 1975, 1989, and 2006 (Figures 2(a), 2(b), and 2(c)). The basic marsh landscape in purple changed into farm landscapes in yellow as the present basic landscapes. A very substantial change of the landscapes occurred in the study area, from the 67.1% of the marsh wetland area in 1975 to 73.1% farmland area in 2006. In 1989, the typical marsh wetland loss was 47.4% compared with 1975, and the loss was 89.8% in 2006. The marsh landscape shrank in the HNNR, with a few odd marsh wetlands in the farm areas.

In the past 30 years, a large loss of rivers and ponds occurred during the progression of marsh loss. The landscape in blue lost 53%, and the natural forest loss was 58.2% since 1975. During the progression of the basic natural landscape of the marsh wetlands changing into an agricultural landscape, the dry farmland landscape changed to an increasing number of paddy fields. No paddy fields existed in the study area in 1975, but this landscape comprised one-third of the study area in 2006. A large amount of dry farmland was replaced by paddy fields with the extensive development of agricultural irrigation, which had a very negative impact on the regional marsh wetlands. The few remaining marsh wetlands degraded into meadows because of the loss of healthy habitats attributable to irrigation activity. Therefore, the area of the meadow landscape has seen a nearly 32.3% increase even after most of the original meadows were reclaimed into croplands in the past 30 years.

3.2. Progression and Characteristics of Encroachment on Marsh Wetlands.
Two matrices of the landscape changes were made for the two periods according to the three landscape maps in 1975, 1989, and 2006 based on interpretations of the satellite images (Table 1, Table 2). From these, we analyzed how the marsh wetlands shrunk while the farm landscapes increased in the study area.

Table 1 shows the apparent loss of marsh wetlands from 1975 to 1989, during which a large amount of marsh wetlands were reclaimed into dry farmland or paddy fields. A 47% loss of the marsh area occurred, and the area of dry farmland increased by 380%, a four-fold increase compared with 1975. In 1989 paddy fields comprised 15% of the study area, while in 1975 almost no paddy fields existed. Twenty percent of the forest area was reclaimed into crop land or for other purposes. The originally existing marsh wetlands were the basic landscape in the study area in 1975, and natural marsh, river, and pond landscapes comprised nearly 90% of the area. Few dry farmlands existed during that time. However, the basic marsh landscape was replaced by a landscape pattern consisting of nearly 40% farmlands in 1989, with dry farmlands being the principle landscape. No significant changes occurred to the other landscapes during this period.

Table 2 shows that the marsh wetlands continued to be lost with a change ratio of over 80%, and the area decreased from 35.3% in 1989 to 6.9% in 2006. At the same time, the other natural landscapes, such as river, pond, and forest, also continuously decreased, with an average loss ratio of 50%. The progression of shrinking natural landscapes coincided with the expansion of farmlands, similar to what happened during the previous period, but some new trends appeared in the change of the landscapes from 1989 to 2006. A substantial change in the farm pattern was a 131% increase in the paddy fields during that period. The dual progression occurred as natural landscapes changed to farm landscapes while dry farmlands were replaced by paddy fields.

3.3. Impacts on the Marsh Wetland Habitat due to the Intensified Agriculture Development.
The uniformity of the changing landscapes in the study area includes the three surrounding farms that experienced a rapid change from a basic marsh landscape to an agricultural landscape, although they experienced different agricultural progressions and retain different landscape structures as a result of regional development (Figure 3). We concluded that the impacts on the natural wetland habitat caused by marsh reclamation have two characteristics. First, it reduced the area of the marsh wetland habitat directly. Wetland habitats for wildlife and plants were lost largely because of the rapid decrease in the marsh wetlands in the study area. The remaining marsh wetlands became fragmented from a landscape perspective. Second, reclamation weakened the ecological function of the remaining marsh wetlands as habitats. The remaining marsh wetlands lost their healthy habitats because environmental flow was cut or reduced as a result of agricultural irrigation systems that were strengthened continuously on the neighboring farms.

Well known as a natural "gene bank" of most wildlife on the Sanjiang Plain in China, the HNNR was established in 1984 and was upgraded to a national reserve in 1996. In 2002, it was listed in the Ramsar Convention as an international wetland reserve [42]. This reserve is a location of the original typical marsh wetland on the Sanjiang Plain.

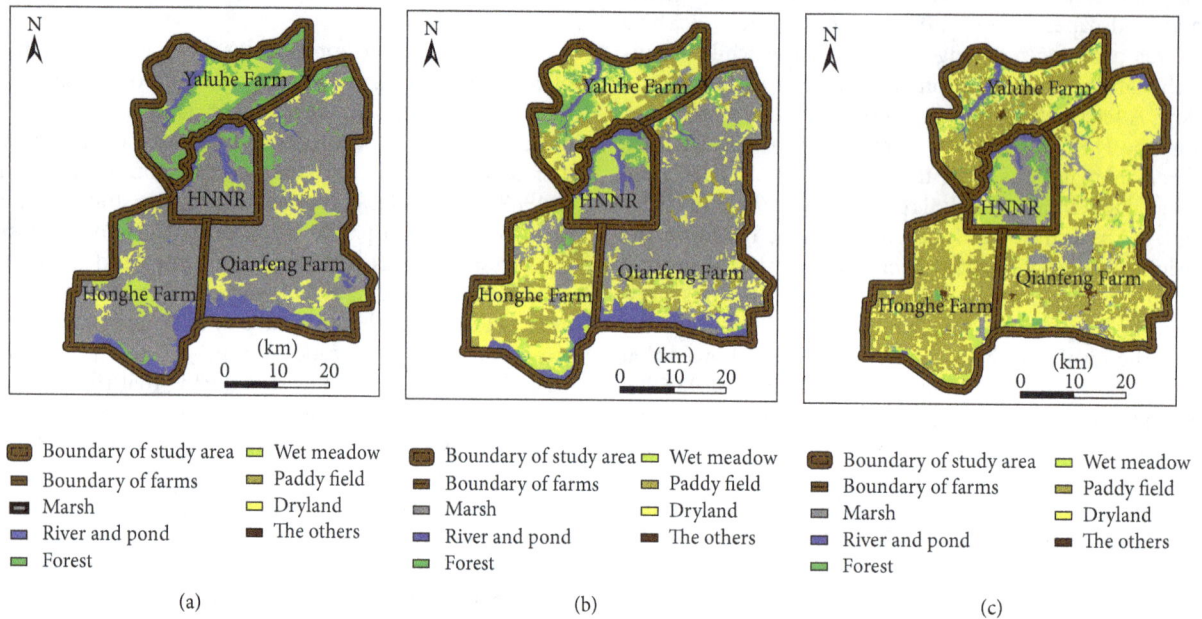

FIGURE 2: Changes of the wetland landscape within the past 30 years.

TABLE 1: Transformation matrix of landscape and land use within the study area during the period from 1975 to 1989 (unit: km²).

| 1975 | 1989 | | | | | | | | |
	Marsh	River and pool	Forest	Meadow	Paddy field	Dry farmland	Other types	Total	Proportion (%)
Marsh	759.87	16.18	72.20	137.31	252.9	383.30	0	1621.74	67.10
River and pool	9.20	206.93	7.39	19.29	5.61	7.75	0	256.17	10.60
Forest	24.79	5.45	86.11	14.14	44.27	62.22	0	236.96	9.81
Meadow	36.37	9.95	22.05	72.68	15.74	33.12	0	189.90	7.86
Paddy field	0	0	0	0	0	0	0	0	0
Dry farmland	22.70	0.04	1.85	3.41	32.65	51.38	0	112.03	4.64
Other types	0	0	0	0	0	0	0	0	0
Total	852.92	238.54	189.60	246.82	351.16	537.76	0	2416.80	
Proportion (%)	35.29	9.87	7.85	10.21	14.53	22.25	0		100
Variation rate (%)	−47.41	−6.88	−19.99	+29.97	/	+380.01	0		

TABLE 2: Transformation matrix of landscape and land use within the study area during the period from 1989 to 2006 (unit: km²).

| 1989 | 2006 | | | | | | | | |
	Marsh	River and pool	Forest	Meadow	Paddy field	Dry farmland	Other types	Total	Proportion (%)
Marsh	118.68	12.61	23.89	99.06	143.93	453.49	1.26	852.92	35.29
River and pool	22.83	86.06	8.17	45.62	9.96	64.85	1.05	238.54	9.87
Forest	6.05	10.84	39.84	15.12	31.37	86.31	0.08	189.60	7.85
Meadow	13.34	6.74	11.04	61.37	61.14	93.02	0.17	246.82	10.21
Paddy field	0.87	1.06	6.37	14.98	233.82	93.13	0.93	351.16	14.53
Dry farmland	3.76	3.19	9.71	15.13	332.45	162.5	11.02	537.76	22.25
Other types	0	0	0	0	0	0	0	0	0
Total	165.53	120.50	99.02	251.28	812.67	953.30	14.50	2416.80	
Proportion (%)	6.85	4.99	4.10	10.40	33.63	39.44	0.60		100
Variation rate (%)	−80.59	−49.48	−47.77	+18.07	+131.42	+77.27	/		

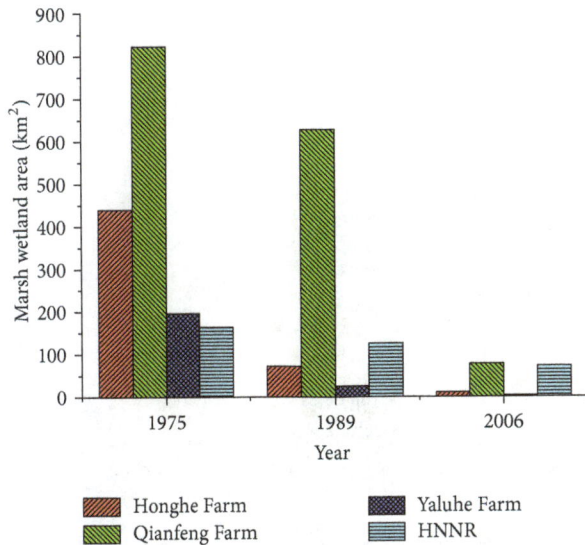

FIGURE 3: Loss of the marsh wetland of 4 units within the study area in the past 30 years.

Compared with the other three farms that have experienced extensive disturbances, the HNNR maintains a basic marsh landscape with less human disturbance. However, its marsh area decreased since the 1980s. Rapidly developing irrigation projects in the surrounding farms cut the water sources to the marsh ecosystem in the HNNR. Therefore, 30% of the marsh wetlands in the HNNR degraded into meadow wetlands [43]. From this research, we can conclude that the establishment of this natural reserve protected the remaining marsh wetlands with the intensified regional agricultural development. Because of the limited reserve area of the HNNR, however, our further analysis showed that marsh wetland ecosystems in the HNNR have been indirectly influenced by the agricultural activity of the surrounding farms that have changed the landscape pattern of the HNNR.

3.4. Damage to the Natural Wetland Ecosystem Caused by Local Agricultural Development. Wetlands are well-known habitats of water fowl. The HNNR is a transfer location in East Asia for rare water fowl, such as Grus japonica and Ciconia boyciana, which have first-order protection status in China [14]. Over 23 species of Grus japonica and 400 species of Ciconia boyciana were listed in the study area in the early 1970s [44], but only three species of Grus japonica and five species of Ciconia boyciana were recorded during an uninterrupted observation period in the study area between 2003 and 2004. This represents a nearly 90% loss since the 1970s [36]. For most water fowl, the increasing farmlands and paddy fields cannot replace their natural habitat. The shrinking natural marsh wetlands have an obvious negative impact on the existence of these water fowl [2, 45].

Damage to the wetland habitat for wildlife and plants has also resulted in the loss of rare plant species. Over 50 wetland plant species are listed as endangered at the national level in the region. Both Dysophylla yatabeana and D. fauriei

are now extinct, although they were very common species 30 years ago. The damage to natural habitats harms wetland plants from both biological and ecological perspectives [43]. Carex lasiocarpa is a representative species of the local marsh wetland ecosystem. It was recorded in the 1970s as a robust and large plant with an average height of 73.7 cm. However, its height has decreased to an average of 40.5 cm, 33.2 cm shorter than 30 years earlier, according to a field survey conducted between 2003 and 2004 [36]. Its average biomass decreased from $653 \, g/m^2$ to $403 \, g/m^2$ (a 30% decrease) in the past 50 years [46]. With regard to plant composition, the richness of the species of wetland plants has also decreased because of decrease in the quality of the wetland habitats. Currently, there is an average of 6.7 species per square meter, a reduction of one species compared with 30 years ago [47]. The biodiversity of the natural ecosystems has definitely been damaged in the region because of the large amount of marsh wetland degradation into meadows. Our future research will precisely assess the weakness of ecological function due to the agriculture development.

3.5. Discussion. Marsh wetlands were widespread on the Sanjiang Plain before the 1980s. The growing season is very short (only 4 months) on the Sanjiang Plain. Most of the area in this region is flooded year round because of the extremely cold and moist climate. The rough natural conditions result in few permanent residents in this region. Therefore, the Sanjiang Plain is well known as "The Big Wild" because of its unique natural marsh landscape [9]. With the increasing Chinese population, the country is seriously challenged by the increasing demand for grain. Grain production is a priority for the Chinese government. Therefore, a series of agricultural policies were made to encourage marsh reclamation and the expansion of farmlands for the purpose of agricultural development [13]. The Sanjiang Plain became the highest priority for reclamation because of its abundance of wild land [10]. Within our study area, the Qianfeng Farm was established in 1969. The Yaluhe Farm was established in 1977, and the Honghe Farm was established in 1980. The purpose of these established farms was to reclaim marsh wetlands. However, encroachment on the marsh wetlands was not excessive because of the lower productivity during that time. Local farmers were not willing to produce more grain because of socialist equalitarianism [48–50], and people were busy engaging in various political movements throughout China during that time.

The initial stage of Chinese reform and open policy occurred from 1978 to 1983. During that time, China implemented successful reform of socialistic economic institutions throughout its widespread countryside. Under the reformation rubric, some local farms on the Sanjiang Plain were selected by the central government for pilot projects of modern agricultural farming. The farmers achieved efficient grain production while continuing to reclaim marsh wetlands under reclamation leadership in Jianshanjiang, although this did not reach a climax of regional marsh reclamation [40, 50]. The progression of encroachment on marsh wetlands accelerated on the Qianfeng Farm because of the widespread

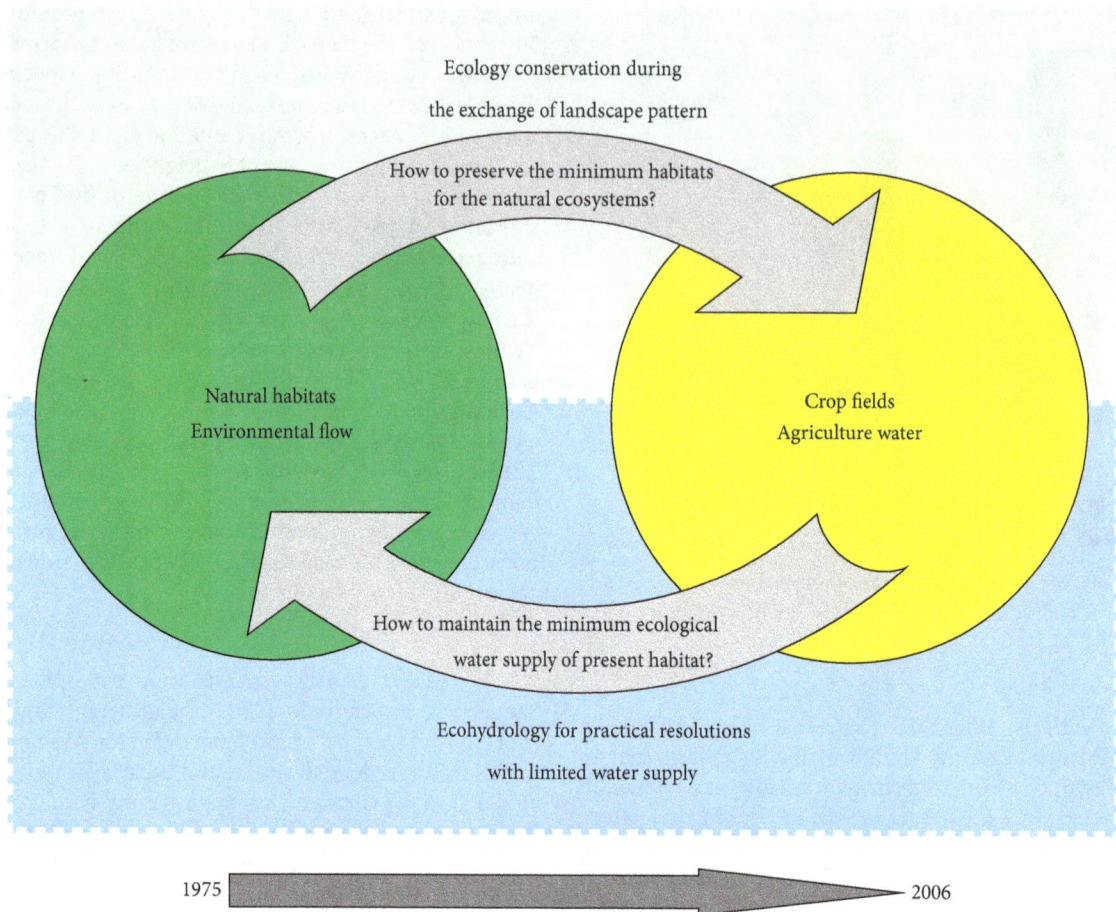

FIGURE 4: Two key issues in wetland eco-hydrology during the regional process of marsh wetland reclamation.

policy of organizing family farms encouraged by the parent body after 1985 [50]. Following the Qianfeng Farm, the Yaluhe Farm, which was previously a socialist institution, was divided into many small family farms in 1988, and this policy was followed by the Honghe Farm in 1993 [48]. With the new policy, farmers were actively involved in running their family farms. They made investments in various agricultural equipments to expand their own production capacities. Farming efficiency was improved so much that grain production increased during this period [13, 49] by somehow successfully reclaiming marsh wetlands to expand the farmland owned by the family. Encroachments on marsh wetlands most rapidly occurred on the Sanjiang Plain (Figure 3).

The Government of Heilongjiang province published the Regulation of Wetland Protection in Heilongjiang Province on June 20, 2003. It was the first regional regulation on wetland protection by a local government in China. The regulation declared the prohibition of all activities that encroach on wetlands [36]. However, number 1 document from the central government that encouraged an increase in the income of farmers at the national level was published in 2004. The document suggested subsidizing farmers by reducing their agricultural tax [9]. This policy stimulated the farmers' will to increase grain production. Local farmers attempted

to reclaim the marsh wetlands to expand their farmland to maximize grain production, even through various illegal means that were against the Wetland Protection Regulation [16, 36]. The technical means of reclaiming marsh wetlands improved substantially during that period, and the modern agricultural facilities helped farmers reduce the cost of marsh reclamation [43]. Marsh reclamation also took disadvantage of both global warming and regional aridity [51]. The gradual illegal encroachment on the few remaining marsh wetlands has not been suspended in the study area, although the reclamation of marsh wetlands has been ceased on a large scale.

Marsh reclamation causes obvious negative impacts to wetland ecosystems. Wetlands, the natural habitats of most wildlife and plants, are well known as the "gene bank of wildlife." Wetlands have significant value for biodiversity in most ecosystems [52–54]. Extensive alterations of both regional hydrology and ecological patterns have occurred at a large scale on the Sanjiang Plain. Marsh reclamation has caused an irreversible and rapid change from a natural ecosystem to an agricultural ecosystem at the regional level. As a consequence of the change, irrigation water has replaced the previous natural environmental flow. However, little research has scientifically assessed the huge disturbance and ecological impact [55, 56]. The challenges include resolving two key scientific issues at the regional level. The first issue is how to preserve a minimum of natural habitats during the rapid progression of ecosystem reductions. The second issue is how to maintain a minimum amount of environmental flow for the remaining natural ecosystems confronted by the increased demand of irrigation water (Figure 4).

4. Conclusions

(1) Intensified agriculture development has changed a unique natural marsh landscape into an agricultural landscape during the past 30 years in the study area. The reclamation process of marsh wetlands accelerated in response to various national policies that demanded grain production beginning in the 1980s. Ninety percent of the natural marsh wetland area was lost in the study area from 1975 to 2006 while most dry farmland has been replaced by paddy fields.

(2) Attempt of current Chinese institution for preserving the regional natural wetlands has achieved limited success. A few wetlands remain healthy because of the establishment of the HNNR, although their ecological quality has declined because of increased threats to the remaining wetland habitats. Irrigation expansion plays a key role in such threats.

(3) The sustainability of the natural wetland ecosystems is being threatened by continuous reduction in the wetland habitats number and decline in the ecological quality due to the intensified agriculture development. In the future, it is a big challenge to preserve a minimum of natural habitats during the rapid progression of natural ecosystem reductions while natural resource administrators attempt to maintain a reasonable amount of environmental flow for the remaining natural ecosystems confronted by the increased demand of irrigation water.

Acknowledgments

This study was funded by the National Natural Science Foundation of China (NSFC 41171415 and NSFC 41001050). The research also received support from the Project of the National Basis Research Program of China (2009CB421103), and the special S&T Project on Treatment and Control of Water Pollution (2012ZX07201004). The authors would like to thank Professor Wei Ji from Missouri University for his valuable suggestions during the preparation of the paper. The authors also thank the Sanjiang Marsh Wetland Experimental Station, Chinese Academy of Sciences, and Honghe National Natural Reserve for their help with the field work and ecohydrological monitoring.

References

[1] M. C. Thoms, "Floodplain-river ecosystems: lateral connections and the implications of human interference," *Geomorphology*, vol. 56, no. 3-4, pp. 335–349, 2003.

[2] M. D. Bryant, R. T. Edwards, and R. D. Woodsmith, "An approach to effectiveness monitoring of floodplain channel aquatic habitat: salmonid relationships," *Landscape and Urban Planning*, vol. 72, no. 1–3, pp. 157–176, 2005.

[3] J. H. Bai, R. Xiao, K. J. Zhang, H. F. Gao, B. S. Cui, and X. H. Liu, "Soil organic carbon as affected by land use in young and old reclaimed regions of a coastal estuary wetland, China," *Soil Use and Management*, vol. 29, no. 1, pp. 57–64, 2013.

[4] L. B. Huang, J. Bai, B. Chen, K. J. Zhang, C. Huang, and P. P. Liu, "Two-decade wetland cultivation and its effects on soil properties in salt marshes in the Yellow River Delta, China," *Ecological Informatics*, vol. 10, pp. 49–55, 2012.

[5] T. Nakayama, "Shrinkage of shrub forest and recovery of mire ecosystem by river restoration in northern Japan," *Forest Ecology and Management*, vol. 256, no. 11, pp. 1927–1938, 2008.

[6] H. Y. Liu, X. G. Lü, S. K. Zhang, and Q. Yang, "Fragmentation process of wetland landscape in watersheds of Sanjiang Plain, China," *Chinese Journal of Applied Ecology*, vol. 16, no. 2, pp. 289–295, 2005.

[7] M. S. Wondzell, A. M. Hemstrom, and A. P. Bisson, "Simulating riparian vegetation and aquatic habitat dynamics in response to natural and anthropogenic disturbance regimes in the Upper Grande Ronde River, Oregon, USA," *Landscape and Urban Planning*, vol. 80, no. 3, pp. 249–267, 2007.

[8] Y. X. Yin, Y. P. Xu, and Y. Chen, "Relationship between changes of river-lake networks and water levels in typical regions of Taihu Lake Basin, China," *Chinese Geographical Science*, vol. 22, no. 6, pp. 673–682, 2012.

[9] X. T. Liu and X. H. Ma, *Natural Environmental Changes and Ecological Protection in the Sanjiang Plain*, Sciences Press, Beijing, China, 2002.

[10] K. S. Song, D. W. Liu, Z. M. Wang et al., "Land use change in Sanjiang Plain and its driving forces analysis since 1954," *Acta Geographica Sinica*, vol. 63, no. 1, pp. 93–109, 2007.

[11] J. Y. Liu, M. L. Liu, H. Q. Tian et al., "Spatial and temporal patterns of China's cropland during 1990–2000: an analysis based on Landsat TM data," *Remote Sensing of Environment*, vol. 98, no. 4, pp. 442–456, 2005.

[12] J. Y. Liu, Q. Zhang, and Y. F. Hu, "Regional differences of China's urban expansion from late 20th to early 21st century based on remote sensing information," *Chinese Geographical Science*, vol. 22, no. 1, pp. 1–14, 2012.

[13] H. Y. Liu, S. K. Zhang, Z. F. Li, X. G. Lu, and Q. Yang, "Impacts on wetlands of large-scale land-use changes by agricultural development: the Small Sanjiang Plain, China," *Ambio*, vol. 33, no. 6, pp. 306–310, 2004.

[14] D. M. Zhou and H. L. Gong, *Hydro-Ecological Modelling of the Honghe National Nature Reserve*, Chinese Environmental Scientific Press, Beijing, China, 2007.

[15] Z. Q. Luan, W. Deng, and J. H. Bai, "Protection of Honghe National Nature Reserve wetland habitat," *Water and Soil Conservation Research*, vol. 10, pp. 154–157, 2003.

[16] D. M. Zhou, H. L. Gong, Z. Q. Luan, J. M. Hu, and F. L. Wu, "Spatial pattern of water controlled wetland communities on the Sanjiang Floodplain, Northeast China," *Community Ecology*, vol. 7, no. 2, pp. 223–234, 2006.

[17] B. L. Wen, X. T. Liu, X. J. Li, F. Y. Yang, and X. Y. Li, "Restoration and rational use of degraded saline reed wetlands: a case study in western Songnen Plain, China," *Chinese Geographical Science*, vol. 22, no. 2, pp. 167–177, 2012.

[18] J. H. Bai, Z. F. Yang, B. S. Cui, H. F. Gao, and Q. Y. Ding, "Some heavy metals distribution in wetland soils under different land use types along a typical plateau lake, China," *Soil and Tillage Research*, vol. 106, no. 2, pp. 344–348, 2010.

[19] R. Xiao, J. H. Bai, H. G. Zhang, H. F. Gao, X. H. Liu, and W. Andreas, "Changes of P, Ca, Al and Fe contents in fringe marshes along a pedogenic chronosequence in the Pearl River estuary, South China," *Continental Shelf Research*, vol. 31, no. 6, pp. 739–747, 2011.

[20] J. H. Bai, R. Xiao, B. S. Cui et al., "Assessment of heavy metal pollution in wetland soils from the young and old reclaimed regions in the Pearl River Estuary, South China," *Environmental Pollution*, vol. 159, no. 3, pp. 817–824, 2011.

[21] X. L. Wang, Y. M. Hu, and R. C. Bu, "Analysis of wetland landscape changes in Liaohe delta," *Scientia Geographica Sinica*, vol. 16, pp. 260–265, 1996.

[22] C. L. Yi, S. M. Cai, J. L. Huang, and R. R. Li, "Classification of wetlands and their distribution of the Jianghan-Dongting Plain, central China," *Journal of Basic Science and Engineering*, vol. 6, pp. 19–25, 1998.

[23] J. L. Huang, "The area change and succession of Dongtinghu wetland," *Geographical Research*, vol. 18, pp. 297–304, 1999.

[24] G. L. Huang, J. J. Zhang, and Y. X. Li, "Wetland classification and actuality analysis of Liaohe Delta," *Forest Resources Management*, vol. 4, pp. 51–56, 2000.

[25] H. Y. Liu, X. G. Lu, and Z. Q. Liu, "Deltaic wetlands in Bohai Sea: resources and development," *Journal of Natural Resources*, vol. 16, pp. 101–106, 2001.

[26] Z. F. Zhang, H. L. Gong, W. Zhao, R. H. Fu, and T. L. Zhang, "Research on dynamic change in wetland resource in Peking Widgeon-lake based on 3S techniques," *Remote Sensing Technology and Application*, vol. 18, pp. 291–296, 2003.

[27] T. Zhang, A. X. Mei, and Y. L. Cai, "Application of Spot Remote Sensing image in landscape classification of Chongming Dongtan," *Urban Environment & Urban Ecology*, vol. 17, pp. 45–47, 2004.

[28] G. W. Yong, C. C. Shi, and P. F. Qiu, "Monitoring on desertification trends of the grassland and shrinking of the wetland in Ruoergai Plateau in north-west Sichuan by means of Remote Sensing," *Journal of Mountain Science*, vol. 21, pp. 758–762, 2003.

[29] F. Xiao and S. M. Cai, "Studies on the Honghu wetland changes," *Journal of Central China Normal University*, vol. 37, pp. 266–268, 2003.

[30] J. M. Bian and N. F. Lin, "Application of the 3S technology on the landscape evolution in the wetland of lower reach of Huolin River Basin," *Journal of Jilin University*, vol. 35, pp. 221–225, 2005.

[31] H. Y. Liu, S. K. Zhang, and X. G. Lu, "Processes of wetland landscape changes in Naoli River Basin since 1980s," *Journal of Natural Resources*, vol. 17, pp. 698–705, 2002.

[32] A. H. Wang, S. Q. Zhang, and Y. F. He, "Study on dynamic change of mire in Sanjiang Plain based on RS and GIS," *Scientia Geographica Sinica*, vol. 22, pp. 636–640, 2002.

[33] W. Hou, S. W. Zhang, Y. Z. Zhang, and W. H. Kuang, "Analysis on the shrinking process of wetland in Naoli river basin of Sanjiang Plain since the 1950s and its driving forces," *Journal of Natural Resources*, vol. 6, pp. 725–731, 2004.

[34] B. Schröder and R. Seppelt, "Analysis of pattern-process interactions based on landscape models-Overview, general concepts, and methodological issues," *Ecological Modelling*, vol. 199, no. 4, pp. 505–516, 2006.

[35] P. Treitz and J. Rogan, "Remote sensing for mapping and monitoring land-cover and land-use change-an introduction," *Progress in Planning*, vol. 61, no. 4, pp. 269–279, 2004.

[36] K. Y. Zhao, Y. J. Luo, J. M. Hu, D. M. Zhou, and X. L. Zhou, "A study of current status and conservation of threatened wetland ecological environment in Sanjiang Plain," *Journal of Natural Resources*, vol. 23, pp. 790–796, 2008.

[37] M. Santelmann, K. Freemark, J. Sifneos, and D. White, "Assessing effects of alternative agricultural practices on wildlife habitat in Iowa, USA," *Agriculture, Ecosystems and Environment*, vol. 113, no. 1–4, pp. 243–253, 2006.

[38] A. Bär and J. Löffler, "Ecological process indicators used for nature protection scenarios in agricultural landscapes of SW Norway," *Ecological Indicators*, vol. 7, no. 2, pp. 396–411, 2007.

[39] J. Álvarez-Rogel, F. J. Jiménez-Cárceles, M. J. Roca, and R. Ortiz, "Changes in soils and vegetation in a Mediterranean coastal salt marsh impacted by human activities," *Estuarine, Coastal and Shelf Science*, vol. 73, no. 3-4, pp. 510–526, 2007.

[40] Editorial committee for publication of historical records on Honghe Farm in Heilongjiang Province, Statistical Yearbook of Honghe Farm, 1980–1984, 1986.

[41] H. Y. Zhang, D. M. Zhou, and Y. H. Wang, "The changing process of wetland landscape in Honghe National Nature Reserve and surrounding farms in Sanjiang Plain," *Remote Sensing Technology and Application*, vol. 24, pp. 57–62, 2009.

[42] The List of Wetlands of International Importance, http://www.ramsar.org/pdf/sitelist_order.pdf.

[43] D. M. Zhou, H. L. Gong, Y. Y. Wang, S. Khan, and K. Y. Zhao, "Driving forces for the marsh wetland degradation in the Honghe National Nature Reserve in Sanjiang Plain, Northeast China," *Environmental Modeling and Assessment*, vol. 14, no. 1, pp. 101–111, 2009.

[44] K. Y. Zhao, *Mires of China*, Sciences Press, Beijing, China, 1999.

[45] G. F. Wilhere, M. J. Linders, and B. L. Cosentino, "Defining alternative futures and projecting their effects on the spatial distribution of wildlife habitats," *Landscape and Urban Planning*, vol. 79, no. 3-4, pp. 385–400, 2007.

[46] W. Deng, P. Y. Zhang, and B. Zhang, *Development Report in Northeast China*, Sciences Press, Beijing, China, 2004.

[47] Y. H. Ji, X. G. Lu, Q. Yang, and K. Y. Zhao, "The succession character of Carex lasiocarpa community in the Sanjiang Plain," *Wetland Science*, vol. 2, pp. 140–144, 2004.

[48] Editorial committee for publication of historical records on Honghe Farm in Heilongjiang Province, Statistical Yearbook of Honghe Farm, 1985–2002, 2005.

[49] Bureau of historical records of general administration of agricultural reclamation in Heilongjiang Province, Statistical Yearbook of Honghe Farm, 2003–2006, 2006.

[50] Bureau of historical records of general administration of agricultural reclamation in Heilongjiang Province, Statistical Yearbook of Qianfeng Farm, 1968–2000, 2004.

[51] M.-H. Yan, W. Deng, and X.-H. Ma, "Climate variation in the sanjiang plain disturbed by large scale reclamation during the last 45 years," *Acta Geographica Sinica*, vol. 56, pp. 159–170, 2001.

[52] E. K. Antwi, R. Krawczynski, and G. Wiegleb, "Detecting the effect of disturbance on habitat diversity and land cover change in a post-mining area using GIS," *Landscape and Urban Planning*, vol. 87, no. 1, pp. 22–32, 2008.

[53] C. Boutin, A. Baril, and P. A. Martin, "Plant diversity in crop fields and woody hedgerows of organic and conventional farms in contrasting landscapes," *Agriculture, Ecosystems and Environment*, vol. 123, no. 1-3, pp. 185–193, 2008.

[54] T. G. O'Connor and P. Kuyler, "Impact of land use on the biodiversity integrity of the moist sub-biome of the grassland biome, South Africa," *Journal of Environmental Management*, vol. 90, no. 1, pp. 384–395, 2009.

[55] M. F. Carreño, M. A. Esteve, J. Martinez, J. A. Palazón, and M. T. Pardo, "Habitat changes in coastal wetlands associated to hydrological changes in the watershed," *Estuarine, Coastal and Shelf Science*, vol. 77, no. 3, pp. 475–483, 2008.

[56] T. S. Seilheimer, T. P. Mahoney, and P. Chow-Fraser, "Comparative study of ecological indices for assessing human-induced disturbance in coastal wetlands of the Laurentian Great Lakes," *Ecological Indicators*, vol. 9, no. 1, pp. 81–91, 2009.

Permissions

The contributors of this book come from diverse backgrounds, making this book a truly international effort. This book will bring forth new frontiers with its revolutionizing research information and detailed analysis of the nascent developments around the world.

We would like to thank all the contributing authors for lending their expertise to make the book truly unique. They have played a crucial role in the development of this book. Without their invaluable contributions this book wouldn't have been possible. They have made vital efforts to compile up to date information on the varied aspects of this subject to make this book a valuable addition to the collection of many professionals and students.

This book was conceptualized with the vision of imparting up-to-date information and advanced data in this field. To ensure the same, a matchless editorial board was set up. Every individual on the board went through rigorous rounds of assessment to prove their worth. After which they invested a large part of their time researching and compiling the most relevant data for our readers. Conferences and sessions were held from time to time between the editorial board and the contributing authors to present the data in the most comprehensible form. The editorial team has worked tirelessly to provide valuable and valid information to help people across the globe.

Every chapter published in this book has been scrutinized by our experts. Their significance has been extensively debated. The topics covered herein carry significant findings which will fuel the growth of the discipline. They may even be implemented as practical applications or may be referred to as a beginning point for another development. Chapters in this book were first published by Hindawi Publishing Corporation; hereby published with permission under the Creative Commons Attribution License or equivalent.

The editorial board has been involved in producing this book since its inception. They have spent rigorous hours researching and exploring the diverse topics which have resulted in the successful publishing of this book. They have passed on their knowledge of decades through this book. To expedite this challenging task, the publisher supported the team at every step. A small team of assistant editors was also appointed to further simplify the editing procedure and attain best results for the readers.

Our editorial team has been hand-picked from every corner of the world. Their multi-ethnicity adds dynamic inputs to the discussions which result in innovative outcomes. These outcomes are then further discussed with the researchers and contributors who give their valuable feedback and opinion regarding the same. The feedback is then collaborated with the researches and they are edited in a comprehensive manner to aid the understanding of the subject.

Apart from the editorial board, the designing team has also invested a significant amount of their time in understanding the subject and creating the most relevant covers. They scrutinized every image to scout for the most suitable representation of the subject and create an appropriate cover for the book.

The publishing team has been involved in this book since its early stages. They were actively engaged in every process, be it collecting the data, connecting with the contributors or procuring relevant information. The team has been an ardent support to the editorial, designing and production team. Their endless efforts to recruit the best for this project, has resulted in the accomplishment of this book. They are a veteran in the field of academics and their pool of knowledge is as vast as their experience in printing. Their expertise and guidance has proved useful at every step. Their uncompromising quality standards have made this book an exceptional effort. Their encouragement from time to time has been an inspiration for everyone.

The publisher and the editorial board hope that this book will prove to be a valuable piece of knowledge for researchers, students, practitioners and scholars across the globe.

List of Contributors

John Gichuki, Reuben Omondi, Priscillar Boera and Tsuma Jembe
Kenya Marine and Fisheries Research Institute, P.O. Box 1881, Kisumu 40100, Kenya

John Gichuki
Big Valley Rancheria Band of Pomo Indians, 2726 Mission Rancheria Road, Lake Port, CA 95453-9637, USA

Tom Okorut and Ally Said Matano
Lake Victoria Basin Commission, P.O. Box 1510, Kisumu 40100, Kenya

Ayub Ofulla
Maseno University, P.O. Box Private Bag Maseno, Kenya

M. Shuhaimi-Othman, R. Nur-Amalina and Y. Nadzifah
School of Environmental and Natural Resource Sciences, Faculty of Science and Technology, National University of Malaysia (UKM), Selangor, 43600 Bangi, Malaysia

Dimitrios E. Bakaloudis
Laboratory of Wildlife Ecology and Management, Department of Forestry and Natural Environment Management, Technological Educational Institute of Kavala, 1st km Drama-Mikrohori, 661 00 Drama, Greece

Dimitrios E. Bakaloudis, Christos G. Vlachos, Malamati A. Papakosta, Vasileios A. Bontzorlos and Evangelos N. Chatzinikos
Department of Wildlife and Freshwater Fisheries, Faculty of Forestry and Natural Environment, Aristotle University of Thessaloniki, P.O. Box 241, 540 06 Thessaloniki, Greece

Vasileios A. Bontzorlos,
Hunting Confederation of Greece, 8 Fokionos Street, 105 63 Athens, Greece

Evangelos N. Chatzinikos
4th Hunting Federation of Sterea Hellas, 8 Fokionos Street, 105 63 Athens, Greece

Sérgio P. Ávila and Antonio M. de Frias Martins
Departamento de Biologia, Universidade dos Açores, 9501-801 Ponta Delgada, Açores, Portugal
CIBIO-Açores, Universidade dos Açores, 9501-801 Ponta Delgada, Açores, Portugal

Sérgio P. Ávila
MPB-Marine Palaeo Biogeography Working Group of the University of the Azores, Rua da Mãe de Deus, 9501-801 Ponta Delgada, Açores, Portugal

Jeroen Goud
National Museum of Natural History, Invertebrates, Naturalis Darwinweg, Leiden, P.O. Box 9517, 2300 RA Leiden, The Netherlands

Yichun Xie
Department of Geography and Geology, Eastern Michigan University, Ypsilanti, MI 48197, USA

Zongyao Sha
International School of Software, Wuhan University, Wuhan 430079, China

Ulysses Paulino de Albuquerque, Elcida de Lima Araújo, Ana Carla Asfora El-Deir, Geraldo Jorge Barbosa deMoura, Glauco Alves Pereira, Joabe Gomes de Melo, Marcelo Alves Ramos, Maria Jesus Nogueira Rodal, Nicola Schiel and SeverinoMendes de Azevedo-J´unior
Departamento de Biologia, Universidade Federal Rural de Pernambuco, Rua Dom Manoel de Medeiros, s/n, Dois Irm~os, 52171-900 Recife, PE, Brazil

André Luiz Alves de Lima
Unidade Acadêmica de Serra Talhada (UAST), Universidade Federal Rural de Pernambuco (UFRPE), Fazenda Saco s/n, 56.900-000, Serra Talhada, PE, Brazil

Antonio Souto and Bruna Martins Bezerra
Departamento de Zoologia, Centro de Ciências Biológicas Universidade Federal de Pernambuco (UFPE), Avenida Professor Moraes Rego, 1235-Cidade Universit´aria, 50670-901 Recife, PE, Brazil

Elba Maria Nogueira Ferraz
Direção de Ensino/Gerência de Pesquisa e Pós-Graduação-Cidade Universit´aria, Instituto Federal de Pernambuco-Reitoria, Campus Recife, Avenida Professor Luis Freire 500, 50740-540 Recife, PE, Brazil

Eliza Maria Xavier Freire
Laboratório de Herpetologia, Departamento de Botânica, Ecologia e Zoologia, Centro de Biociências, Universidade Federal do Rio Grande do Norte, Lagoa Nova, 59072-900 Natal, RN, Brazil

Everardo Valadares de Sá Barreto Sampaio
Departamento de Energia Nuclear, Centro de Tecnologia, Universidade Federal de Pernambuco (UFPE), Avenida Professor Lu´ıs Freire 1000, Cidade Universit´aria, 50740-540 Recife, PE, Brazil

Flor Maria Guedes Las-Casas
Programa de Pós-Graduação em Ecologia e Recursos Naturais, Centro de Ciências Biológicas e da Saúde, Departamento de Ecologia e Biologia Evolutiva, Universidade Federal de São Carlos, Rodovia Washington Luiz Km 235, 13565-905 São Carlos, SP, Brazil

Rachel Maria de Lyra-Neves and Wallace Rodrigues Telino Júnior
Unidade Acadêmica de Garanhuns, Universidade Federal Rural de Pernambuco, Avenida Bom Pastor, s/n, Boa Vista, Heliópolis, 55296-901 Garanhuns, PE, Brazil

Rômulo Romeu Nóbrega Alves
Departamento de Biologia, Universidade Estadual da Paraíba, Avenida das Baraúnas 351, Bodocongó, 58109-753 Campina Grande, PB, Brazil

William Severi
Departamento de Pesca e Aquicultura, Universidade Federal Rural de Pernambuco, Rua Dom Manoel de Medeiros, s/n, Dois Irm~aos, 52171-900 Recife, PE, Brazil

Huan Yu, Zheng-Wei He, Cheng-Jiang Zhang, Chao-Xu Xia and Xuan-Qiong Li
College of Earth Sciences, Chengdu University of Technology, Chengdu 610059, China

Shi-Jun Ni and Bo Kong
Institute of Mountain Hazards and Environment, Chinese Academy of Sciences, Chengdu 610041, China

Shu-Qing Zhang and Xin Pan
Northeast Institute of Geography and Agroecology, Chinese Academy of Sciences, Changchun 130012, China

Qinggai Wang, Shibei Li, Peng Jia, Changjun Qi and Feng Ding
Appraisal Center for Environment and Engineering, Ministry of Environmental Protection, Beijing 100012, China

Lanlan Guo, Yi Chen and Zhao Zhang
State Key Laboratory of Earth Surface Processes and Resource Ecology, Academy of Disaster Reduction and Emergency Management, Beijing Normal University, Beijing 100875, China

Lanlan Guo
Key Laboratory of Environmental Change and Natural Disaster, MOE, Beijing Normal University, Beijing 100875, China

Takehiko Fukushima
Graduate School of Life and Environmental Science, University of Tsukuba, Tsukuba 305-8572, Japan

Yu Qin and Shuhua Yi
State Key Laboratory of Cryospheric Sciences, Cold and Arid Regions Environmental and Engineering Research Institute, Chinese Academy of Sciences, 320 DonggangWest Road, Lanzhou 730000, China

Wassie Anteneh
Department of Biology, College of Science, Bahir Dar University, P.O. Box 79, Bahir Dar, Ethiopia

Eshete Dejen
FAO-Sub Regional Office for Eastern Africa, P.O. Box 5536, Addis Ababa, Ethiopia

Abebe Getahun
Fisheries and Aquatic Science Stream, Faculty of Life Sciences, Addis Ababa University, P.O. Box 1176, Addis Ababa, Ethiopia

Ramón H. Barraza-Guardado, Ramón Casillas-Hernández and Cuauhtemoc Ibarra-Gámez
Instituto Tecnol'ogico de Sonora (ITSON), 85000 Ciudad Obregón, SON, Mexico

Ramón H. Barraza-Guardado and Marco A. López-Torres
Departamento de Investigaciones Cient'ıficas y Tecnológicas de la Universidad de Sonora (DICTUS), 83000 Hermosillo, SON, Mexico

José A. Arreola-Lizárraga and Francisco Magallón-Barrajas
Centro de Investigaciones Biol'ogicas del Noroeste, S.C. (CIBNOR, S.C.), 85454 Guaymas, SON, Mexico

Anselmo Miranda-Baeza
Universidad Estatal de Sonora (US), 85800 Navojoa, SON, Mexico

Juan Huang, Jinyan Zhan, Haiming Yan and Feng Wu
State Key Laboratory ofWater Environment Simulation, School of Environment, Beijing Normal University, Beijing 100875, China

Xiangzheng Deng
Institute of Geographic Science and Natural Resource Research, CAS, Beijing 100101, China
Center for Chinese Agricultural Policy, CAS, Beijing 100101, China

Deborah Gnana Selvam, K. M. Mujeeb Rahiman and A. A. Mohamed Hatha
Department of Marine Biology, Microbiology, and Biochemistry, Cochin University of Science and Technology, Lakeside Campus, Cochin 682 016, India

Randall W. Myster
Biology Department, Oklahoma State University, Oklahoma City, OK 73107, USA

Marcel Martinez-Porchas
Departamento de Tecnología de Alimentos de Origen Animal, Centro de Investigación en Alimentación y Desarrollo, Km. 0.7 Carretera a La Victoria, Hermosillo, SON, Mexico

Luis R. Martinez-Cordova
Departamento de Investigaciones Científicas y Tecnológicas de la Universidad de Sonora, Boulevard Luis Donaldo Colosio s/n, 83000 Hermosillo, SON, Mexico

Dongping Meng and Min Li
Institute of Loess Plateau, Shanxi University, Taiyuan 030006, China

Jin-Tun Zhang
College of Life Sciences, Beijing Normal University, Beijing 100875, China

Zhaoqing Luan, Zhongxin Wang, Dandan Yan, Guihua Liu and Yingying Xu
Key Laboratory of Wetland Ecology and Environment Science, Northeast Institute of Geography and Agroecology, Chinese Academy of Sciences, 4888 Shengbei Road, Changchun 130102, China

Zhongxin Wang, Dandan Yan and Guihua Liu
University of Chinese Academy of Sciences, 19A Yuquan Road, Beijing 100049, China

Zhaoqing Luan and Demin Zhou
Key Laboratory of Wetland Ecology and Environment, Northeast Institute of Geography and Agricultural Ecology, Chinese Academy of Sciences, 4888 Shengbei Street, Changchun 130102, China

Demin Zhou
College of Resources, Environment and Tourism, Capital Normal University, Beijing 100048, China